Pavol Ivanyi (Ed.)

# MHC + X

## Complex Formation and Antibody Induction

With 63 Figures and 47 Tables

Springer-Verlag Berlin Heidelberg New York
London Paris Tokyo

Pavol Ivanyi, MD PhD

Central Laboratory of the Netherlands
Red Cross Blood Transfusion Service and
Laboratory for Experimental and Clinical Immunology
University of Amsterdam
1006 AD Amsterdam, The Netherlands

---

Proceedings of the Workshop-Symposium meeting held in Ommen/Amsterdam, 7–9 October 1987

---

*Figure on the front cover:* A blackboard drawing of Dr. J.L.Strominger based on the data of Bjorkman, P.J.Saper, M.A.Samraoui, B., Bennett, W.S., Strominger, J.L., and Wiley, D.C. (Nature, 329: 506, 1987)

ISBN-13:978-3-642-74028-2 e-ISBN 13:978-3-642-74026-8
DOI: 10.1007/978-3-642-74026-8

Library of Congress Cataloging-in-Publication Data
MHC + X: complex formation and antibody induction/P. Ivanyi, ed. p. cm. "Proceedings of the workshop-symposium meeting held in Ommen/Amsterdam, 7–9 October 1987"-T.P. verso. Includes bibliographies and index.
ISBN-13:978-3-642-74028-2 (U.S.: alk. paper)
1. Major histocompatibility complex-Congresses, 2. MHC antibodies-Congresses, 3. Immunoglobulins-Synthesis-Congresses. 4. Immune response-Regulation-Congresses. I. Ivanyi, P. (Pavol). 1930– . II. Title: MCH plus X.
OR184.3.M53 1988 616.07'-dc19 88-26562 CIP



2123/3145-543210 – Printed on acid-free paper

# Preface

After the discovery of the function of MHC molecules, namely to provide the context for T cell recognition of foreign antigens, in 1974 Zinkernagel and Doherty made the first drawing of MHC+X (Fig. 1 from Zinkernagel and Doherty, Nature, 251: 547, 1974). Over the next 18 years a very large number of similar drawings ensued, some of real artistic beauty. One side of the problem, the nature of the T cell receptor, was unraveled; however, we still do not know exactly what kind of a structure the T cell receptor recognizes, although in 1987 we learned so much about the structure of MHC molecules and antigen presentation. In schematic presentations no one is now placing the foreign antigen beside the MHC molecule, but rather on top of it, as pointed out by J. L. Strominger at the MHC+X meeting in Paris. The complex of MHC and antigen is named MHC+X, but the precise meaning of this formula remains a "problem perplex", as illustrated in these proceedings by Peter Parham.

When planning the Ommen/Amsterdam meeting at the beginning of 1987, its major aim was seen as to discuss the question of whether MHC+X can induce antibodies and, consequently, their specificity. In other terms, whether – in analogy to antigen specific MHC restricted T cells – MHC restricted antibodies also exist. There must be no more hesitation about the existence of *a new phenomenon,* namely that *injections of syngeneic virus infected cells in some situations can induce MHC (class 1) specific antibodies.* What is disputed and remains unclear is the mechanism by which these antibodies are triggered and, especially, what is their specificity. For our meeting the problem was formulated as "MHC specific antibodies induced by foreign antigen." We made every effort to bring together researchers who have tackled these questions and contributed to a solution of the problem. A reputable if not complete representation was achieved, and I believe that this series of articles will guide interested immunologists through the data scattered in the literature.

Three other topics were included in the program: alloreactive CTL induced by virus injections, naturally occurring MHC specific antibodies, and the antigenic determinants formed by the complex of $\beta_2$-microglobulin and the class I heavy chain. It is probably the first time that data from various laboratories on these topics, appear in one volume.

During the preparation of the meeting it became obvious that very important new information is accruing on the nature of MHC + X. A few presentations highlight recent developments on the topic of peptide binding to MHC molecules. However, in this volume the term MHC + X is extended by the inclusion of information other than that concerning only the binding of peptides into the MHC groove. These data show that the MHC molecule possibly interacts with other ligands as well (endorphins, volatiles, insulin receptor). Thus while MHC + peptide data are represented to a rather limited extent, the other "nonimmunological" functions of the MHC molecules are brought fully to the reader's attention.

It was very interesting to organize this type of meeting at a time when so much news is accumulating on MHC function.

I am thankful to the Research Director of the CLB (Prof. Dr. V.P. Eijsvoogel) and the Head of the Department of Immunohematology (Prof. Dr. C.P.Engelfriet) for encouragement and help in organizing the meeting. The assistance of the local organizing committee (Femia Kievits, Marlies Lutkie, T.J.F.Buunen, L.P. de Waal, and O.Weiler) was invaluable, and the secretarial assistance of Wanda Winkel was most helpful. Encouragement, advice, and support from Springer-Verlag were of the utmost importance to the preparation of the meeting and the proceedings.

Amsterdam, January 1988 PAVOL IVANYI

# Acknowledgments

This meeting was organized and financially supported by the Dr. Karl Landsteiner Foundation of the Central Laboratory of the Netherlands Red Cross Blood Transfusion Service, Amsterdam.

Financial support was kindly provided by:

- Abbot Diagnostics Division, Amstelveen
- Ahrend B.V., Amsterdam
- Amstelstad B.V., Zwanenburg
- Beun de Ronde B.V., Abcoude
- Clean Air Techniek, Woerden
- Costar Europe Ltd., Badhoevedorp
- Depex, De Bilt
- Geostick Nederlandse Etikettenfabrik B.V., Uithoorn
- Gibco Europe B.V., Breda
- Harlan CPB, Indianapolis
- Koningin Wilhelmina Fonds, Amsterdam
- Linde Verpakkingsindustrie B.V., Denekamp
- F. Meinders, Amsterdam
- Nedalco, Bergen op Zoom
- Nier Stichting Nederland, Bussum
- Omnilabo Nederland B.V., Tiel
- Pharmacia Nederland B.V., Woerden
- Retsch N.V., Ochten
- Springer Verlag, Heidelberg
- Nederlandse Organisatie voor Zuiver Wetenschappelijk Onderzoek (ZWO), Den Haag
- Zedek B.V., Deventer

# Table of Contents

# List of Participants and Contributors

(Current addresses as far as known)

J. P. ABASTADO
National Institute of Health, Building 6, 311 Bethesda, MD 20892, USA

I. ALGARRA
Servicio Analisis Clinicos, C. S. Virgen de las Nieves, Granada 18014, Spain

H. ALLEN
Biogen Research Corporation, Cambridge, MA 02142, USA

F. AMEGLIO
Regina Elena Institute for Cancer Research, Rome, Italy

E. J. BAAS
The Netherlands Cancer Institute, Department of Cellular Biochemistry, Plesmanlaan 121, 1066 CX Amsterdam, The Netherlands

M. BARAD
Centre d'Immunologie INSERM-CNRS, 70 Route Leon-Lachamp, 132888 Marseille 2, Cedex 9, France

B. H. BARBER
Department of Immunology, Faculty of Medicine, University of Toronto, Medical Sciences Building, Toronto, Ontario M5S 1AB, Canada

C. J. BEUKELMANN
Department of Microbiology, University of Utrecht, Utrecht, The Netherlands

J. Boerenkamp
Central Laboratory of the Netherlands Red Cross Blood Transfusion Service, and Laboratory for Experimental and Clinical Immunology, Plesmanlaan 125, 1066 CX Amsterdam, The Netherlands

R. E. Brown
Department of Psychology, Dalhousie University, Halifax, Nova Scotia B3H 4J1, Canada

S. Buus
Institute for Experimental Immunology, None Alle 71, 2100 Copenhagen 0, Denmark

A. Caballero
Servicio Analisis Clinicos, C.S. Virgen de las Nieves, Granada 18014, Spain

R. Černý-Provaznik
Neurologische Klinik und Polyklinik der Technischen Universität München, Möhlstrasse 28, 8000 München 80, Federal Republic of Germany

F. H. J. Claas
Department of Immunohematology and Blood Bank, University Hospital, Rijnsburgerweg 10, 2333 AA Leiden, The Netherlands

L. Corbo
Istituto di Scienze Biochimiche, II Facoltà di Medicina e Chirurgia, Napoli, Italy

M. Cramer
Institute for Genetics, University of Köln, Weyertal 121, 5000 Köln 41, Federal Republic of Germany

M. de Felice
Istituto di Scienze Biochimiche, II Facoltà di Medicina e Chirurgia, Napoli, Italy

G. G. de Lange
Central Laboratory of the Netherlands Red Cross Blood Transfusion Service, Plesmanlaan 125, 1066 CX Amsterdam, The Netherlands

L. P. de Waal
Central Laboratory of the Netherlands Red Cross Blood Transfusion Service Plesmanlaan 125, 1066 CX Amsterdam, The Netherlands

L. Degos
Laboratory of Mouse Immunogenetics, U93 INSERM Hôpital Saint Louis, 2, Place du Dr. A. Fournier, 75010 Paris, France

C. Delgado
Servicio Analisis Clinicos, C. S. Virgen de las Nieves, Granada 18014, Spain

P. Demant
The Netherlands Cancer Institute, Plesmanlaan 121, 1066 CX Amsterdam, The Netherlands

P. J. F. M. Derhaag
The Netherlands Cancer Institute, Department of Cellular Biochemistry, Plesmanlaan 121, 1066 CX Amsterdam, The Netherlands

A. Dolei
Institute of Microbiology, University of Sassari, Sassari, Italy

V. P. Eijsvoogel
Central Laboratory of the Netherlands Red Cross Blood Transfusion Service, Plesmanlaan 125, 1066 CX Amsterdam, The Netherlands

C. P. Engelfriet
Central Laboratory of the Netherlands Red Cross Blood Transfusion Service, Plesmanlaan 125, 1066 CX Amsterdam, The Netherlands

T. E. W. Feltkamp
Central Laboratory of the Netherlands Red Cross Blood Transfusion Service, Plesmanlaan 125, 1066 CX Amsterdam, The Netherlands

S. Ferrone
Department of Microbiology and Immunology, New York Medical College, Valhalla, NY 10595, USA

L. Flaherty
Wadworth Center for Laboratories and Research, New York State Department of Health, Albany, NY 12201, USA

A. Garrido
Servicio Analisis Clinicos, C. S. Virgen de las Nieves, Granada 18014, Spain

F. Garrido
Servicio Analisis Clinicos, C.S. Virgen de las Nieves,
Granada 18014, Spain

A. Gillet
Centre d'Immunologie INSERM-CNRS, 70 Route Leon-Lachamp,
132888 Marseille 2, Cedex 9, France

R.S. Goodenow
University of California, Department of Genetics, 345 Mulford
Hall, Berkeley, CA 94720, USA

E. Goulmy
Department of Immunohematology and Blood Bank, Univesity
Hospital, Rijsnburgerweg 10, 2333 AA Leiden, The Netherlands

H.M. Grey
Cytel Cooperation, 11099 North Torrey Pines Road, Suite 100,
La Jolla, CA 92037, USA

P.D. Griffiths
Department of Virology, Royal Free Hospital School of Medicine,
Pond Street, London NW3 2QG, United Kingdom

J.E. Grundy
Virology Unit, Royal Free Hospital School of Medicine, Pond
Street, London NW3 2QG, United Kingdom

E. Günther
Zentrum für Hygiene und Human Genetik, Humangenetik der
Universität, Gosslerstrasse 12, 3400 Göttingen, Federal Republic
of Germany

R. Hakem
Centre d'Immunologie INSERM-CNRS, 70 Route Leon-Lachamp,
132888 Marseille 2, Cedex 9, France

T.H. Hansen
Department of Genetics, Washington University School of
Medicine, St. Louis, MO 63110, USA

E. Heber-Katz
The Wistar Institute, Philadelphia PA, USA

D. Ivanyi
Netherlands Cancer Institute, Plesmanlaan 121,
1066 CX Amsterdam, The Netherlands

P. Ivanyi
Central Laboratory of the Netherlands Red Cross Blood Transfusion Service, Plesmanlaan 125, 1066 CX Amsterdam, The Netherlands

C. Jaulin
Département d'Immunologie, Institut Pasteur, 25 Rue du Dr. Roux, 75724 Paris, France

F. Kievits
Central Laboratory of the Netherlands Red Cross Blood Transfusion Service, Plesmanlaan 125, 1066 CX Amsterdam, The Netherlands

P. Kourilsky
Département d'Immunologie, Institut Pasteur, 25 Rue du Dr. Roux, 75724 Paris, France

W. Kuon
Institute for Immunology, 8000 München 2, Federal Republic of Germany

K. Kuypers
Central Laboratory of the Netherlands Red Cross Blood Transfusion Service, Plesmanlaan 125, 1066 CX Amsterdam, The Netherlands

D. R. Lee
Department of Genetics, Washington University School of Medicine, St. Louis, MO 63110, USA

A. van Leeuwen
Department of Immunohematology and Blood Bank, University Hospital, Rijnsburgerweg 10, 2333 AA Leiden, The Netherlands

F. A. Lemmonnier
Centre d'Immunologie, INSERM-CNRS, 70 Route Leon-Lachamp, 132888 Marseille 2, France

R. Linsk
Department of Genetics, University of California, 345 Mulford Hall, Berkeley, CA 94720, USA

J. Maryanski
Ludwig Institute for Cancer Research, Lausanne Branch, Chemin des Boveresses, 1066 Epalinges, Switzerland

P. Matzinger
Basel Institute of Immunology, Grenzacherstrasse 487,
4085 Basel, Switzerland

S. Mayer
Laboratoire d'Histocompatibilité, Centre de Transfusion Sanguine, 10 Rue Spielmann, 67085 Strasbourg Cedex, France

J. A. McKeating
Department of Virology, Royal Free Hospital School of Medicine, Pond Street, London NW3 2QG, United Kingdom

C. J. M. Melief
The Netherlands Cancer Institute, Plesmanlaan 121,
1066 CX Amsterdam, The Netherlands

R. Mierau
Institute for Genetics, University of Köln, Weyertal 121,
5000 Köln 41, Federal Republic of Germany

Y. Mitsuishi
Laboratoire d'Histocompatibilité, Centre de Transfusion Sanguine, 10 Rue Spielmann, 67085 Strasbourg Cedex, France

M. Mommaas-Kienhuis
Department of Dermatology, University Hospital Leiden, Leiden, The Netherlands

M. L. Moule
Banting and Best Department of Medical Research, University of Toronto, Toronto, Ontario M5G 1L6, Canada

N. B. Myers
Department of Genetics, Washington University School of Medicine, Saint Louis, MO 63110, USA

A. Opolski
U93 INSERM, Hopital St. Louis, 2, Place Dr. Fournier, 75475 Paris, France

M. Oudshoorn-Snoek
The Netherlands Cancer Institute, Plesmanlaan 121,
1066 CX Amsterdam, The Netherlands

P. Parham
Department of Cell Biology, Stanford University School of Medicine, Stanford CA 94305, USA

B. PERARNEAU
Centre d'Immunologie INSERM-CNRS, 70 Route Leon-Lachamp, 132888 Marseille 2, Cedex 9, France

M. PÉREZ
Servicio Analisis Clinicos, C.S. Virgen de las Nieves, Granada 18014, Spain

B. PERNIS
College of Physicians and Surgeons, Columbia University, Department of Microbiology, 701 W 168 Street, NY 10032, USA

P.A. PETERSON
Div. Mol. Immunogenetics, Department Imm. Scripps Res. Institute, 10666 North Torrey Pines Road, La Jolla, CA 92037, USA

M.L. PHILLIPS
University of Toronto, Charles H. Best Institute, Banting and Best Department of Medical Research, 112 College Street, Toronto, Ontario M5G 1L6, Canada

M. PLA
U93 INSERM, Hôpital St. Louis 2, Place Dr Fournier, 75475 Paris, France

H.L. PLOEGH
The Netherlands Cancer Institute, Plesmanlaan 121, 1066 CX Amsterdam, The Netherlands

S. RAGHOEBIER
Wadworth Center for Laboratories and Research, New York State Department of Health, Albany, NY 12201, USA

H.G. RAMMENSEE
Basel Institute of Immunology, Grenzachterstrasse 487, 4085 Basel, Switzerland

P.J. ROBINSON
Institute of Immunology, German Cancer Research Center, P.O. Box 101949, 6900 Heidelberg, Federal Republic of Germany

A. ROCCA
U93 INSERM, Hôpital St. Louis 2, Place Dr Fournier, 75475 Paris, France

B. J. Roser
Department of Immunology, AFRC Institute of Animal Physiology and Genetics Research, Babraham Research Station, Babraham Hall, Cambridge, CB2 4AT, United Kingdom

F. Rotteveel
Central Laboratory of the Netherlands Red Cross Blood Transfusion Service, Plesmanlaan 125, 1066 CX Amsterdam, The Netherlands

W. Schmidt
Institut für Zellbiologie (Tumorforschung), Universität Essen (GH), Hufelandstrasse 55, 4300 Essen 1, Federal Republic of Germany

H. Schreiber
Department of Pathology, University of Chicago, 5841 So. Maryland Avenue, P.O. Box 414, Chicago, IL 60637, USA

A. Sette
Cytel Cooperation, 11099 North Torrey Pines Road, Suite 100, La Jolla, CA 92037, USA

P. B. Singh
AFRC Inst. of Animal Physiology and Genetic Research/Immunology Dept., Babraham, Cambridge CB2 4AT, United Kingdom

M. H. Smith
Department of Immunology, Medical Sciences Building, University of Toronto, Toronto, Canada M5S 1A8

E. S. Song
Department of Genetics, University of California, 345 Mulford Hall, Berkeley, CA 94720, USA

R. Sorrentino
Department of Cell Biology and Development, University La Sapienza, Rome, Italy

N. Tanigaki
Department of Molecular Immunology, Roswell Park Memorial Institute, Buffalo, NY, USA

W. L. Tamminen
University of Toronto, Department of Immunology, Medical Sciences Building, Toronto, Ontario M5S 1A8, Canada

U. THEOPOLD
Max-Planck-Institut für Immunbiologie, Stübeweg 51,
7800 Freiburg, Federal Republic of Germany

M. M. TONGIO
Institute d'Hématologie, Centre de Transfusion Sanguine,
67085 Strasbourg Cedex, France

R. M. TOSI
Laboratorio di Biologia Cellulare, Via G. Romagnosi 18/A,
00196 Roma, Italy

M. C. TURCO
Istituto di Scienze Biochimiche, II Facoltà di Medicina
e Chirurgia, Napoli, Italy

A. URLACHER
Laboratoire d'Histocompatibilité, Centre de Transfusion Sanguine, 10 Rue Spielmann, 67085 Strasbourg Cedex, France

G. VAN BLEEK
Central Laboratory of the Netherlands Red Cross Blood
Transfusion Service, Plesmanlaan 125, 1066 CX Amsterdam,
The Netherlands

F. P. M. J. VAN BREE
Department of Immunohaematology and Blood Bank, University
Hospital, Rijnsburgerweg 10, 2333 AA Leiden, The Netherlands

P. C. VAN MOURIK
Central Laboratory of the Netherlands Red Cross Blood Transfusion Service and Laboratory for Experimental and Clinical Immunology, Plesmanlaan 125, 1066 CX Amsterdam, The Netherlands

J. M. VAN REE
Rudolf Magnus Institute of Pharmacology, University of Utrecht,
Utrecht, The Netherlands

J. J. VAN ROOD
Department of Immunohematology and Blood Bank, University
Hospital, Rijnsburgerweg 10, 2333 AA Leiden, The Netherlands

G. VAN SEVENTER
Central Laboratory of the Netherlands Red Cross Blood
Transfusion Service, Plesmanlaan 125, 1066 CX Amsterdam,
The Netherlands

E. VAN TWUYVER
Central Laboratory of the Netherlands Red Cross Blood Transfusion Service, Plesmanlaan 125, 1066 CX Amsterdam, The Netherlands

S. VENUTA
Istituto di Scienze Biochimiche, II Facoltà di Medicina e Chirurgia, Napoli, Italy

W. VERDUYN
Department of Immunohaematology and Bloodbank, University Hospital Leiden, Leiden, The Netherlands

W. M. A. VERHOEVEN
Department of Biological Psychiatry, University of Utrecht, Utrecht, The Netherlands

P. L. WARD
Department of Pathology, University of Chicago, 5841 So. Maryland Ave., P. O. Box 414, Chicago, IL 60637, USA

E. WEISS
Institute for Immunology, 8000 München 2, Federal Republic of Germany

R. M. WELSH
Department of Pathology, University of Massachusetts Medical School, 55 Lake Avenue North, Worcester, MA 01655, USA

D. B. WILLIAMS
Department of Biochemistry, Sciences Building, University of Toronto, Toronto, Canada M5S 1A8

H. YANG
Virology Department, Pharmaceutical Research and Development Division, Bristol-Myers Company, 5 Research Parkway, P. O. Box 5100, Wallingford, CT 06492, USA

C. C. YIP
Banting and Best Department of Medical Research, University of Toronto, Toronto, Ontario M5G 1L6, Canada

# Induction of Allospecific Cytotoxic T Cells by Virus Infections In Vivo

HYEKYUNG YANG and RAYMOND M. WELSH

Department of Pathology, University of Massachusetts Medical School, Worcester, MA 01655, USA

## Introduction

Rejection of implanted tissue grafts is due to T cell responses against allogeneic histocompatibility antigens. Graft rejection was for some time thought to be mediated by helper T cells, but more recent evidence indicates that cytotoxic T cells infiltrate grafts and probably play a major role in rejection (Ascher et al. 1981; Tyler et al. 1984). Graft rejection in transplant patients sometimes occurs following infections with viruses, most notably human cytomegalovirus (Simmons et al. 1974; May et al. 1978) (HCMV) and influenza virus (Briggs et al. 1972), and it has been suggested that these virus infections might help to stimulate immune responsiveness against the foreign tissue (Gaston and Waer 1985). Recently, several T cell lines bearing cross-reactivity between virus-modified syngeneic MHC antigens and allogeneic MHC antigens have been isolated, leading to the hypothesis that virus-specific T cells bearing cross-reactivity with alloantigens may provide the mechanism for the virus-induced graft rejection (Gaston and Waer 1985; Braciale et al. 1981; Sheil et al. 1987). This hypothesis invokes specific cross-reactive clones as mediators of graft rejection, in contrast to a more generalized less specific biological response modifier effect.

Indeed, viruses are potent biological response modifiers and can stimulate various components of the host response in a relatively nonspecific manner. Virus-induced interferon (IFN) stimulates the activation of natural killer (NK) cells and macrophages (Welsh 1984), and viruses have even been used as adjuvants to stimulate specific immune responses against autologous tumor tissue (Wallack 1982). Some virus infections are capable of polyclonal B cell stimulation, an event mediated either by direct virus-induced transformation of B cells, by mitogen-like effects of viral glycoproteins, or, perhaps, by bystander stimulation of B cells via the release of B cell growth and differentiation factors by highly stimulated virus-specific helper T cell populations (Rosen et al. 1977; Goodman-Snitkoff et al. 1981; Armstrong et al. 1981; Ahmed and Oldstone 1984).

If virus infections were capable of polyclonal cytotoxic T-lymphocyte (CTL) stimulation, one might expect to see allospecific CTLs generated during acute infection. This is because precursors of allospecific CTLs are present at high frequency, as can be readily shown in mixed lymphocyte cultures (MacDonald et al. 1980). Generation of allospecific CTLs could, however, be masked by the high levels of NK cell activity stimulated during infection. In this report we show that the depletion of NK cells in virus-infected mice reveals the presence of allospecific CTLs which appear to be distinct from the virus-specific CTLs concomitantly generated. This suggests that another mechanism by which virus infections could result in allograft rejection is by the polyclonal stimulation of CTLs, some of which recognize alloantigens displayed on the allograft.

## Cytotoxic Cell Response to Infection

Infection of C57BL/6 mice with lymphocytic choriomeningitis virus (LCMV) results in a biphasic cytotoxic cell response. The first response consists of nonspecific, highly activated NK cells, which peak in activity and number from 2 to 5 days postinfection. The second peak is composed of virus-specific, MCH-restricted CTLs, peaking at days 7–9 postinfection (Welsh 1978; Yang and Welsh 1986). A significant amount of cytotoxic activity against uninfected syngeneic, allogeneic, and xenogeneic cell lines remains detectable at 7–9 days postinfection and on first impression would seem to be attributed to NK cells. However, if NK cells are depleted in vivo by intravenous injections of antibody to asialo GM1 (a neutral glycosphingolipid found in high concentrations on NK cells), the cytotoxicity against syngeneic and xenogeneic targets is completely eliminated, whereas the killing against allogeneic targets is only partially reduced (Table 1). Killing against all types of target cells is eliminated by this antibody at day 3 postinfection, coincidental with the peak in NK cell activity. The remaining alloreactive killer cells on day 7 postinfection have a typical CTL antigenic phenotype, i.e., thy $1.2^+$, Lyt $2.2^+$, $L3T4^-$, asialo $GM1^{+/-}$ (Yang and Welsh 1986).

**Table 1.** Cytotoxic cells generated during LCMV infection[a]

| Days postinfection | Anti-asialo GM1 | % Lysis at E/T = 100 | | | |
|---|---|---|---|---|---|
| | | $H\text{-}2^b$ | $H\text{-}2^b$ + LCMV | $H\text{-}2^k$ | $H\text{-}2^d$ |
| 0 | – | <1 | 1.6 | 1.0 | <1 |
| 3 | – | 24 | 22 | 27 | 23 |
| 7 | – | 7.8 | 58 | 26 | 31 |
| 3 | + | <1 | <1 | <1 | <1 |
| 7 | + | 2.4 | 66 | 22 | 32 |

[a] C57BL/6 mice were inoculated intraperitoneally with $8 \times 10^4$ PFU of LCMV. Spleen leukocytes were tested in 4–6-h $^{51}Cr$ release cytotoxicity assays of $H\text{-}2^b$ (MC57G), $H\text{-}2^k$ (L-929), and $H\text{-}2^d$ (P-815) cell targets. Antiserum to asialo GM1 was injected into mice to deplete NK cell activity

## Target Specificities of the Alloreactive CTLs

The CTLs from the NK cell-depleted mice lysed LCMV-infected syngeneic targets and over ten allogeneic targets derived from H-$2^a$, H-$2^k$, and H-$2^d$ mice. They did not lyse xenogeneic targets or uninfected syngeneic targets (Table 1). Work with cloned T cell lines has indicated that some CTLs recognize both allogeneic and virus-modified syngeneic determinants (Gaston and Waer 1985; Braciale et al. 1981; Sheil et al. 1987). Cold target competition assays were thus run to determine whether the virus-specific and allospecific CTL functions were mediated by the same cells (Yang and Welsh 1986). Unlabeled LCMV-infected syngeneic (H-$2^b$) cells competed well against themselves but not against radiolabeled allogeneic H-$2^k$ and H-$2^d$ targets. The allogeneic targets failed to compete against the virus-infected targets or against targets bearing a heterologous allotype. H-$2^d$ targets did, however, compete well against different target cells also expressing H-$2^d$. These data suggest that the LCMV-specific CTLs were distinct from the allospecific CTLs and that the allospecific CTL population was composed of subpopulations of cells with distinct allospecificities.

## Generation of Allospecific CTLs Is Mediated by Several Viruses But Is Mouse Strain Dependent

To determine how universal the induction of allospecific CTLs by virus infections was, C57BL/6 mice were also infected with murine CMV (MCMV), vaccinia virus (VV), or Pichinde virus (PV). All three viruses stimulated allospecific CTLs at levels similar to that induced by LCMV. This argues against cross-reactivity between alloantigens and virus-modified self and indicates that the generation of the allospecific CTLs is a property common to many virus infections.

The ability of LCMV to stimulate allospecific CTLs was examined in several other strains of mice, including C3H/St, BALB/c, SJL, SWR/J, DBA/2, and CBA/J. Compared to the C57BL/6 mice, relatively low levels of allospecific CTLs were generated, even though high levels of LCMV-specific CTLs were generated. This indicates that the amount of virus-induced allospecific CTL generation is dependent on the genetic background of the host. This might relate to the fact only some patients reject allografts after virus infection.

## Stimulation of "Memory" CTLs

The stimulation of apparently non-cross-reactive allospecific CTLs during virus infection suggested that viruses might be polyclonal CTL inducers. This being the case, one virus infection might be able to stimulate CTLs against a non-cross-reactive virus which previously had infected the host. We found that MCMV, VV, and PV each stimulated LCMV-specific CTLs in mice which previously had been immunized to LCMV. Cold target competition assays indicated no cross-reactivities at the CTL level between these viruses. None of these viruses, however, stimulated VV-specific CTLs in VV-immune mice, indicating that conditions had to be right for non-cross-reactive CTL stimulation to be detectable. This may relate to the facts that high numbers of CTL precursors are found in LCMV-immune mice (Moskophidis et al. 1987) and that the LCMV infection is never completely cleared from an infected mouse, but the VV infection is (treatment of LCMV-immune mice with antilymphocyte serum will stimulate LCMV synthesis) (Volkert and Lundstedt 1968).

## Mechanism of Allospecific CTL Stimulation

Efforts were made to examine the mechanism behind the allospecific CTL stimulation in concordance with the LCMV-specific CTL stimulation in C57BL/6 mice. Splenocytes from NK cell depleted, 5-day LCMV-infected mice displayed little or no cytotoxicity when directly isolated from mice, but, after culture for 2 days in vitro, they lysed cells in the typical virus-specific and allospecific pattern. Depletion of helper T cells with antibody to L3T4 and complement abrogated the generation of either CTL type. However, addition of culture fluid from unfractioned day 5 cells incubated for 2 days in vitro restored the ability of the L3T4$^{+}$ cell-depleted cultures to generate CTLs. Antibody to LCMV blocked the ability of day 5 cells to generate into CTLs, but this ability likewise was restored by supplementation with culture fluid from responding cells. The likely explanation for these results is that helper T cells responding to viral antigens release lymphokines which stimulate the generation of both virus-specific and allospecific CTL populations. These T cells first had to be primed in vivo to respond to these lymphokines, however, as spleen cells from control or 3-day infected mice did not develop into allospecific or LCMV-specific CTLs after exposure to the lymphokine-containing culture fluids in vitro.

## Discussion

These data indicate that many viruses can stimulate allospecific CTL generation in a manner dependent upon the genetic background of the host. This likely is an indicator of polyclonal CTL stimulation, as virus infections also stimulate CTLs directed against an unrelated virus in mice previously immunized against that unrelated virus. Our data do not rule out the possibility that a very low frequency of cross-reactive CTL clones are induced, but cold target competition assays support the concept that this is primarily a nonspecific polyclonal stimulation not due to the generation of cross-reactive clones. It is difficult to rule out that there may be some very low affinity cross-reactivity that, though being impossible to detect in competition assays, may nevertheless stimulate receptors sufficiently to render cells responsive to growth and differentiation factors.

Current concepts of CTL stimulation suggest that the T cell receptor must be stimulated or cross-linked before cells become responsive to growth factors (Weiss et al. 1986). In our system there appears to be an in vivo conditioning of allospecific

CTLs to become responsive to lymphokines, as culture fluids which stimulated CTLs isolated 5 days postinfection did not stimulate CTLs isolated 3 days postinfection. Whether there is a preferential expansion of CTL clones with allospecificities over CTLs with other specificities is not known.

The potential significance of this virus-induced allospecific CTL stimulation in the rejection of allografts is obvious. Virus infections also stimulate IFN, which upregulates MHC antigen expression (Bukowski and Welsh 1986). This combined stimulation of allospecific CTLs and upregulation of their target antigen may prove lethal for the allograft. Whether there is a human corollary to this phenomenon is under investigation, and recent work suggests that Epstein-Barr virus may indeed be capable of polyclonal allospecific CTL stimulation in man (Thompkinson and Sullivan, personal communication).

*Acknowledgments.* This work was supported by U.S. PH.S.Grants AM35506, AI17672, and CA34461. We thank Patricia Small and Linda Colby for technical assistance and Dottie Walsh for preparation of the manuscript.

## References

Ahmed R, Oldstone MBA (1984) Mechanisms and biological implications of virus-induced polyclonal B-cell activation. In: Notkins AL, Oldstone MBA (eds) Concepts in viral pathogenesis. Springer, Berlin Heidelberg New York, pp 231–238

Armstrong RB, Butchko GM, Kiley SC, Phelan MA, Ennis F (1981) Mitogenicity of influenza hemagglutinin glycoproteins and influenza viruses bearing H2 hemagglutinin. Infect Immun 34: 140–143

Ascher NL, Chen S, Hoffman R, Simmons RL (1981) Maturation of cytotoxic effector cells at the site of allograft rejection. Transplant Proc 13: 1105–1107

Braciale TJ, Andrew ME, Braciale VL (1981) Simultaneous expression of H-2 restricted and alloreactive recognition by a cloned line of influenza virus-specific cytotoxic T lymphocytes. J Exp Med 153: 1371–1376

Briggs JD, Timbury MC, Paton AM, Bell PR (1972) Viral infection and renal transplant rejection. Br Med J 4: 520–522

Bukowski JF, Welsh RM (1986) Enhanced susceptibility to cytotoxic T lymphocytes of target cells isolated from virus-infected or interferon-treated mice. J Virol 59: 735–739

Gaston JSH, Waer M (1985) Virus-specific MHC-restricted T lymphocytes may initiale allograft rejection. Immunol Today 6: 237–239

Goodman-Snitkoff G, Mannino RJ, McSharry JJ (1981) The glycoprotein isolated from vesicular stomatitis virus is mitogenic for mouse B lymphocytes. J Exp Med 13: 1489–1502

MacDonald RH, Cerottini J-C, Ryser J-E, Maryanski JL, Taswell C, Widmer MB, Brunner KT (1980) Quantitation and cloning of cytolytic T lymphocytes and their precursors. Immunol Rev 51: 93–123

May AG, Betts RF, Freeman RB, Andrus CH (1978) An analysis of cytomegalovirus infection and HLA antigen matching on the outcome of renal transplantation. Ann Surg 187: 110–117

Moskophidis D, Assmann-Wischer V, Simon MM, Lehmann-Grube F (1987) The immune response of the mouse to lymphocytic choriomeningitis virus. V. High numbers of cytolytic T lymphocytes are generated in the spleen during acute infection. Eur J Immunol 17: 937–942

Rosen A, Gergely P, Jondal M, Klein G, Britton S (1977) Polyclonal Ig production after Epstein-Barr virus infection of human lymphocytes in vitro. Nature 267: 52–54
Sheil JM, Bevan MJ, LeFrancois L (1987) Characterization of dual-reactive H-$2K^b$-restricted anti-vesicular stomatitis virus and alloreactive cytotoxic T cells. J Immunol 138: 3654–3660
Simmons RL, Lopez C, Balfour M, Kalis J, Rattazzi LC, Najarian JS (1974) Clinical virological correlations in renal transplant patients. Ann Surg 180: 623–634
Tyler JD, Galli SJ, Snider ME (1984) Cloned Lyt-$2^+$ cytolytic T lymphocytes destroy allogeneic tissue in vivo. J Exp Med 159: 234–243
Volkert M, Lundstedt C (1968) The provocation of latent lymphocytic choriomeningitis virus in mice by treatment with anti-lymphocyte serum. J Exp Med 127: 327–339
Wallack MK (1982) Specific tumor immunity produced by the injection of vaccinia viral oncolysates. J Surg Res 33: 11–16
Weiss A, Imboden J, Hardy K, Manger B, Terhorst C, Stobo J (1986) The role of the T3/antigen receptor complex in T cell activation. Annu Rev Immunol 4: 593–620
Welsh RM (1978) Cytotoxic cells induced during lymphocytic choriomeningitis virus infection of mice. I. Characterization of natural killer cell induction. J Exp Med 148: 163
Welsh RM (1984) Natural killer cells and interferon. CRC Crit Rev Immunol 5: 55–93
Yang H, Welsh RM (1986) Induction of alloreactive cytotoxic T cells by acute virus infection of mice. J Immunol 136: 1186–1193

# Naturally Occurring H-2 Specific Antibodies

P. Ivanyi, R. Černý-Provaznik[1], and P. C. van Mourik

Central Laboratory of the Netherlands Red Cross Blood Transfusion Service and Laboratory for Experimental and Clinical Immunology. University of Amsterdam, Amsterdam, The Netherlands

[1] Present address: Neurologische Klinik, Technische Universität, Mohlstrasse 28, 8000 München, Federal Republic of Germany

## H-2 Antibodies Induced by Injections of Syngeneic Cells

The serendipity of discovery of "naturally occurring" H-2 specific antibodies (NA H-2 Abs) started in our laboratory in the late 1970s. For the induction of antibodies against the H-2$L^d$ molecule, we immunized BALB/c-$L^d$-loss mutant mice (BALB/cBy, H-$2^{dm2}$) with lymphocytes of the coisogenic strain BALB/cBy H-$2^d$ (Ivanyi et al. 1979); we shall designate this immunization as dm2-anti-d. Twenty-two mice were immunized and individual sera (Ivanyi et al. 1978) analyzed separately for cytotoxic antibodies against a panel of lymphocytes from congeneic strains with various H-2 haplotypes. After the first 4–8 weeks of immunization, anti-$L^d$ antibodies – as expected – could be detected (H-$2^d$, H-$2^q$, and H-$2^b$ cells were lysed). However, after continuous immunization, in most sera antibodies appeared which surprisingly also lysed H-$2^k$ ($K^k$) and H-$2^r$ cells. These unexpected antibodies exerted the reactivity pattern of a short public antibody, namely anti-H-2.25. We shall designate these antibodies here as anti-$K^k$.

It has not been clear how this finding should be interpreted. One possibility was that the immunogen, i.e., the $L^d$ molecule, induced anti-$L^d$ antibodies that cross-react with $K^k$. This would be a new aspect for the specificity of anti-$L^d$ antibodies, although in principle not unexpected, because cross-reactions among H-2K, D, and L antigens can and do exist. An alternative was that we are facing a new phenomenon, because $L^d$ was *not* the trigger for the induction of anti-$K^k$ antibodies and thus we have no clear explanation of how and why the anti-$K^k$ antibodies appeared in the sera. A third possibility was a technical mistake in the immunization experiments, most probably due to injections of lymphocytes of unknown origin.

The reactivity pattern of the individual sera strongly suggested that the observation could not be explained by a technical mishap (Ivanyi et al. 1979). Of chief importance in this respect was the fact that a few individual mouse sera reacted only with $K^k$. Moreover, the presence of anti-$K^k$ antibodies in some dm2-anti-d (anti-$L^d$) sera was confirmed by independent work in other laboratories (Hansen et al. 1979; McKenzie, personal communication).

The critical experiment was performed with regard to the question of whether donor cells (H-$2^d$) can absorb all antibodies from the respective sera, i.e., whether H-$2^d$ cells can absorb not only the anti-$L^d$ antibodies but also the anti-$K^k$ antibodies. These experiments were performed independently in both laboratories, though on

**Table 1.** Absorption experiments on dm2-anti-d sera[a]

| Target cells | dm2-anti-d sera used for absorptions | | | | | | | | |
|---|---|---|---|---|---|---|---|---|---|
| | Serum no. 15 | | | Serum no. 1 | | | Serum no. 18 | | |
| | Not abs. | Absorbed by BALB/c | Absorbed by B10.BR | Not abs. | Absorbed by BALB/c | Absorbed by B10.BR | Not abs. | Absorbed by BALB/c | Absorbed by B10.BR |
| BALB/c H-$2^d$ | 120 | 0 | 32 | 16 | 0 | 4 | 0 | 0 | 0 |
| B10.BR H-$2^k$ | 16 | 4 | 0 | 32 | 8 | 0 | 16 | 4 | 0 |

[a] The sera were obtained from individual BALB/c-H-$2^{dm2}$ mice referred to in Iványi et al. (1979). Titers indicate the highest serum dilution in which more than 50% of target cells have been lysed. The lymphocytotoxic test was used; absorption techniques have been described in Iványi et al. (1979). 100–150 × $10^6$ cells have been used for absorption of 50 μl serum

a limited scale (Ivanyi et al. 1979; Hansen et al. 1979). The results showed that with a high dosis of donor cells (BALB/c, H-$2^d$) all antibody activity from the anti-$L^d$ sera, including the anti-$K^k$ antibodies, could be absorbed.

For reasons we do not understand (we were not able to repeat the experiments because of the small amounts of individual sera), these absorption results could not be confirmed later by an extended systematic approach with a number of other sera. As Table 1 shows, H-$2^d$ cells could absorb the anti-$L^d$ but not the anti-$K^k$ or H-$2^r$ (anti-H-2.25) antibodies. Vice versa, H-$2^k$ cells could not absorb the anti-$L^d$ antibodies. Thus, most of our peculiar dm2-anti-d sera, reactive with H-$2^k$ and H-$2^r$ cells, did not contain anti-$L^d$ antibodies cross-reactive with H-2.$K^k$. Rather, two different populations of antibodies were present, namely (a) anti-$L^d$ antibodies, induced by the $L^d$ molecule difference between dm2 and d mice, and (b) anti-$K^k$ antibodies that illustrate a new phenomenon. Probably, in our initial absorption experiments (Ivanyi et al. 1979; Hansen et al. 1979), sera with relatively low titers were used or, alternatively, some anti-$L^d$ antibodies really cross-react with $K^k$ molecules. However, the absorption experiments illustrated in Table 1 were performed with various sera and various doses of cells and allowed us to conclude that H-$2^d$ cells are unable to absorb the anti-$K^k$ antibodies from dm2-anti-d sera, although the latter have been induced by injections of H-$2^d$ cells. In other terms, the $L^d$ alloantigen difference did *not induce* the anti-$K^k$ antibodies in the dm2-anti-d sera.

What have been the consequences of these data? Our interpretation implied that the anti-$K^k$ antibodies in the dm2-anti-d sera occurred *without allostimulation,* because the anti-$L^d$ difference is the only allo-difference between H-$2^{dm2}$ and H-$2^d$ individuals. *Hence injections of H-$2^d$ cells into H-$2^d$ (syngeneic) mice might provide the same result.* Therefore, we started our first syngeneic immunization by injections of donor BALB/cBy H-$2^d$ lymphocytes into BALB/cBy H-$2^d$ recipients, all experiments animals being raised in our own Animal House Breeding Unit. The result of this experiment was really the induction of H-2 specific antibodies by injections of syngeneic cells (Ivanyi et al. 1980). These observations in BALB/c mice were confirmed by analogous experiments in C57B1/6 (H-$2^b$) mice (Ivanyi et al. 1982). In the BALB/c *syngeneic-immune sera,* anti-$K^k$ antibodies were predominant. In C57B1/6 syngeneic-immune sera, anti-H-$2^d$ antibodies were the most frequent. However, individual sera exerted a broad range of variations. Independent experiments with syngeneic immunization in another institute gave the same result (Capkova et al. 1983). Taken together, these results show that it is possible to induce H-2 specific antibodies by injections of syngeneic lymphocytes. To the best of our knowledge, this was the first demonstration of induction of H-2 specific antibodies by syngeneic lymphocytes. It is not clear whether induction of H-2 specific antibodies by injections of syngeneic tumor cells with modified H-2 antigen expression ("alien" H-2 antigens) is a related phenomenon (Meschini and Parmiani 1978; Garrido et al. 1979; Risser and Grunwald 1981; Schmidt and Alonzo 1983; Philipps et al. 1985; Schmidt 1987, and data in these Proceedings).

However, as Capkova et al. (1983) and we (Ivanyi et al. 1979, 1980) reported, and as we could see later in various further experiments, induction of H-2 specific antibodies by injections ofn syngeneic cells is an *irregular* phenomenon. Only a fraction of the injected mice develop H-2 specific antibodies and the percentage of responding mice can vary from 0% to 50% in groups of 10–20 mice injected with aliquots of the same donor cell suspension; similarly, the antibody titers and specificities in the sera of individual mice are very variable. The reasons for these variations are unknown and, therefore, we cannot monitor the induction of H-2 specific antibodies upon injections of "normal" syngeneic cells (lymphocytes). We presume that unknown factors in the endemic virus status of the donor and/or recipient mice play an important role. At the time of the experiments referred to above, we worked in an "open" animal house with accumulation of various virus infections by introducing strains of various origin.

At present, we work with SPF mice. Syngeneic injections with normal cells are considered as negative controls, with only negligible H-2 antibody formation (Kievits et al. 1987). Syngeneic immunization with Sendai virus in vitro infected cells induced H-2 specific antibodies in about 80% of C57B1/6 mice. We consider the experiments with SV-infected cells (Kievits et al. 1987) in C57B1/6 mice, and in a series of other strains (Kievits et al., in preparation), as a *model* for the explanation of our previous syngeneic immunizations. In other words, under some – not defined – circumstances of virus infections or the virological status in the animal house, injections of syngeneic cells can induce H-2 specific antibodies. This happens irregularly – and cannot be monitored – under "natural" circumstances. In experiments with injections of syngeneic in vitro SV-infected cells, H-2 specific antibodies are induced regularly. Experiences of a similar nature, but with other viruses, are reported in various papers in these Proceedings.

## Detection of Naturally Occurring H-2 Specific Antibodies

An extension of the above interpretation of the experimental data would be the induction of H-2 specific antibodies by injections of syngeneic cells from aged mice into young mice, because old mice frequently express MuLV determinants on cell membranes. Experiments of this nature were initiated and performed in cooperation with Drs. Ada Kruisbeek and J. Radl from the Institute of Experimental Gerontology (Rijswijk), where a limited number of unique aged mouse strains are available. In the

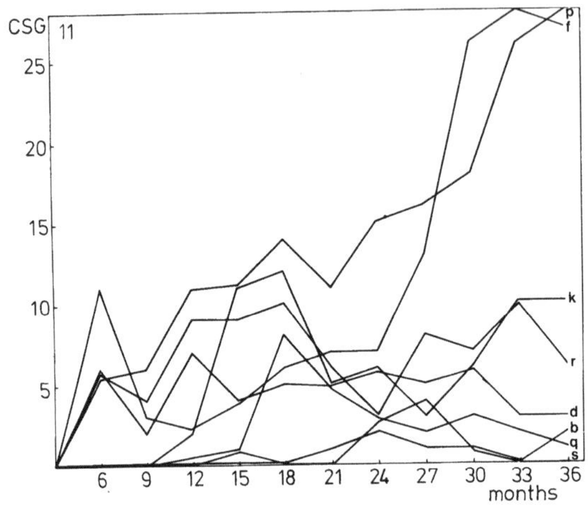


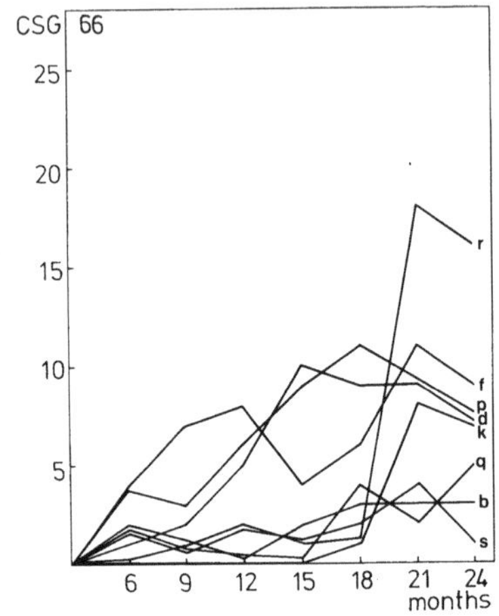


**Fig. 1.** *(left)* Time occurrence and H-2 specificity of antibodies in samples of serum from one individual C57B1/KaLwRij (H-$2^b$) mice (no. 11). All serum samples were tested against lymph node cells with haplotypes H-$2^{b,d,f,k,p,q,r,s}$ on the B10 background. CSG values not higher than 5 are too weak reactions and give poor reproducibility (Černý-Provaznik et al. 1985 a)

**Fig. 2.** *(right)* Time occurrence and H-2 specificity of antibodies in serum of mouse no. 66 (see also Fig. 1)

context of these experiments, we have observed in control bleedings that nonimmunized virgin aged mice frequently contain natural H-2 specific antibodies (Ivanyi et al. 1982). They are designated as "naturally occurring," because the aged mice have not been used in any experiments and, although not raised under strict SPF conditions, their high age illustrates their good health status. NA H-2 Abs were found in about 25% of aged mice, but can be detected – albeit much less frequently – also in young adult mice (Černý-Provaznik et al. 1985 a). With age, the frequeny, the cross-reactivity, and the titer of the antibodies increase. The titers have been low or medium but, in exceptional sera, high or very high titers of NA H-2 Abs have been found (Černý-Provaznik et al. 1985 a, b). Concerning all the above irregularities in the induction of H-2 antibodies by injections of syngeneic cells and the occurrence of NA H-2 Abs, the old-versus-young syngeneic immunization experiments might still be feasible, but the results difficult to evaluate on a statistical basis.

These efforts resulted (about 35 years after the discovery of alloimmune H-2 antibodies) in the observation that naturally occurring H-2 specific antibodies can also be detected (Ivanyi et al. 1982). Similar observations were reported by Wolosin and Greenberg (1981) and Lynes et al. (1984).

The results of a systematic lifelong study regarding the occurrence of NA H-2 Abs have been published (Černý-Provaznik et al. 1985 a). Here we show examples of the development of NA H-2 Abs in four individual mice (Figs. 1, 2). The figures illustrate a relatively sharp onset of specific antibody production in some mice. Under the given circumstances of life history and environmental conditions (Černý-Provaznik et al. 1985 a), such a picture is difficult to accommodate with the effect of nonspecific polyclonal stimulation, a randomly outgrowing B cell clone, or cross-stimulation by shared antigen epitopes of environmental agents. More probably, such findings show the effect of specific stimulation by modified self-H-2 exposure due to MuLV activation and representation at the cell membrane.

The frequency of NA H-2 Abs in individual mice of one highly inbred strain is age, strain, and environment dependent (Černý-Provaznik et al. 1985a). The role of environment is indicated by the sharp differences of the NA H-2 Abs specificities detected in C57B1 (H-$2^b$) mice by Lynes et al. (1984) and Ivanyi et al. (1982) and Černý-Provaznik et al. (1985a). However, many more studies in various strains, maintained under the same conditions and examined in lifelong studies, are needed.

Our efforts to provide an explanation for the origin of NA H-2 Abs have not been conclusive. However, we accumulated the following major observations:

1. Only class I specific, i.e., no class II specific, NA H-2 Abs are detected.
2. The NA H-2 Abs are mostly IgM.
3. The specificity of the antibodies is mostly against public H-2 determinants.
4. NA H-2 Abs are only exceptionally autoreactive.
5. The antibody specificities in sera of individual mice from the same highly inbred strain are very diverse, and the percentage of antibody specificities against various H-2 haplotypes can change over time in various groups of mice (e.g., in C57B1 sera, anti-H-$2^k$ antibodies were most frequent in 1980, but anti-H-$2^f$ antibodies were most frequent in 1982, $P = 0.001$) (Černý-Provaznik et al. 1985a).
6. With a highly sensitive cytotoxic test, NA H-2 Abs of low titers could be detected in 85% of aged mice (Černý-Provaznik and Ivanyi 1985).
7. Monoclonal antibodies (NA H-2 McAbs) have been isolated from mice that have never been immunized, including a first example of an autoreactive NA H-2 McAb (Černý-Provaznik et al. 1985b, c, 1986b).
8. An exceptional group of BALB/c mice has been found where all mice in one cage produced the same (autoreactive) NA H-2 Ab (Černý-Provaznik et al. 1986a).
9. None of the various in vivo or in vitro approaches for polyclonal stimulation (including situations resulting in immunopathology, such as GvH or NZB mice) resulted in induction of NA H-2 Abs. Groups of deceased mice (e.g., TBC) do not have a predictable increase of NA H-2 Abs (Černý-Provaznik et al. 1985a and unpublished observations).
10. There has been no correlation between the presence of NA H-2 Abs and other naturally occurring antibodies or benign Ig paraproteinemias (Černý-Provaznik et al. 1985a).

All in all, in our interpretation, NA H-2 Abs are not due to polyclonal stimulation or cross-stimulation by environmental agents (i.e., cross-stimulation by mimicry epitopes or microbes and viruses) and they are also not randomly proliferating B cell clones (i.e., proliferating without specific stimulation) (Černý-Provaznik et al. 1985a; Ivanyi et al. 1986).

We suggest that murine viruses (endogenous and exogenous) induce modifications in the exposure of self-MHC class I structures and that this modified self-MHC exposure triggers the production of NA H-2 Abs. The modified self-MHC is presumably presented only on a limited number of cells, or exposure is time limited, or the modified cells are eliminated ("immune surveillance") and, therefore, autoreactivity is not detected by the NA H-2 Abs. Hence, the antibodies are usually not autoreactive, although they have been "auto"-triggered.

The experimental model for this explanation are the data which show that injections of syngeneic virus-infected cells can regularly induce H-2 specific antibodies (see Kievits et al. 1987 and several papers in this volume). Therefore, we also think that better understanding of the origin and biological (and pathological) significance of NA H-2 Abs will progress in parallel with the analysis of the mechanism responsible for the induction of H-2 specific antibodies by injections of syngeneic virus-infected cells.


## Summary

Naturally occurring H-2 specific antibodies have been detected in about 25% of aged C57B1 (H-$2^b$) mice and occur less frequently in young individuals. Such antibodies have also been found in sera from various other strains but the occurrence is irregular. The natural H-2 antibodies are mostly IgM and directed against public H-2 determinants. Autoreactivity was found only exceptionally. With the lymphocytotoxic test on spleen cells, we have found only class I specific antibodies. The serum antibody titers are usually low or medium, but exceptional individual mice have high antibody titers. A few natural H-2 antibodies have been isolated in monoclonal form, including one autoreactive monoclonal antibody. The origin of these antibodies is unclear. It has not been possible to induce H-2 antibodies by nonspecific or polyclonal stimulation. Our results suggest that naturally occurring H-2 specific antibodies are induced by the occurrence of a modified exposure of self-MHC class I molecules. Most probably, such self-MHC modifications are due to self-structure alterations induced by viruses.


*Acknowledgments.* This investigation was supported in part by the Foundation for Medical Research MEDIGON, which is subsidized by the Netherlands Organization for the Advancement of Pure Research (ZWO) (grant 509-087) and in part by the Deutsche Forschungsgemeinschaft (grant no. 5/1–2).

## References


Capkova J, Hasek M, Kousalova M, Holan V, Boubelik M (1983) Alloreactive cytotoxic anti-H-2 antibodies in sera of the syngeneically immunized C57Bl/10ScSnPh, BALB/cJPh, and B10.A/SnPh mice. Folia Biol (Praha) 29: 200–210

Černý R, Iványi P (1985) Frequency of naturally occurring H-2-specific antibodies in mouse sera monitored by "superreactive" rabbit complement. Tissue Antigens 26: 259–261

Černý-Provaznik R, Radl J, Leupers T, van Mourik P, Brondijk R, de Greeve P, Iványi P (1985 a) Anti-MHC immunity detected prior to intentional alloimmunization. I. Naturally occurring H-2-specific antibodies in C57B1/KaLwRij (H-$2^b$) mice. Nat Immun Cell Growth Regul 4: 138–159

Černý-Provaznik R, van Mourik P, Radl J, Leupers T, Limpens JC, Iványi P (1985b) Anti-MHC immunity detected prior to intentional alloimmunization. II. Monoclonal H-2-specific antibodies obtained from a normal C57B1/KaLwRij (H-$2^b$) mouse. Nat Immun Cell Growth Regul 4: 160–168

Černý-Provaznik R, van Mourik P, Limpens J, Leupers T, Iványi P (1985c) Anti-MHC immunity detected prior to intentional alloimmunization. III. Natural autoreactive H-2-specific antibodies. Immunogenetics 21: 491–504

Černý-Provaznik R, Kloosterman T, Verkerk D, Vroom TM, Iványi P (1986a) Natural autoreactive H-2-specific serum antibodies in a group of BALB/cBy (H-$2^d$ mice). Tissue Antigens 27: 106–111

Černý-Provaznik R, Kloosterman T, Verkerk D, van Mourik P, Reboul M, Frangoulis B, Pla M, Ivanyi P (1986b) Anti-MHC immunity detected prior to intentional alloimmunization. IV. Natural monoclonal H-2-specific antibodies. J Immunogenet 13: 287–297

Garrido F, Perez M, Torres MD, Garcia-Olivares E, Iványi P, Schirmacher V (1979) A syngeneic anti-tumor serum recognizing a complex H-2 alloantigen. Immunobiology 156: 110–120

Hansen TH, Iványi P, Levy RB, Sachs DH (1979) Cross-reactivity among the products of three non-allelic H-2 loci. Transplantation 28: 339–342

Iványi P, van den Berg-Loonen EM, de Greeve P (1978) Individual mice of one inbred strain produce anti-H-2 and anti-HLA antibodies of different specificities. Tissue Antigens 12: 32–38

Iványi P, Melief CJM, de Greeve P, van Mourik P (1979) Individual mice recognize the complex nature of H-2 antigens; unexpected reaction (anti-$K^k$) in anti-BALB/c-H-$2^d$ sera produced in the BALB/c-H-$2^{db}$ mutant. Transplant Proc 9: 642–646

Iványi P, van Mourik P, Breuning MH, Melief CJM (1980) Anti-H-2 antibodies induced by syngeneic immunization. Immunogenetics 10: 319–332

Iványi P, van Mourik P, Breuning MH, Kruisbeek AH, Kröse CJM (1982) Natural anti-H-2 antibodies in sera of aged mice. Immunogenetics 15: 95–102

Iványi P, Černý-Provaznik R, Radl J (1986) Naturally occurring antibodies against antigens of the major histocompatibility complex (H-2) in ageing mice. In: Facchini A, Haaijman JJ, Labo G (eds) Immunoregulation in aging. Eurage, Paris, pp 307–314

Kievits F, Rocca A, Opolski A, Limpens J, Leupers T, Kloosterman T, Boerenkamp WJ, Pla M, Iványi P (1987) Induction of H-2-specific antibodies is triggered by injections with syngeneic Sendai virus-coated cells. Eur J Immunol 17: 27–35

Lynes MA, Flaherty L, Michaelson J, Collins JJ, Richnik AM (1984) Allogeneic reactivity in normal mouse serum. J Immunogenet 11: 189–196

Meschini A, Parmiani G (1978) Anti-H-2 alloantibodies elicited by syngeneic immunizations with a chemically induced fibrosarcoma. Immunogenetics 6: 117–127

Philipps C, McMillan M, Flood PM, Murphy DB, Forman J, Lancki D, Womack JE, Goodenow RS, Schreiber H (1985) Identification of a unique tumor-specific antigen as a novel class-I major histocompatibility molecule. Proc Natl Acad Sci USA 82: 5140–5144

Risser R, Grunwald DJ (1981) Production of anti-self-H-2 antibodies by hybrid mice immune to a viral tumor. Nature 289: 563–568

Schmidt W (1987) H-2-specific antibodies induced by injection of syngeneic leukemia cells. Immunogenetics 25: 215–221

Schmidt W, Alonzo A (1983) Biologic significance of autoantibodies against H-2 antigens in syngeneic antitumor sera. Transplant Proc 15: 2118–2120

Wolosin LB, Greenberg AH (1981) Genetics of natural antitumor antibody production: antibodies to MHC-linked determinants detected in the serum of unstimulated mice. J Immunol 126: 1456–1459

# Naturally Occurring Anti-MHC Class I Antibodies in the Rat

EBERHARD GÜNTHER

Abteilung Immungenetik der Universität, Gosslerstrasse 12d, 3400 Göttingen, Federal Republic of Germany

Immunity to major histocompatibility complex (MHC) determined antigens is usually studied after deliberate immunization or pregnancy, but can also be detected serologically in the absence of prior intentional immunization, notably against class I antigens. Our studies in the rat, a preliminary report of which has been published previously (Günther et al. 1983), were initiated by the observation that rat sera to be used as normal sera in hemagglutination assays occasionally agglutinated the red blood cells (RBCs).

## Sera, Assay, Strains

The sera were obtained from untreated rats of 17 different inbred strains (see Table 1) and assayed by hemagglutination using dextran as a developing agent (Stock and Günther 1982), a procedure much more sensitive and reliable when typing class I antigens than are similar methods in the mouse. Screening included a panel of RBCs representing 9 of the established 11 independent class I MHC haplotypes of the rat, i.e., the RT1 a, b, c, d, f, k, l, n, and u haplotypes (Butcher and Howard 1986). Serum donors were 2–18 months old and of male or female sex. They had been bred under specified pathogen-free or conventional conditions or had been transferred from the former to the latter environment (breeding was at either the Max-Planck-Institut für Immunbiologie, Freiburg, FRG, or the Abteilung Immungenetik, Universität Göttingen, FRG). No influence of age, sex, and hygienic breeding conditions was detectable, so the data were pooled for each strain.

## Frequency and Specificity of Hemagglutinins

Of 369 rats tested, 69 showed hemagglutinin activity against one or more of the RBCs included in the panel (Table 1), most often against RBCs of the a, f, and k RT1 haplotypes, which are known from immune sera to be highly cross-reactive with each other, and rarely with RBCs of the b and l RT1 haplotypes. The titer, being low (<1:160) compared to "immune" antibodies, and the reaction pattern sometimes

**Table 1.** Anti-RT1 hemagglutinin activity in normal sera from various inbred rat strains

| Group | Strain | RT1 haplotype | Positive/ tested | Reaction pattern (RT1 haplotypes) |
|---|---|---|---|---|
| 1 | AVN | a | 0/ 7 | |
| 2 | LEW.1A[a] | a | 1/19 | f |
| 3 | LEW.1AR2[a] | ar2 | 1/ 4 | |
| 4 | DA | av1 | 0/11 | |
| 5 | COP | av2 | 1/ 8 | d |
| 6 | BUF | b | 8/30 | a, k, u, a+d |
| 7 | PVG/C | c | 12/33 | a, d, d+n, a+d, a+f, a+d+f+l+n+u |
| 8 | LEW.1D[a] | d | 1/17 | k |
| 9 | LEW.1F[a] | f | 12/53 | n, a+k, a+k+n, k+n |
| 10 | LEW[a] | l | 0/20 | |
| 11 | AS | l | 2/ 6 | |
| 12 | LEW.1N[a] | n | 2/20 | a |
| 13 | LEW.1W[a] | u | 2/34 | a, d |
| 14 | LEP | u | 5/16 | a, d, a+d+f |
| 15 | LOU/C | u | 6/40 | a, d, a+d+l |
| 16 | WF | u | 16/46 | a, c, d, f, k, a+d, a+f, a+k, a+d+f, a+d+k |
| 17 | LEW.1WR2[a] | wr2 | 0/5 | |
| | | | 69:369=19% | |

[a] RT1-congeneic strains

showed considerable fluctuation when the same rat was tested repeatedly. In no case could autoreactivity be observed. By using RBCs carrying RT1 recombinant haplotypes the sera could be shown to react with gene products determined by the RT1. A region, which codes for the classical class I antigens, and not by the B/D region encoding class II antigens or the C region, which determines class I-like antigens resembling in several respects H-2 Qa molecules (Stock and Günther 1982). Thus there is no evidence that the target antigens and – as the cross-reaction patterns reveal – the antigenic determinants themselves differ between natural and "immune" antibodies.

## Genetics of the Natural Antibody Response

Table 1 clearly shows that the fraction of positive sera varies between the strains. A comparison between the RT1-congeneic strains LEW or LEW.1W and LEW.1F indicates that the RT1 system is involved in controlling the response with respect to the fraction of positive rats ($P < 0.05$). On the other hand the comparison between the LEW.1W and WF strains which carry phenotypically identical $RT1^u$ haplotypes, but

**Table 2.** Effect of syngeneic immunization on the occurrence of anti-RT1 alloantibodies

| Exp. | Strain | Treatment[a] (6×) | Positive/ treated | Reaction pattern (number of preceding treatments) |
|---|---|---|---|---|
| 1 | LEW | PBS | 2/5 | d, a+d+f (6) |
| | | LEW | 1/5 | a+u (2) |
| 2 | LEW.1D | PBS | 2/5 | a (6) |
| | | LEW.1D | 1/5 | a (6) |
| 3 | LEW.1D | PBS | 0/5 | |
| | | LEW.1D | 0/5 | |
| 4 | LEW.1N | PBS | 0/4 | a+d, a+d+f+k, a+b+d+f+k, |
| | | LEW.1N | 5/5 | a+b+d+f+u (2–6) |
| 5 | LEW.1N | PBS | 2/5 | d (4–5) |
| | | LEW.1N | 5/5 | a, d, f, k, a+d, a+f (2–3) |

[a] Phosphate-buffered saline (PBS) or $50 \times 10^6$ syngeneic spleen cells in PBS i.p.

differ with respect to the genetic background, also shows clear differences in the frequency of positive sera ($P<0.01$). The rare occurrence of natural anti-RT1.A antibodies in sera from congenic rats carrying RT1.A$^a$ or RT1.A$^u$ is in accord with data reported by Klöting and Stark (1985). These authors observed high proportions of positive sera in diabetes- and non-diabetes-prone BB rats. Among the strains tested here, BUF is susceptible to spontaneous autoimmune thyroiditis and LOU develops myeloma-like tumors; the other strains, however, seem "normal," so that the production of natural anti-MHC antibodies does not appear to be restricted to immunologically anomalous rat strains.

## Syngeneic Immunization

Several experiments have been performed in which rats were treated in parallel either with about $50 \times 10^6$ syngeneic spleen cells each month or with buffer and tested for natural anti-RT1.A antibodies 1–2 weeks later. Table 2 shows that in two experiments a clear conversion-like reaction had occurred, indicating the potential effectiveness of syngeneic immunization. It should be added that other experiments of this type were less conclusive since the control groups developed anti-RT1·A activity also rather frequently.

## Effects of Germ-Free Status and Polyclonal Stimulation

Two out of 13 sera from germ-free DA, BDIX, BN, LEW, F344, OM, and SPRD rats, made available by Dr. Hedrich, Zentralinstitut für Versuchstierzucht, Hannover, FRG, agglutinated RBCs of the a, f, and k RT1 haplotypes. In order to evaluate the possible effect of mitogenic stimulation, WF and LEW.1A rats were injected with different doses of *S. minnesota* lipoprotein. Whereas an increased anti-sheep RBC response indicated the effectiveness of the stimulation, no clear induction or amplification of anti-RT1 antibodies was detectable.

## Concluding Remarks

About 20% of rats are found to develop anti-MHC alloantibodies as far as is detectable by hemagglutination. The target antigens and determinants do not appear to be different from those involved in inducing antibodies by deliberate immunization. MHC as well as non-MHC genes seem to determine the frequency of positive sera. It is noteworthy that the immune response to certain RT1.A antigens is MHC-controlled; PVG/C rats, however, have been found to be weak responders to RT1.$A^a$ determinants (Butcher and Howard 1986), in contrast to the findings shown in Table 1. The MHC appears also to be involved by shaping the pattern of the response because autoreactivity is avoided. The data do not contribute clearly to a decision between the various hypotheses put forward so far in order to explain the phenomenon of natural anti-MHC immunity, such as mimicry by environmental antigens or modified self, polyclonal stimulation, or a physiologically existing anti-alloantigen response.

## References

Butcher GW, Howard JC (1986) The MHC of the laboratory rat, *Rattus norvegicus*. In: Weir DM (ed) Handbook of experimental immunology, vol 3. Blackwell, Oxford, pp 101.1–101.18

Günther E, Elsässer-Beile U, Hedrich HJ (1983) Occurrence of antibodies against RT1.A antigens in normal rat sera. Transplant Proc 15: 1560–1561

Klöting I, Stark O (1985) Occurrence of antibodies against gene products of the major histocompatibility complex in normal sera of diabetes-prone BB rats. Folia Biol (Praha) 31: 34–37

Stock W, Günther E (1982) Serologic and cellular characterization of products of a new major histocompatibility gene region, RT1.C, of the rat; possible homology to mouse H-2Qa. J Immunol 128: 1923–1928

# Naturally Occurring HLA Antibodies

M. M. Tongio, Y. Mitsuishi, A. Urlacher, and S. Mayer

Laboratoire d'Histocompatibilité, Centre de Transfusion Sanguine, 10 Rue Spielmann, 67085 Strasbourg Cedex, France

## Introduction

The occurrence of natural HLA antibodies has long been considered a very rare feature. In fact, up until 1985 only three cases had been reported: the first by Collins et al. (1973), who found a very strong natural anti-HLA-B8 antibody in a patient affected with thyroiditis; the second and third by Lepage et al. (1976) and Lenhardt et al. (1982), who found, respectively, a natural anti-HLA-A2 and a natural anti-HLA-B8 antibody. However, both these latter sera were particular in that they reacted only with cells which were homozygous for the HLA-A2 and HLA-B8 antigens respectively. The fact that natural HLA antibodies are more frequent than expected was revealed by chance in our laboratory when we searched for new class II antibodies. A summary of this work, published elsewhere (Tongio et al. 1985), is given here.

## Natural Anti-HLA Antibodies in Normal Blood Bank Donors

Screening of 5242 blood bank donor sera (4640 female and 602 male) was performed with chronic lymphocytic leukemia (CLL) lymphocytes which express DR, DQ, and DP antigens. Parallel screening with peripheral blood lymphocytes (PBLs) was carried out in order to retain only those sera which were positive with CLL lymphocytes but negative with PBLs. Of the 5242 sera tested, 5% of the female sera and 7% of the male sera met these criteria. Specificity studies were performed in 53 of these cases for which enough serum was available (42 female and 11 male sera). Twenty-one of them contained antibodies directed against class I specificities. HLA typing was performed in 11 of these donors and their families, who had agreed to give blood for this purpose. For these cases the class I specificity of the antibodies was confirmed by a blocking reaction with an anti-$\beta$2-microglobulin antibody. In five female cases, the husband and at least one of the children carried the class I antigen against which the antibodies were directed. These sera were considered to contain weak class I alloantibodies. In the other six cases no cause of alloimmunization could be found, since two sera came from nontransfused men, one from a woman who had never been transfused or pregnant, and three from women who had had children but whose husband and children did not carry the antigen against which the antibody's specificity was directed. In order to make sure that these antibodies did not develop against the noninherited class I antigen carried by his mother when the donor was a fetus, maternal typing was performed in two cases. None of the mothers tested carried the HLA antigen against which the donor's specificity was directed. Because of this, these antibodies were called natural antibodies.

The study of the immunoglobulin class to which the antibodies belonged showed that the first five cases which were considered to contain alloantibodies were IgGs, and that the six natural antibodies were IgMs. Since these two kinds of antibody, the IgG alloantibodies and the IgM natural antibodies, had the same kind of reactivity (positive reactivity with CLL lymphocytes but negative reactivity with PBLs), it was decided to compare their behavior.

It was first checked whether B cell varieties other than CLL lymphocytes were able to react with them in the lymphocytotoxicity technique. Three types of cell were tested: normal B-lymphocytes, cord B-lymphocytes, and splenocytes. The results showed that the natural HLA antibodies were cytotoxic with all these cell varieties provided the cells carried the class I antigen corresponding to the antibodies' specificity. This was not the case for some of the alloimmune antibodies since three of the five sera in this category were negative or nearly negative with normal B-lymphocytes whereas they reacted specifically with splenocytes.

It was then determined whether, despite negative reactivity in the lymphocytotoxicity technique with PBLs, the antigens corresponding to the natural and to the weak alloimmune antibodies were nevertheless expressed on T-lymphocytes. The indirect immunofluorescence technique and absorption techniques were used to study this question.

The indirect immunofluorescence technique showed that T-lymphocytes carrying the relevant antigen were stained, allowing one to conclude that the antigens corresponding to the natural antibodies as well as those corresponding to the weak alloimmune antibodies were indeed present on T-lymphocytes.

Absorption techniques confirmed their presence, since the natural antibodies could be absorbed by PBLs as well as by CLL lymphocytes carrying the relevant antigen. Nevertheless, when comparing the amount of lymphocytes necessary to achieve the absorption of the natural antibodies with that necessary for the absorption of the immune antibodies, it was seen that a mean of 10 times more cells was necessary for most of the natural antibodies to be absorbed than for the immune antibodies to be absorbed. Absorptions were also carried out with lymphocytes not carrying the relevant antigen, nor antigens cross-reacting with the relevant antigen. It was seen that even when a very large amount of lymphocytes was used for the absorption, the natural antibodies could not be absorbed, whereas the alloimmune antibodies could be absorbed, provided the number of lymphocytes was 10 times higher than that necessary for the absorption with lymphocytes carrying the relevant antigen.

Finally, the persistence of the six natural antibodies was studied over 2–4 years; it was found that in four cases the antibody reactivity remained stable, in one case it had disappeared 3 years after the first testing, and in one case which was an anti-HLA-B13 at the first blood donation, the antibody became an anti-HLA-B13+B7. But in this case the mother had had a pregnancy between the two blood donations, which suggests that the additional B7 antibody might have been the consequence of an alloimmunization, since the husband carries the B7 antigen.

In summary, after this work we came to the following conclusions:

1. Natural antibodies may be present in about 1% of normal individuals.
2. They belong to the IgM immunoglobulin class.
3. Most of them are not detected, since they are weak and react negatively in the standard lymphocytotoxicity technique with PBLs, but can be found by using B-lymphocytes as target cells. However, since the three natural antibodies reported in the literature were detected with PBLs, it is also possible to detect some strong natural antibodies with PBLs.
4. Their absorption with lymphocytes carrying the antigen corresponding to the antibody's specificity can be achieved by using 5–10 times more lymphocytes than is necessary to absorb immune antibodies having the same serological pattern.
5. These antibodies cannot be absorbed by lymphocytes not carrying the antigen corresponding to the antibody's specificity.
6. Their persistence seems to be rather stable in normal individuals.

## Natural HLA Antibodies in Alloimmunization Situations

According to Ivanyi et al. (1982) and Černý-Provaznik (1985), the most probable hypothesis to explain the natural MHC antibodies' occurrence is that these antibodies are induced against a self-MHC + X combination and that this combination cross-reacts with allo-MHC determinants. However, not only self-MHC antigens but also allo-MHC antigens combined with an X antigen could be responsible for the appearance of natural antibodies. Two cases found in our laboratory serve to illustrate this point. The first comes from a male post-renal-transplant patient who had had five transfusions, the last one 14 days before the transplant. He received no transfusion during the transplantation. No anti-HLA, anti-DR, or anti-monocytic antibodies could be detected prior to transplantation. His serum was also negative with CLL cells. There was a good compatibility between donor and recipient, since the recipient's HLA type was HLA-A2; B7, w41; Cw7; DR5, w6 and the donor's HLA-A2, A3; B7, W47; Cw6,w7; DR5,W6. The cross-match was negative. On postoperative day 13, the patient was restarted on hemodialysis on account of an increase in urea; on postoperative day 21, a *partial* anti-HLA-A1 + A11 antibody belonging to the IgM immunoglobulin class was detected using the lymphocytotoxicity technique with PBLs at +37 °C. This antibody was specific, since it reacted positively with kidney donor PBLs (which had been frozen at the time of transplantation), although these did not carry the HLA-A1 or HLA-A11 antigens. The reactivity of this antibody was blocked by a turkey anti-$\beta^2$-microglobulin serum, confirming its class I specificity. After postoperative day 37, the antibody became an HLA-A1 alone, which finally disappeared at day 45. At the same time, cold strong IgM lymphocytotoxic antibodies detected at +15 °C appeared, which were reactive with the patient's autocells as well as all the allocells tested.

Since the anti-HLA-A1 + A11 antibody reacted positively with kidney donor PBLs which carried the HLA-A3 antigen incompatible with the recipient, and since there are known cross-reactions between HLA-A3, HLA-A1, and HLA-A11, an absorption test was carried out with HLA-A1 and HLA-A3 CLL cells. No absorption took place even when $2 \times 10^9$ CLL cells were used to absorb 1 ml of serum. In the same manner, an absorption test was performed with HLA-A3 homozygote platelets; there too, no absorption was observed. Nevertheless, in testing the eluate it was interesting to note that a positive reactivity was observed with all the lymphocytes tested, independently of their HLA type.

From a clinical point of view, this antibody, which was probably responsible for the renal dysfunction as long as it was present, nevertheless did not destroy the kidney, since after postoperative day 90 hemodialysis was stopped and the urea remained stable at 0.93 g/liter, and 6 years after transplantation the patient has not been restarted on hemodialysis.

In conclusion, this antibody, which was a class I HLA antibody belonging to the IgM class, appeared 21 days after transplantation and disturbed the renal function during the time it was present; its HLA specificity did not correspond to an antigen carried by the kidney donor but nevertheless reacted positively with his lymphocytes. We think this antibody could be a natural HLA antibody directed against a donor's allo-MHC antigen combined with an unknown X antigen.

The second case was found even earlier, in 1973, when we performed extensive studies of the HLA fetomaternal immunization. HLA screening was performed in a woman during the 6th month of her first pregnancy and an anti-B17+B13+B37 antibody was found. The husband carried none of these antigens. The antibody, which belonged to the IgM class, was tested during a second pregnancy and its specificity had not changed although a few of the positive reactions became negative. When tested with the husband's lymphocytes and with the second child's lymphocytes, all the sera from the first and the second pregnancies were negative. When tested with the first child's lymphocytes, the serum collected during the first pregnancy was positive, but the sera collected during the second pregnancy were negative. Thus, one can suppose that in this case the natural HLA antibody was directed against an allo-MHC antigen of the first child combined with an X antigen. Nevertheless, one would also expect that the sera collected during the second pregnancy, which had the same specificity (although a little narrower) as the serum collected during the first pregnancy, would also be positive with the first child's lymphocytes, which was not the case.

Thus, although these two cases furnished some complementary information about the appearance of natural HLA antibodies, no definite conclusions can be reached.

We are now proceeding with the work by trying to:

1. Identify the hypothesized foreign X antigen. By supposing that in some cases the foreign X antigen could be a viral agent, we plan to compare the titer of viral antibodies in sera containing natural antibodies so that in sera containing weak alloantibodies. Up to now, EBV antibodies have been studied with no differences found between the natural antibody and the weak alloantibody sera.
2. Determine whether the blood donors who have natural antibodies also have T-lymphocytes sensitized against the relevant allospecificity. This work has just begun and up to now the tests performed have been negative.

*Acknowledgments.* This work was performed with the financial help of ARC (Association pour la Recherche contre le Cancer) grant Nr. 6240.

## References

Černý-Provaznik R, van Mourik P, Limpens J, Leupers T, Ivanyi P (1985) Anti-MHC immunity detected prior to intentional alloimmunization. III. Natural autoreactive H-2 specific antibodies. Immunogenetics 21: 491–504

Collins ZV, Arnold PF, Peetoom F, Smith GS, Walford RL (1973) A naturally occurring monospecific anti-HL-A8 isoantibody. Tissue Antigens 3: 358–363

Ivanyi P, van Mourik P, Breuning M, Kruisbeek AM, Kröse CJM (1982) Natural H-2-specific antibodies in sera of aged mice. Immunogenetics 15: 95–102

Lenhard V, Diehm H, Römer W, Rauterberger EW, Kownatziki E, Roelcke D (1982) A "spontaneous" cold reactive IgM antibody with anti-HLA-B8 specificity in a patient with multiple sclerosis. Immunobiology 160: 382–391

Lepage V, Degos L, Dausset J (1976) An natural anti-HLA-A2 antibody reacting with homozygous cells. Tissue Antigens 8: 139–142

Tongio MM, Falkenrodt A, Mitsuishi Y, Urlacher A, Bergerat JP, North ML, Mayer S (1985) Natural HLA antibodies. Tissue Antigens 26: 271–285

# Anti-Class I and Anti-Class II Antibodies with Monomorphic Reaction Patterns in Non-alloimmunized Human Subjects

Franco Ameglio[1], Rosa Sorrentino[2], Antonina Dolei[3], Nobuyuki Tanigaki[4], and Roberto Tosi[5]

[1] Regina Elena Institute for Cancer Research, Rome, Italy
[2] Department of Cell Biology and Development, University La Sapienza, Rome, Italy
[3] Institute of Microbiology, University of Sassari, Sassari, Italy
[4] Department of Molecular Immunology, Roswell Park Memorial Institute, Buffalo, NY, USA
[5] Institute of Cell Biology, CNR, Via Romagnosi 18A, 00196, Rome, Italy

## Introduction

In a previous study we demonstrated the occurrence of anti-HLA class II antibodies in a group of untreated Hodgkin's disease patients, none of whom had an history of pregnancy or blood transfusions (Romagnani et al. 1985). This reactivity was associated with a significant suppression of autologous and allogeneic MLR, was not directed against polymorphic DR determinants, and was strongly reduced after chemotherapy.

More recently, several reports have appeared on the occurrence, both in experimental animals (Černý-Provaznik et al. 1985, 1986; Opolski et al. 1986; Kievits et al. 1987) and in human subjects (Okudaira et al. 1982b; Tongio et al. 1985), of anti-MHC antibodies without previous alloimmunization. This information stimulated us to perform a more extensive search for these antibodies in a variety of pathological conditions using our established methodology based on purified radiolabeled HLA molecules. Surprisingly, we have observed the presence of a non-alloantigenically induced HLA reactivity in two apparently unrelated groups of subjecs: active drug abusers and patients infected by *Echinococcus granulosus*. The present paper summarizes data that have been the subject of separate reports (Ameglio et al. 1987a, b, 1988).

## Methodology

A brief description of the experimental conditions is necessary for an understanding of the data. More details have been reported elsewhere (Tanigaki and Tosi 1982). The sera to be tested are incubated in a fixed aliquot (5 μl) with a radiolabeled preparation of either class I or class II HLA molecules. These are purified from B-lymphoblastoid cell lines by a combination of affinity chromatography and gel filtration and are labeled with $^{125}$I by the chloramine-T method. After overnight incuba-

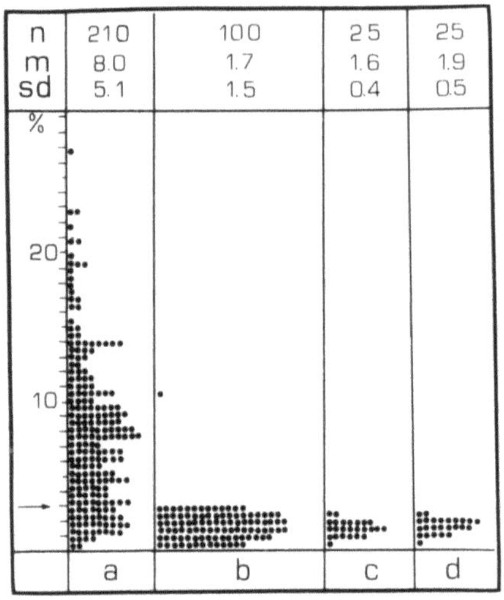


**Fig. 1a–d.** Reactivity with HLA class II molecules of sera. **a** Heroin addicts, **b** normal male blood donors, **c** anti-HIV positive, **d** HBsAg-positive

tion the immune complexes are precipitated by the addition of a sheep anti-human immunoglobulin antiserum, washed, and counted. The results are usually expressed as the proportion of counts that can be precipitated by an excess of anti-HLA xeno-antiserum. The immune complexes are also bound by Sepharose-conjugated protein A and analyzed by SDS-PAGE.

## Anti-HLA Reactivity in Sera from Heroin Addicts

Sera from 210 active heroin addicts obtained from the Center for Assistance to Drug Abusers at the San Camillo Hospital in Rome were studied. Most of the addicts (about 80%) were males, and their mean age was about 28 years. Because of their irregular and various habits, no stratification for drug quantity, quality, or frequency could be made. The sera were tested by the double precipitation assay for their ability to bind a radiolabeled class II preparation (from the LG10 cell line: DR7; w53; DQw2). A high proportion of them gave a significant binding (Fig. 1 a) when compared to sera of 100 normal male blood donors (Fig. 1 b). Similar data were obtained by testing the sera with a purified class I preparation (from the RM2 cell line: A11, 24; B28, 44; C–). In each serum the anti-class I and the anti-class II binding activity were parallel. About 80% of heroin addict sera possessed a binding activity above that of the controls. There was no relationship with the presence of anti-HIV antibodies (although 52% of the sera were anti-HIV-positive) or with the presence of hepatitis virus (21% of the sera were HBsAg-positive). This was an important point and it was more directly checked by examining a group of sera of non-drug abusers, anti-HIV-positive homosexuals (Fig. 1 c) and HBsAg-positive patients (Fig. 1 d). Neither of these groups showed a significant HLA-binding activity.

## Anti-HLA Reactivity of Sera from Hydatidosis Patients

Sera from 32 untreated hydatidosis patients with surgically confirmed diagnoses were tested. Twenty-five of them were male and only three were women with two or more pregnancies. In about 80% of cases the cysts were localized in the liver. Of the 32 patients, 31 were found to possess an anti-HLA reactivity (against both class I and class II molecules) above the controls. No correlation was found with the presence of antibodies to *Echinococcus granulosus* (as determined by an IHA test). Patients with *Plasmodium falciparum* malaria, acute urinary schistosomiasis, or invasive amebiasis (20 patients for each group, all Italians infected in Africa) were also tested and found to be negative for anti-HLA reactivity.

## Specificity Controls

The radiolabeled preparations used as antigen were only 40%–60% pure. To exclude the possibility that contaminants in the labeled preparations were involved in the reaction, three positive and three negative heroin addict sera were incubated with class I or class II preparation and the resultant immune complexes were recovered by protein A-Sepharose and analyzed by SDS-PAGE. The patterns obtained (Fig. 2) conformed to the known dimeric structure of HLA molecules.

Several $^{125}$I-labeled molecules, including BSA, $\beta_2$-microglobulin, and four different glycoproteins $\alpha$-fetoprotein, luteinizing hormone, follicle stimulating hormone, and $\beta$-human chorionic gonadotropin) were tested with the same sera. No difference was found between heroin addict and control sera for binding of these molecules (Table 1). Similar results were obtained with hydatidosis sera.

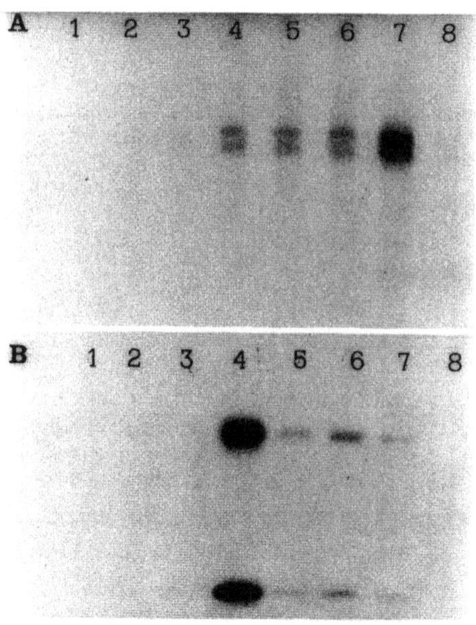


**Fig. 2A, B.** SDS-PAGE analysis of HLA molecules bound by heroin addict. **A** Class II preparation KT13; *lanes 1–3:* negative sera; *4–6:* positive sera; *7:* rabbit anti human class II serum; *8:* normal rabbit serum. **B** Class I preparation RM2; *lanes 1–3:* negative sera; *5–7:* positive sera; *4:* rabbit anti human class I serum; *8:* normal rabbit serum

**Table 1.** Reactivity of heroin addicts' sera against different antigens

| | | Labeled antigens | | | | | | |
|---|---|---|---|---|---|---|---|---|
| | | KT13 class II | BSA | $\beta_2$-m | AFP | LH | FSH | BHCG |
| No. | | 210 | 60 | 60 | 60 | 60 | 60 | 60 |
| Addicts' | CPM – | 3400[a] | 564 | 998 | 1579 | 1222 | 1105 | 1358 |
| | – SD | 1866 | 39 | 168 | 211 | 144 | 129 | 156 |
| No. | | 100 | 60 | 60 | 60 | 60 | 60 | 60 |
| Controls' | CPM – mean | 1220 | 551 | 1022 | 1412 | 1208 | 1078 | 1277 |
| | – SD | 200 | 45 | 159 | 223 | 163 | 112 | 139 |

[a] $P < 0.001$ by Student's *t*-test

## Lack of Allospecificity

As shown in Fig. 3, the hydatidosis sera react with four class II and two class I preparations of different phenotypes (class II: LG10: DR7; LG38: DR5; RM5: DR3; KT13: DR4; class I: RM2: A11, 24; B28, 44; C–; RPMI1788: A2; B7; C–). By pairwise comparisons, a very high correlation (*r* above 0.80) was always found between the reactivities against the different antigens. The same results were obtained with heroin addict sera.

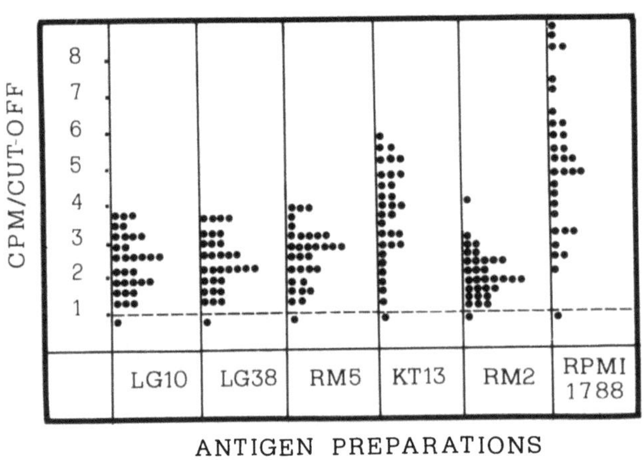


**Fig. 3.** Distribution of reactivities of sera from patients with hydatidosis against different antigenic preparations

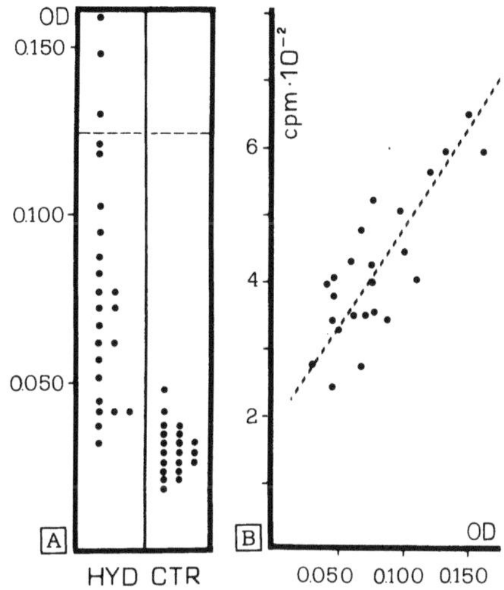


**Fig. 4. A** Distribution of levels *(OD)* of antibody to HIV of 24 serum samples from hydatidosis patients *(HYD)* and 20 normal controls *(CTR)*. The *dotted line* represents the cut-off threshold calculated according to the manufacturers' instructions.
**B** Correlation between the OD levels obtained with an ELISA test and cpm representing the reactivity against HLA class I molecules

## Interference in Anti-HIV Screening Test

An anti-HIV screening test (Abbott HTLV-III EIA kit) was performed on 24 hydatidosis sera. As shown in Fig. 4a, the OD values observed were significantly higher as compared with controls. Their distribution correlation ($r = 0.79$) with anti-HLA class I reactivity (Fig. 4b). Three sera (12.5%) were above the threshold for positivity in two repeated sets of analyses, but these positive results were not confirmed by using recombinant antigens (Abbott env/core recombinant, *E. coli* EIA HTLV-III confirmatory test). These results are of obvious importance for the interpretation of positivities in screening tests of sera from heroin addicts.

## Possible Biological Significance

The first problem to discuss is how the described anti-HLA reactivity is generated in these two apparently unrelated situations. A common underlying factor might be a continuous solicitation of the immune system. In heroin addicts this could be due to daily injections of mixtures of drugs and various additives under nonaseptic conditions, in hydatidosis patients to the release of parasitic antigens. Such continuous stimulation might enhance the synthesis and perhaps the shedding and elimination of antigen presenting molecules. According to a plausible model (Grabar 1983) an autoimmune response could follow as a consequence of this overproduction. If this is true the same phenomenon should be observed in other conditions involving continuous antigenic stimulation. In the mouse, autoreactive antibodies have been observed consistently after repeated injections of syngeneic virus-coated lymphocytes (Opolski et al. 1986; Kievits et al. 1987), although they possessed some alloreactive preference. In man, the appearance of anti-class II antibodies, devoid of alloantigenic specificity, in normal subjects after immunization with tetanus toxoid or *Candida albicans* antigens has been reported (Okudaira et al. 1982a). It remains to be explained why other parasitic diseases were negative.

A second, related problem is that of the consequences of the presence of anti-HLA reactivity in these subjects. Several disturbances of immune parameters have been described in heroin addicts, who are generally considered as immunologically hyporeactive (Nair et al. 1986). Impaired cellular immunity and hypergammaglobulinemia have been described in hydatidosis (Riley et al. 1986). An attractive hypothesis would be that these immunological alterations are related to the presence of autoreactive HLA antibodies, maybe by some interference with antigen presentation. The effect of these sera on in vitro lymphocyte responses is under study.

Even though the biological importance of this reactivity is blurred, its practical importance as a source of false-positive results in anti-HIV screening tests must be underlined. Due to the lack of allospecificity, high OD values are expected when the virus preparation is contaminated with HLA regardless of the phenotype of the cell in which the virus was grown, and thus the recurrence of antibodies may be more widespread than the interference by alloantibodies previously decribed by Kuhn et al. (1985). Considering the high frequency of anti-HLA reactivity in drug addicts, we regard it as mandatory to confirm all positivities in this group at risk with more specific tests.

*Acknowledgments.* This investigation was supported by NIH grant A1-20251, by the Regione Sardegna 37/138, by a NATO grant to R.S., and by Progetto Bilaterale CNR 8600265.04. We thank Dora Centis and Eleuteria Lancia for technical assistance.

## References

Ameglio F, Saba F, Bitti A, Aceti A, Tanigaki N, Sorrentino R, Dolei A, Tosi R (1987a) Anti-HLA class I and II reactivity in sera from patients with hydatidosis. J Infect Dis 156: 673–676

Ameglio F, Dolei A, Benedetto A, Sorrentino R, Tanigaki N, Tosi R (1987b) Antibodies reactive with non-polymorphic epitopes on HLA molecules interfere in screening tests for human immunodeficiency virus. J Infect Dis 156: 1034–1035

Ameglio F, Benedetto A, Marotta P, Centis D, Sorrentino R, Tanigaki N, Tosi R (1988) A high proportion of sera of heroin addicts possesses anti-HLA class I and class II reactivity. Clin Immunol Immunopathol 46: 328–334

Černý-Provaznik R, Radl J, van Mourik P, Leupers T, Brondijk B, de Greeve P, Ivanyi P (1985) Anti-MHC immunity detected prior to intentional alloimmunization. Nat Immun Cell Growth Regul 4: 138–159

Černý-Provaznik R, Kloosterman T, Verkerk D, Vroom TM, Ivanyi P (1986) Natural autoreactive H-2-specific serum antibodies in a group of BALB/cBy (H-$2^d$) mice. Tissue Antigens 27: 106–111

Grabar P (1983) Autoantibodies and the physiological role of immunoglobulins. Immunol Today 4: 337–341

Kievits F, Rocca A, Opolski A, Limpens J, Leupers T, Kloosterman T, Boerenkamp WJ, Pla M, Ivanyi P (1987) Induction of H-2 specific antibodies by injections of syngeneic Sendai virus-coated cells. Eur J Immunol 17: 27–35

Kuhn P, Seidl S, Holzberger G (1985) HLA DR4 antibodies cause positive HTLV-III antibody ELISA results. Lancet 1: 1222–23

Nair MPN, Lating TJ, Schwartz SA (1986) Decreased natural and antibody-dependent cellular cytotoxic activities in intravenous drug abusers. Clin Immunol Immunopathol 38: 68–78

Okudaira K, Goodwin JS, Williams RC (1982a) Anti-Ia antibody in sera of normal subjects after in vivo antigenic stimulation. J Exp Med 156: 255–267

Okudaira K, Searles RP, Jan JL, Goodwin JS, Williams RC (1982b) Anti Ia reactivity in sera from patients with systemic lupus erythematosus. J Clin Invest 69: 17–30

Opolski A, Kievits F, Ivanyi P (1986) Polymorphic and autoreactive H-2-specific monoclonal antibody isolated after injection of syngeneic Sendai virus-coated lymphocytes. Immunogenetics 24: 402–408

Riley EM, Dixon JB, Jenkins P, Ross G (1986) *Echinococcus granulosus* infection in mice: host responses during primary and secondary infections. Parasitology 92: 391–403

Romagnani S, Almerigogna F, Giudizi MG, Biagiotti R, Centis D, Alessi A, Ricci M, Tosi R (1985) Anti-Ia reactivity in sera of untreated patients with active Hodgkin's disease. Clin Immunol Immunopathol 34: 1–10

Tanigaki N, Tosi R (1982) Assessment of the specificity of human alloantisera and alloantigens by the use of radioiodinated human Ia antigens. Tissue Antigens 20: 1–21

Tongio MM, Folkenrodt A, Mitsuishi Y, Urlacher A, Bergerat JP, North ML, Mayer S (1985) Natural HLA antibodies. Tissue Antigens 26: 271–285

# Antisera to "Modifying Factor" in Patients with Ankylosing Spondylitis: The Present Situation

A. van Leeuwen[1], F. P. M. J. van Bree[1], C. J. Beukelman[2], and J. J. van Rood[1]

[1] Department of Immunohaematology and Bloodbank, University Hospital of Leiden, Rijnsburgerweg 10, 2333 AA Leiden, The Netherlands
[2] Department of Microbiology, University of Utrecht, Utrecht, The Netherlands

It is now 14 years since the first reports appeared on the close association between HLA-B27 and ankylosing spondylitis (AS) (Brewerton et al. 1973; Schlosstein et al. 1973) and 8 years since the Sydney group (Seager et al. 1979) published evidence for a "modifying factor" on the surface of lymphocytes of AS patients. Using sera of rabbits immunized with enteric organisms, such as *Klebsiella, Salmonella, Shigella,* or *E. coli,* Geczy and his colleagues were able to distinguish lymphocytes of patients suffering from AS and healthy, HLA-B27 positive, control lymphocytes. During the past 8 years many investigators have tried to reproduce these findings, however mainly without success. Table 1 provides an overview of the confirmatory and nonconfirmatory data.

It is remarkable that confirmatory reports are based on experiments carried out in the laboratory in Sydney, using reagents supplied by Geczy, but that on return to their own laboratory these findings could not be reproduced.

**Table 1.** Confirmatory and nonconfirmatory data regarding the detection of "modifying factor" in AS patients

| Author | Confirmation |
|---|---|
| Kinsella et al. (1983) | No |
| Beaulieu et al. (1983) | No |
| Archer et al. (1985) | Yes, London patients tested in Sydney |
| Struthers (1985) | Yes, in 1984 Birmingham patients tested in Sydney<br>No, in 1985 same patients tested in Birmingham |
| Van Rood et al. (1985) | Yes, Dutch patients tested in Sydney (three different shipments) |
| Geczy et al. (1986a) | Yes, Dutch patients tested in Leiden by Geczy |
| McGuigan et al. (1985) | Yes, New Zealand patients tested in Sydney |
| Singh et al. (1986) | No |
| Ivanyi (1987, personal communication) | No, Dutch patients tested in Sydney |

We decided to study the reproducibility of the "Geczy factor" using a different approach. We collected 26 Dutch AS patients and 19 healthy Dutch controls, selected for age and sex (AS is more frequent in male than in female patients) and shipped the cryopreserved cells in three shipments under code to Sydney where they were tested for the HLA-B27-associated determinant (Van Rood et al. 1985). The result of these studies was that Geczy's antibacterial rabbit sera were able to lyse specifically HLA-B27+/AS+ cells in a blind set-up. The cells used in this study were also tested for HLA-B27 subtypes W, K, and C (Breuning et al. 1985; Breur-Vriesendorp 1985). Seven cells (out of 26) of AS patients failed to react with the "Geczy factor." Three of these seven cells were of the HLA-B27K type, three of HLA-B27W type, and one was HLA-B27 negative. The three HLA-B27K cells were repeated when Geczy visited Leiden (see below) and they appeared to be positive in that study.

After the promising findings of the collaborative study in Sydney a follow-up was organized (Van Rood 1986). It was decided to invite Geczy to come to Leiden and to test with his own $^{51}Cr$ release lymphocytotoxicity assay coded fresh and cryopreserved lymphocytes from Dutch AS patients and controls. The results of this blind study showed that Geczy was able to reproduce his finding outside Sydney (Geczy et al. 1986a). Forty-five samples were tested, 24 AS+ patients (23 were HLA-B27+ and one was HLA-B27-) and 21 HLA-B27+ healthy controls, again selected for age and sex (Table 2). During Geczy's stay in Leiden he used his own antibacterial rabbit sera.

Next we decided to attempt to raise "active" antisera to "cross-reactive" bacteria. The group in Utrecht (head Prof. J.M.N.Willers) had already made various unsuccessful attempts at immunizing a great number of Dutch rabbits. It was suggested that the rabbits used in Australia were high responders for the "modifying factor." Geczy et al. (1986b) describe the "cross-reactive" antibody production not only in rabbits, but also in guinea pigs and mice. They show that during the last years over 50 animals have been immunized with either "cross-reactive" or "non-cross-reactive" bacteria and only three animals have been encountered whose naturally occurring serum cytotoxins gave unacceptable cytotoxicity values when tested on a panel of human lymphocytes. Therefore Geczy shipped several rabbits to Holland. Twelve Australian animals were immunized with enterobacteria directly isolated from the stools of Dutch HLA-B27 positive AS patients (Beukelman et al. 1987). One remarkable observation was the low cytotoxic activity in the Australian rabbits before immunization compared to Dutch rabbits. At least one serum, obtained from rabbit no. 16, had discriminating activity using a modified "micro" cytotoxicity assay. The

**Table 2.** Data on 45 samples of Dutch patients and normal controls, tested in $^{51}Cr$ release assay

| | | AS+ | AS− |
|---|---|---|---|
| Rabbit antisera | + | 22 | 0 |
| | − | 2[a] | 21 |

[a] One patient was HLA-B27+; one was HLA-B27-

**Table 3.** Techniques used for testing the antibacterial rabbit sera

| | Sydney method | Leiden method |
|---|---|---|
| Antiserum | 100 μl, undiluted | 10 μl, undiluted |
| Cell suspension | 100 μl, $^{51}$Cr labeled 2–5 × 10$^{6}$/ml | 10 μl, C-FDA labeled 5 × 10$^{6}$/ml |
| Incubation | 30 min 20 °C | 30 min 20 °C |
| Rabbit complement | 100 μl | 10 μl |
| Incubation | 60 min 20 °C | 60 min 20 °C |
| Stain | | 10 μl propidium Iodine 12.6 μg/ml |
| Incubation | | 5 min 20 °C |
| Reading | Supernatant is counted in gamma scintillation spectrometer | Dead cells are counted in automated fluorescence microscope |

test was modified in such a way that the respective volumes were comparable with the $^{51}$Cr release assay as described by Geczy (Table 3). Unfortunately only serum no. 16, collected 3 days after boosting, reacted positively. Serum collected 14 days after boosting appeared to be unreactive. The authors (Beukelman et al. 1987) suggest that the cytolytic activity depends either on antibodies with a short half-life, such as IgM, or on an antibody-independent principle, e.g., an acute phase reactant. The Utrecht group will continue the research into raising antibaterial rabbit sera.

Finally, the production of monoclonal antibodies directed against the "modifying factor" has to be mentioned (Møving, 1986, personal communication): this work is in progress.

In conclusion, 8 years after Geczy's first publication we have not been able to confirm his results unequivocally but continue attempting to do so.

*Acknowledgments.* This work was supported in part by grants from the Dutch Foundation for Medical Healthy Research (Medigon), which is subsidized by the Dutch Organization for the Advancement of Pure Research (ZWO) and the J. A. Cohen Institute for Radiopathology and Radiation Protection (IRS). The authors would like to thank Ingrid Curiël for typing the manuscript.

## References

Archer JR, Stubb MM, Currey HLF, Geczy AF (1985) Antiserum to *Klebsiella* K43 BTS 1 specifically lyses lymphocytes of HLA-B27 positive patients with ankylosing spondylitis from a London population. Lancet 1: 344–345

Beaulieu AD, Rousseau F, Israel-Assayag E (1983) *Klebsiella* related antigens in ankylosing spondylitis. J Rheumatol 10: 102–105

Beukelman CJ, van Leeuwen A, Aerts PC, van Kregten E, van Bree S, van Rood JJ, Willers JMN, van Dijk H (1988) Rabbit antisera to Enterobacteriaceae isolated from HLA-B27 positive patients with ankylosing spondylitis (AS) are lytic for the mononuclear cells of AS-patient. Br J Rheumatol (1988) Vol. 27, suppl. 2: 58–60

Breuning MH, Lucas CJ, Breur BS, Engelsma MY, de Lange GG, Dekker AJ, Biddison WE, Ivanyi P (1982) Subtypes of HLA-B27 detected by cytotoxic T lymphocytes and their role in self-recognition. Hum Immunol 5: 259–268

Breur-Vriesendorp BS, Huis B, Dekker AJ, Breuning MH, Ivanyi P (1985) Subtypes of antigen HLA-B27 (B27W and B27K) defined by cytotoxic T lymphocytes: Identification of a third subtype (B27C) prevalent in Oriental populations. In: Ziff M, Cohen SB (eds) The spondylarthropathies. Raven, New York, pp 55–65 (Advances in inflammation research, vol 9)

Brewerton DA, Caffrey M, Hart FD, James DCO, Nicolls A, Sturrock RD (1973) Ankylosing spondylitis and HLA-B27. Lancet 1: 904–907

Geczy AF, van Leeuwen A, van Rood JJ, Ivanyi P, Breur BS, Cats A (1986a) Blind confirmation in Leiden of Geczy factor on the cells of Dutch patients with ankylosing spondylitis. Hum Immunol 17: 239–245

Geczy AF, Upfold LI, Sullivan JS, Prendergast JK, McGuigan LE, Bashir HV, Edmonds JP (1986b) Rabbit antisera to cross-reactive organisms. Where's the problem? Ann Rheum Dis 45: 85–86

Kinsella TD, Fritzler MJ, McNeil DJ (1983) Ankylosing spondylitis. A disease in search of microbes. J Rheumatol 10: 1–4

McGuigan LE, Geczy AF, Prendergast JK, Pedmonds JP, Hart HH, Bashir HV (1985) HLA-B27 associated cross-reactive marker on the cells of New Zealand patients with ankylosing spondylitis. Ann Rheum Dis 45: 144–148

Schlosstein L, Terasaki PI, Bluestone R, Pearson CM (1973) High association of an HLA antigen, W27, with ankylosing spondylitis. N Engl J Med 288: 704–706

Seager K, Bahir HV, Geczy AF, Edmonds J, de Vere-Tyndall A (1979) Evidence for a specific B27-associated cell-surface marker on lymphocytes of patients with ankylosing spondylitis. Nature 277: 68–70

Singh B, Milton JD, Woodrow JC (1986) Ankylosing sondylitis, HLA-B27, and *Klebsiella:* A study of lymphocyte reactivity of anti-*Klebsiella* sera. Ann Rheum Dis 45: 190–197

Struthers GR (1985) Cross-reactivity of anti-*Klebsiella* K43 antiserum and lymphocytes from HLA-B27 positive patients with ankylosing spondylitis. Lancet 1: 764

Van Rood JJ (1986) HLA antigens as carrier molecules. Hum Immunol 16: 246–249

Van Rood JJ, van Leeuwen A, Ivanyi P, Cats A, Breur-Vriesendorp BS, Dekker-Saeys AJ, Kijlstra A, van Kregten E (1985) Blind confirmation of Geczy factor in ankylosing spondylitis. Lancet 2: 943–944

# Recognition by Human B Cells of Foreign Antigen in the Context of the MHC: a Review

E. Goulmy, A. van Leeuwen, and J. J. van Rood

Department of Immunohaematology and Bloodbank, University Hospital of Leiden, Rijnsburgerweg 10, 2333 AA Leiden, The Netherlands

The experimental evidence for the existence of *human* B cells that recognize foreign antigen in the context of the MHC is very restricted, not to say limited to only one example (Van Leeuwen et al. 1979), at least with regard to the known nature of the MHC allele and the antigen involved. On the other hand in exerimental animal models, the number of antigen specific MHC restricted antibodies reported (see the detailed descriptions by different authors in this volume) greatly exceeds one, though it does not reach overwhelming levels. Nevertheless, the independent observations strongly support the notion that MHC restricted antibodies do exist, although with special regard to the human studies more data are needed to strengthen the evidence for the existence of such antibodies in man.

The assumption that some B cells are capable of generating MHC restricted antigen specific antibodies is based on our earlier observation. More than one decade has now elapsed since participation of the HLA molecules in the recognition of foreign antigen by cytotoxic T-lymphocytes (CTLs) was reported (Goulmy et al. 1976). Although such recognition is characteristic for T cells, a careful analysis of the serum of the female patient whose lymphocytes were able to demonstrate HLA-A2 restricted H-Y specific cytotoxic activity revealed B cell mediated activity comparable with the cytotoxic T cell responses (Van Leeuwen et al. 1979). It became apparent that lympholytic antibodies were present in the serum which reacted with some mononuclear cells labeled with anti-Ig. The antibody activity was found in the IgM fraction. The serum was tested against a panel of lymphocytes of healthy male and female donors either positive or negative for the restricting molecule (i.e., HLA-A2). The results demonstrated that all HLA-A2 positive male donors were reacting with this antiserum, whereas the lymphocytes of HLA-A2 negative male or female donors were negative in the complement dependent cytotoxicity assay. Nevertheless, of the HLA-A2 positive female donors, two gave a borderline positive reaction. It should be noted that the lymphocytes of *one* of the latter females were also borderline lysed by the HLA-A2 restricted anti-H-Y CTLs in the CML assay. Comparative analysis between the outcome of the CML assay and the reactions obtained with the antiserum showed striking similarity (Table 1). Because of the almost perfect correlation between the reactivity patterns of the CTLs and the antibody, it is reasonable to assume that the HLA-A2 restricted antibody recognizes the same epitopes on the HLA-A2 and on the H-Y antigen as do the HLA-A2 restricted CTLs obtained from the same individual. Since "variants" exist both at the level of the antigen

**Table 1.** Correlation between the lysis patterns by the CTLs and by the antiserum, both derived from the same individual, tested against several healthy male and female individuals

| Cytotoxic activity as exhibited by | Panel cells tested by CTLs and antibody: | | |
|---|---|---|---|
| | A2+ males[a] ($n = 20$) | A2+ females ($n = 7$) | A2− females, A2− males ($n = 16$) |
| CTLs | 20 | 1[b] | 0 |
| Antibody | 20 | 2 | 0 |

[a] A2+ = HLA-A2 positive; A2− = HLA-A2 negative
[b] Only one of the two HLA-A2+ female cells tested by both CTLs and antibody gave a borderline positive reaction with the H-Y specific CTLs

H-Y and at the level of the restricting molecule HLA-A2, our hypothesis could be evaluated.

To approach the question of whether CTLs and antibody recognize similar structures on the H-Y molecule, we studied the chromosomes and the reaction patterns of a small series of individuals with abnormal sex chromosome constitutions (Goulmy et al. 1983). It should be emphasized that the number of cases which could be studied was limited since the reagents are applicable only to those individuals who carry the appropriate MHC restricting molecule. The picture that arose from the latter study confirmed the earlier observed correlation between the reactivity patterns of CTLs and antibody. Moreover, the combined analysis (each showing the same discriminatory patterns) revealed a correlation with the presence of the cytologically detectable Y-chromosome material but not with the presence of testes. So far we have assumed that the antigen recognized by both the B cells and the T cells is most probably identical with the classical transplantation antigen H-Y. Nevertheless, we have not yet reached the point at which we would be justified in making statements concerning the nature of the H-Y antigen as recognized by CTLs and antibody. Current examination of sex-reversed humans by combined analyses of different sets of Y-DNA probes and H-Y specific CTLs will hopefully resolve this issue (Simpson et al. 1987).

To elaborate on the assumption that HLA-A2 restricted B cells use the same MHC epitopes as HLA-A2 restricted T cells for the recognition of the "MHC + X" complex, functional analyses of an established MHC class I "variant" HLA-A2 molecule were carried out. For this purpose serological HLA-A2 positive lymphocytes from a male donor designated M7 were used. Biddison et al. (1980 a) demonstrated, by the use of HLA-A2 restricted influenza virus immune T cells, a lack of recognition of virus infected HLA-A2 positive lymphocytes from donor M7. Additionally, biochemical analysis by means of isoelectric focusing of the HLA-A2 molecule from donor M7 demonstrated a difference in the HLA-A2 heavy polypeptide chain when compared with the HLA-A2 molecules of other donors (Biddison et al. 1980 b).

The lymphocytes from donor M7 were subsequently analyzed in our laboratory using both the HLA-A2 restricted H-Y antigen specific CTLs and antibody. Table 2 shows that HLA-A2 restricted anti-H-Y CTLs *failed* to lyse the lymphocytes from

**Table 2.** Differential recognition of HLA-A2 variant cells (M7) by HLA-A2 restricted H-Y antigen specific CTLs and antibody.

| Target Cells | HLA-A2 restricted anti-H-Y | |
|---|---|---|
| | CTLs | Antibody |
| HLA-A2+ male cells | 64[a] | 45[a] |
| HLA-A2+ male cells | 67 | 41 |
| HLA-A2+ male *M7* cells | 1 | 36 |
| HLA-A2+ female cells | 3 | 8 |

[a] % cytotoxicity

donor M7. This reaction pattern is comparable to the absence of recognition using HLA-A2 restricted influenza virus specific CTLs (Biddison et al. 1980 a). However, the HLA-A2 restricted anti-H-Y antiserum *did react* with the M7 cells to an extent comparable to that in other HLA-A2 positive male donors. It is clear from these observations not only that alloimmune HLA-A2 specific antisera and alloimmune HLA-A2 specific CTLs might use different epitope(s) on the same HLA-A2 molecule for allorecognition (Horai et al. 1982) but also that "monospecific" anti-H-Y antiserum and "monospecific" anti-H-Y CTLs (both derived from the same individual) apparently use different epitopes on the HLA-A2 molecule for associative recognition of the foreign antigen. Thus, the H-Y specific antibody recognizes self-determinants on the HLA-A2 molecule that are distinct from those that are seen by H-Y specific CTLs.

Lately, additional information has become available on the more precise location of the target structures for CTLs on the HLA-A2 molecule. Combined investigation (resulting from a collaborative effort) of the HLA-A2 variant molecules at the molecular (Ezquerra et al. 1986) and at the functional (Goulmy et al. 1984) level, demonstrated that amino acid changes at position 43 and in the residues 145–157 (i.e., cellular defined subtypes HLA-A2.2 and HLA-A2.3) lead to the loss of epitope(s) necessary for associative recognition of the H-Y antigen by HLA-A2 restricted CTLs. Interestingly enough, a single amino acid change at position 9 in the heavy chain of the HLA-A2 molecule (i.e., cellular defined subtype HLA-A2.4) does not affect the recognition of H-Y by HLA-A2 restricted CTLs.

Obviously, it is of interest to examine the reactivity patterns of the cells of the individuals expressing the cellular defined HLA-A2.3 and HLA-A2.4 subtypes (donor M7, see Table 2, carries the *HLA-A2.2* subtype) with our HLA-A2 restricted H-Y specific antibody. Unfortunately, the amounts of the H-Y specific antibody at our disposal were small; the H-Y antibody was only present in the patient's serum for 1 year after bone marrow transplantation, and the activity declined thereafter.

At present, attempts are being made at the in vitro production of the H-Y antibody by Epstein-Barr virus transformed B cells derived from the original H-Y producing female patient.

*Acknowledgments.* The authors would like to thank Els Blokland and Jos Pool for their technical assistance and Ingrid Curiël for typing the manuscript. This work was supported in part by grants from the Dutch Foundation for Medical Health Research (Medigon), which is subsidized by the Dutch Organization for the Advancement of Pure Research (ZWO), and the J. A. Cohen Institute for Radiopathology and Radiation Protection (IRS).

## References

Biddison WE, Ward FE, Shearer GM, Shaw S (1980 a) The self determinants recognized by human virus immune T cells can be distinguished from the serologically defined HLA antigens. J Immunol 124: 548–552

Biddison WE, Krangel MS, Strominger JL, Ward FE, Shearer GM, Shaw S (1980 b) Virus immune cytotoxic T cells recognized structural differences between serologically indistinguishable HLA-A2. Hum Immunol 1: 225–232

Ezquerra A, Doménech N, van der Poel JJ, Strominger JL, Vega MA, Lopéz de Castro JA (1986) Molecular analysis of an HLA-A2 functional variant C1a defined by cytolytic T lymphocytes. J Immunol 137: 1642–1649

Goulmy E, Termijtelen A, Bradley BA, van Rood JJ (1976) Alloimmunity to human H-Y. Lancet 2: 1206

Goulmy E, van Leeuwen A, Blokland E, Sachs ES, Geraedts JPM (1983) The recognition of abnormal sex chromosome constitution by HLA-restricted anti-H-Y cytotoxic T cells and antibody. Immunogenetics 17: 523–531

Goulmy E, van der Poel JJ, Giphart M, van Rood JJ (1984) Analysis of the functional epitopes on different HLA-A2 molecules. Immunogenetics 20: 13–21

Horai S, van der Poel JJ, Goulmy E (1982) Differential recognition of the serologically defined HLA-A2 antigen by allogeneic cytotoxic T cells. I. Population studies. Immunogenetics 16: 135–142

Simpson E, Chandler P, Goulmy E, Disteche CM, Ferguson-Smith MA, Page D (1987) Separation of the genetic loci for the H-Y antigen and for testis determination on human Y chromosome. Nature 326: 876–878

Van Leeuwen A, Goulmy E, van Rood JJ (1979) Major histocompatibility complex-restricted antibody reactivity mainly, but not exclusively directed against cells from male individuals. J Exp Med 150: 1075–1083

# Searching for MHC-Restricted Antibodies: Antibodies Induced by Injections with Syngeneic Cells Coated with Sendai Virus, Trinitrophenyl, and Xenogeneic $\beta_2$-Microglobulin Are Not Restricted by the Mouse MHC

FEMIA KIEVITS, WALTER J. BOERENKAMP, and PAVOL IVANYI

Central Laboratory of the Netherlands Red Cross Blood Transfusion Service, and Laboratory for Experimental and Clinical Immunology, University of Amsterdam, Amsterdam, The Netherlands

## Introduction

The capacity of the immune system to discriminate among an enormous number of different antigens is mediated by two distinct but interactive subpopulations of lymphocytes, B and T cells. These subpopulations differ profoundly in the way they recognize foreign antigens. B cells use immunoglobulins (Ig) as cell surface receptors for recognition of free antigen whereas the Ig-like receptors on T cells (TcR) recognize foreign antigens presented by their own major histocompatibility complex (MHC) gene products, a phenomenon known as MHC restriction (Zinkernagel and Doherty 1979). Another major difference exists between the structure and recognition of Ig and the TcR. Immunoglobulins are composed of two heavy and light chain disulfide-linked hetrodimers while one set of disulfide-linked $\alpha$- and $\beta$-glycoproteins forms the functional TcR (Kappler et al. 1983). The genomic organization of the genes for B (Tonegawa 1983; Honjo 1983) and T (Gascoigne et al. 1984; Hayday et al. 1985) cell receptors are similar and it is likely that the gene families originate from the same ancestral gene (Hood et al. 1985). Recently, it has been suggested that the T cell $\alpha\beta$ receptor molecule has an antigen specific binding site that is fundamentally not different from the conventional binding site on antibodies (Novotny et al. 1986). The structural basis for the difference in the mode of antigen recognition between B and T cells might therefore be in the structure of the antigen that is recognized. Thus while the epitopes that occupy the binding sites of immunoglobulins are derived entirely from the antigen, the epitopes recognized by T cell receptors may be derived in part from the nominal antigen and in part from the restricting MHC molecule and can be seen as a neodeterminant.

Some papers, however, described, in experiments analogous to the classical demonstrations of T cell restriction by Zinkernagel and Doherty, how B cells are MHC restricted in their interaction with either T cells or macrophages (Gorczynski et al. 1980; Katz et al. 1978; Nisbet-Brown et al. 1981). Better evidence for the possibility that B cell and T cell receptors can see the same antigenic determinants may come from antibodies that are restricted to react with self-MHC plus foreign antigen (MHC+X) rather than to react with self-MHC or X alone.

Despite extensive efforts by several investigators (Kievits et al. 1987 a; Opolski et al. 1986; Tamminen et al. 1987; Pestalozzi et al. 1987; Schmidt 1987) to induce and isolate MHC-restricted antibodies, only one laboratory (Wylie et al. 1982; Froscher and Klinman 1986) has provided evidence for the existence of this type of antibody. Some other data supported the existence of MHC-restricted antibodies, but these data have not been extended further (Van Leeuwen et al. 1979; Levy et al. 1984; Ivanyi et al. 1979). Here we describe three studies on antibody response in mice to immunizations with self-MHC+X. Syngeneic spleen cells were modified either by Sendai virus (SV), trinitrophenyl (TNP), or foreign $\beta_2$-microglobulin ($\beta_2$m) and injected into mice over several weeks. Immune serum from each indivudual mouse was tested for the presence of antibodies that were antigen specific but restricted in their binding by self-MHC molecules. Our results do not support the proposal that MHC-restricted recognition of foreign antigens is a part of the B cell repertoire. Nevertheless, some interesting and unexpecting findings emerged from these experiments.

## Materials and Methods

### *Mice*

All mice were obtained from the Animal House of our Institute and bred under specific pathogen-free conditions.

### *Immunization*

Mice were injected i.p. weekly with $10^7$ spleen cells as described in detail elsewhere (Kievits et al. 1987a). Spleen cells were coated with SV (Kievits et al. 1987b) or treated with TNP (Levy et al. 1980) in the same manner as described for the preparation of target cells in TNP- or SV-specific cytotoxic T-lymphocyte (CTL) assays. Lymphoblasts were prepared by incubation of spleen cells with 5 μg concanavalin A (ConA) per milliliter for 2 days in culture medium supplemented with 10% fetal calf serum (FCS) or 5% syngeneic mouse serum (SMS). For injection, blast cells were washed 3 times with phosphate-buffered saline (PBS).

### *Serology*

The microlymphocytotoxicity test (MLCT) on target cells was performed in medium supplemented with 0.1% BSA, 2% SMS, or 10% FCS, as described previously (Kievits et al. 1987a). For serum absorption experiments with FCS, 1 μl of serum dilutions was mixed with 1 μl of FCS dilutions in Terasaki plates and incubated for 30 min at 20 °C. After this period, 1 μl of cell suspension ($2.5 \times 10^6$/ml) was added to the wells and the routine MLCT was continued. Absorptions of sera with various cell types or with SV were performed as described elsewhere (Kievits and Ivanyi 1987; Opolski et al. 1986). The monoclonal antibodies (McAbs) 28.14.8S (anti-H-$2D^{b,q}L^{d,q}$), 27.11.13 (anti-H-$2D^b$,$D^d$,q), and 36.7.5S (anti-H-$2K^k$) were a gift of Dr. D. Sachs (National Cancer Institute, Immunology Branch, National Institute of Health, Bethesda MD). Fluorescence experiments were performed as described elsewhere (Kievits et al. 1987a) and binding was analyzed on an Epic-C flow cytometer. $\beta_2$m was partially purified from FCS on the principle of molecular sieving and was labeled with $^{125}$I by the Iodogen method (Hudson and Hay 1980). Exchange conditions of murine $\beta_2$m have been described previously (Kievits and Ivanyi 1987). Usually, $2 \times 10^6$ spleen cells were incubated with 10 µl of partially purified bovine $^{125}$I-$\beta_2$m (~500 ng of labeled proteins) for 2 h at 37 °C in a total volume of 200 µl medium containing 0.1% ovalbumin.

## Results

### *Syngeneic Immunization with SV-Coated Cells*

Results of this study have already been published in detail (Kievits et al. 1987 a) and are briefly summarized here. Seventy-one C57B1/6 (B6, H-2$^b$) mice were injected with syngeneic SV-coated (SV+) spleen cells over 16 weeks. Of the 71 injected mice, 51% (36) produced antibodies cytotoxic for normal (SV−) lymphocytes (Table 1). However, when the syngeneic immune sera were tested on a panel of SV+ lymphocytes, 89% (63) of the sera were positive. The reactivity of the cytotoxic

**Table 1.** Reactivity of immune sera from H-2$^b$ (B6, B10) and H-2$^k$ (B10.BR) mice injected with syngeneic SV-coated (SV+), TNP-treated (TNP+), or ConA-stimulated spleen cells tested on a panel of eight H-2 congeneic mouse strains with the B10 background

| Immunization | No. of sera tested | Targets[a] | No. of lymphocytotoxic sera[b] H-2 haplotypes | | | | | | | | Total positive sera |
|---|---|---|---|---|---|---|---|---|---|---|---|
| | | | b | d | f | k | p | q | r | s | |
| B6←B6 SV+[c] | 71 | SV+ | 39 | 23 | 29 | 26 | 32 | 33 | 30 | 13 | 63 |
| | | SV− | 26 | 15 | 16 | 3 | 14 | 24 | 4 | 2 | 36 |
| B6←B6 ConA$^{fcs}$ | 10 | ConA$^{fcs}$ | 7 | 3 | 7 | 6 | 8 | 8 | 5 | 1 | 8 |
| | | ConA$^{sms}$ | 0 | 0 | 0 | 0 | 0 | 0 | 0 | 0 | 0 |
| B6←B6 ConA$^{sms}$ | 10 | ConA$^{fcs}$ | 0 | 0 | 0 | 0 | 0 | 0 | 0 | 0 | 0 |
| | | ConA$^{sms}$ | 0 | 0 | 0 | 0 | 0 | 0 | 0 | 0 | 0 |
| B10←B10 TNP[c] | 20 | TNP+ | 20 | 20 | 20 | 20 | 20 | 20 | 20 | 20 | 20 |
| | | TNP− | 0 | 0 | 0 | 0 | 0 | 0 | 0 | 0 | 0 |
| B10.BR←B10.BR TNP[c] | 10 | TNP+ | 10 | 10 | 10 | 10 | 10 | 10 | 10 | 10 | 10 |
| | | TNP− | 0 | 0 | 0 | 0 | 0 | 0 | 0 | 0 | 0 |

[a] Spleen cells were coated with SV, treated with TNP, or stimulated with ConA in the presence of 10% FCS or 5% SMS

[b] 60%–100% of cells were lysed in a dilution of 1:2 or higher

[c] Immunizations were performed without any contact between the injected cells and exogenous serum components

**Table 2.** Absorption of lymphocytotoxic antibodies with various cell types from serum pools[a] of mice immunized with syngeneic SV-coated, TNP-treated, or ConA-stimulated spleen cells

| Immunization | Absorption[b] by | Targets[b] tested | Cytotoxic titer[c] H-2 haplotypes | | |
|---|---|---|---|---|---|
| | | | b | q | k |
| B6←B6 SV+ | None | SV− | 32 | 64 | 2 |
| | | SV+ | 128 | 256 | 32 |
| | | FLU+ | 128 | 256 | 32 |
| | H-2$^{b}$ SV− | SV− | 2 | 4 | ·[d] |
| | | SV+ | 8 | 32 | · |
| | H-2$^{q}$ SV− | SV− | 0 | 0 | · |
| | | SV+ | 4 | 8 | · |
| | HRBC SV+ | SV+ | 128 | · | · |
| | Daudi SV+ | SV+ | 128 | · | · |
| | SV coated on plastic | SV+ | 128 | · | · |
| B6←B6 ConA$^{fcs}$ | None | SC$^{fcs}$ | 8 | 32 | · |
| | Human cells | SC$^{fcs}$ | 8 | 32 | · |
| | Bovine cells | SC$^{fcs}$ | 0 | 0 | · |
| | H-2$^{b}$SMS | SC$^{fcs}$ | 8 | 32 | · |
| | H-2$^{b}$FCS | SC$^{fcs}$ | 0 | 0 | · |
| | H-2$^{q}$SMS | SC$^{fcs}$ | 8 | 32 | · |
| | H-2$^{q}$FCS | SC$^{fcs}$ | 0 | 0 | · |
| | FCS undiluted | SC$^{fcs}$ | 0 | 0 | · |
| | FCS 1:16 | SC$^{fcs}$ | 4 | 2 | · |
| | FCS 1:128 | SC$^{fcs}$ | 8 | 32 | · |
| B10←B10 TNP[e] | None | TNP+ | 128 | · | 128 |
| | H-2$^{b}$ TNP− | TNP+ | 64 | · | 128 |
| | TNP+ | TNP+ | 0 | · | 0 |
| | H-2$^{k}$ TNP− | TNP+ | 128 | · | 128 |
| | TNP+ | TNP+ | 0 | · | 0 |
| | Human cells | | | | |
| | TNP− | TNP+ | 128 | · | 64 |
| | TNP+ | TNP+ | 4 | · | 2 |
| | Sepharose | TNP+ | 64 | · | 64 |
| | Sepharose TNP | TNP+ | 0 | · | 0 |

[a] Serum pools were aliquots of lymphocytotoxic sera from one immunization group

[b] For absorption and target cells, human, bovine, or mouse lymphocytes, Daudi cells, or human red blood cells (HRBC) were coated with SV or FLU, treated with TNP, or incubated for 2 h at 37 °C in medium suplemented with 10% FCS or 5% NMS

[c] Reciprocal of the highest serum dilution with ≧60% cell lysis

[d] Not tested

[e] Similar results were obtained in absorption experiments with pool serum for the B10.BR←B10.BR TNP immunization group

sera ranged from narrow, reactive with one or two haplotypes, to widely cross-reacting with six, seven or eight haplotypes. Most of the positive sera were reactive with more than one haplotype, with an apparent preference for H-$2^{b,q,f,p}$ haplotypes. Autoreactivity (H-$2^{b}$) was not present in each positive serum and never occurred as the only specificity. Therefore, autoreactivity did not appear as the leading specificity of the cytotoxic sera and can be interpreted as being due to cross-reactions of alloreactive antibodies. This was supported by the finding that the lymphocytotoxicity was better absorbed from the sera by allogeneic cells than by the same quantity of syngeneic cells (Table 2). To determine the specificity of the cytotoxic antibodies in more detail, studies on F2 hybrids and H-2 recombinant strains, as well as capping and immunoprecipitation experiments, were performed (Kievits et al. 1987 a). Taken together, the results show that the lymphocytotoxic antibodies in sera from mice immunized with SV+ syngeneic spleen cells were directed against H-2 class I molecules. No class II specific antibodies were detectable. Moreover, Tables 1 and 2 show an enhanced recognition of public H-2 determinants on SV+ cells. However, the lysis of target cells coated with influenza virus (FLU) showed that the increased recognition of SV+ target cells was not specific for SV (Table 2). Absorption experiments with SV stuck onto solid material showed that the lymphocytotoxic antibodies were not directed against Sendai viral antigens (Table 2). This type of H-2 class I specific lymphocytotoxic antibodies was only present in B6 mice immunized with syngeneic SV+ cells and was not present in mice immunized either with SV or with syngeneic cells.

In conclusion, injections with syngeneic SV+ spleen cells regularly induced H-2 specific lymphocytotoxic antibodies but no MHC-restricted antiviral antibodies.

### *Syngeneic Immunization with ConA Blast Cells Cultured in the Presence of FCS*

We (Kievits and Ivanyi 1987) and other investigators (Kahn-Perles et al. 1987; Jefferies and MacPherson 1987; Mierau et al. 1987) have shown that binding of bovine $\beta_2$m to mouse heavy chains resulted in the formation of serologically detectable new determinants. Here we investigated whether these newly formed determinants are able to trigger the production of MHC-restricted antibodies that would specifically recognize epitopes elicited by association of self-class I heavy chains with bovine $\beta_2$m. C57B1/6 mice were injected with syngeneic ConA-stimulated spleen cells cultured in the presence of 10% FCS (ConA$^{fcs}$). Sera from these mice were tested for lymphocytotoxicity on a panel of ConA$^{fcs}$ blast cells obtained from eight H-2 congeneic mouse strains. The results were compared with the results from sera obtained from mice injected with syngeneic ConA blast cells cultured in the presence of SMS (ConA$^{sms}$). Table 1 shows that eight out of ten sera from mice immunized with syngeneic ConA$^{fcs}$ blasts contained antibodies cytotoxic for allogeneic and syngeneic target cells. In contrast, no cytotoxicity was present in sera from mice injected with syngeneic ConA$^{sms}$ blast cells. Because the eight B10 congeneic strains differ only in their H-2 haplotype, the polymorphic reaction pattern of the induced cytotoxic antibodies suggests a specific recognition of H-2 molecules. This was confirmed by the precipitation of two peptides with apparent relative masses of 46000 and 12000 from H-$2^{b}$ ConA$^{fcs}$ cell lysates. Preclearing of the cell lysates with anti-

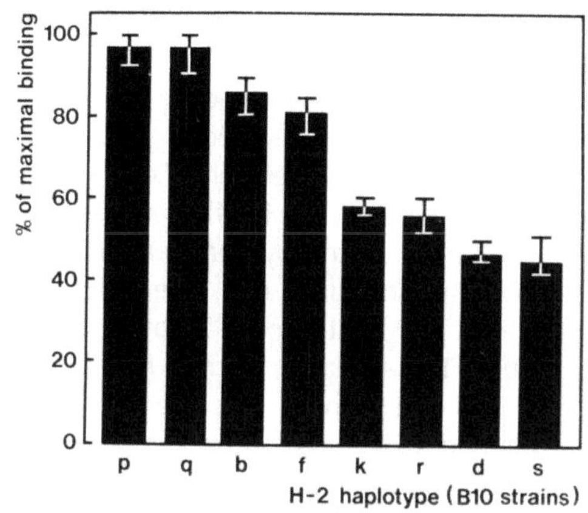


**Fig. 1.** Binding of partially purified bovine $^{125}$I-$\beta_2$m to spleen cells of various B10 congeneic mice, expressed as mean values of three separate experiments each performed in triplicate. The maximal binding value (100%) in each experiment is given to that H-2 haplotype with the highest binding of bovine $^{125}$I-$\beta_2$m. Binding of $\beta_2$m to other H-2 haplotypes is expressed as percentage of maximal binding. *Lines* indicate means of highest and lowest values. Experiments were performed in the presence of 0.1% ovalbumin

H-$2^b$ class I McAbs prevented the precipitation of both peptides, indicating that they were typical class I molecules (results not shown). No cytotoxicity with ConA blasts (Table 1) or precipitation (results not shown) was found in FCS-free conditions. To assign more precisely the class I specificity, absorption experiments were performed (Table 2). Absorption of the immune sera with FCS eliminated the cytotoxic titer to ConA$^{fcs}$ blasts completely, suggesting that the cytotoxic antibodies were directed against FCS specific determinants and not to new determinants formed by the association of self H-$2^b$ heavy chain with bovine $\beta_2$m. Nevertheless, interestingly, the lymphocytotoxic antibodies induced by syngeneic immunization with ConA$^{fcs}$ blasts and directed against bovine $\beta_2$m exhibited an H-2 polymorphic reaction pattern on a panel of various B10 congeneic mice (Table 1). To determine whether this type of polymorphism was due to the capacity of H-2 molecules to associate with bovine $\beta_2$m, lymphocytes of the B10 panel were incubated with $^{125}$I-labeled partially purified bovine $\beta_2$m. Figure 1 shows that the binding of bovine $\beta_2$m is dependent on the H-2 haplotype. The ability to exchange endogenous $\beta_2$m for bovine $\beta_2$m corresponds with the variations in the reaction pattern of the cytotoxic sera. Strains with haplotypes with a high capacity to displace mouse $\beta_2$m and to bind bovine $\beta_2$m (H-$2^{q,p}$) are therefore more intensely recognized by the induced anti-bovine $\beta_2$m antibodies than are strains that displace $\beta_2$m to a lesser extent (H-$2^{d,s}$). Taken together, the results show that syngeneic ConA$^{fcs}$ blasts are not able to trigger the induction of anti-bovine $\beta_2$m MHC-restricted antibodies but trigger the production of bovine $\beta_2$m specific antibodies that bind the free antigen. These antibodies mimic H-2 specific antibodies by precipitation of class I molecules and by a "polymorphic" reaction pattern on a panel of H-2 congeneic target cells. A detailed study of the antibodies showed that the lymphocytotoxicity was directed against bovine $\beta_2$m. The "polymorphic" reaction pattern was due to an H-2 dependent quantitative difference in the capacity to exchange mouse $\beta_2$m for bovine $\beta_2$m.

*Syngeneic Immunization with TNP-Modified Spleen Cells*

Cytotoxic T-lymphocyte responses to self-MHC + TNP have been described for H-$2^b$ mice. The H-$2^b$ + TNP specific CTLs often show cross-reactions with normal (unmodified) H-$2^k$ target cells (Levy et al. 1980). Such cross-reactions have been explained (Levy et al. 1984) as cross-reactions between alloantigens and MHC determinants induced by treatment with TNP (self-MHC + TNP = allo). This explanation was supported by the observation that some anti-H-$2^k$ McAbs were able to bind H-$2^b$ TNP treated cells although they remain negative on normal cells (Levy et al. 1984). The similarity in cross-reactions exhibited by both CTLs and antibodies suggested that T and B cells may recognize the same antigenic (neo)determinants formed by MHC + TNP. Therefore, we investigated whether injections with TNP-modified syngeneic (H-$2^b$) spleen cells could induce anti-TNP antibodies that were restricted by self-MHC (H-$2^b$) and cross-reactive with H-$2^k$ target cells without TNP (H-$2^b$ + TNP = allo H-$2^k$). The humoral immune response to injections of TNP-treated syngeneic spleen cells in B10 (H-$2^b$) was compared to the response in B10.BR (H-$2^k$) mice. After 16 weeks of immunization, sera contained antibodies that were not reacting with normal or ConA-stimulated spleen cells but were cytotoxic for spleen cells treated with TNP. Table 1 shows that the induced lymphocytotoxic antibodies reacted strictly with TNP-treated cells and recognized no H-2 polymorphism because they reacted almost equally with all TNP-treated cells of the panel. To further analyze the specificity of the syngeneic antibodies, absorption experiments with normal or TNP-modified syngeneic and allogeneic spleen cells and with TNP bound to Sepharose were performed. From the experiments (Table 2) it became clear that the syngeneic cytotoxic antibodies recognized only TNP specific determinants and were not directed against neodeterminants induced by treatment with TNP. This implies that if neodeterminants of self-MHC + TNP were formed (and recognized by T cells), they were not seen by B cells. However, Table 3 shows that treatment of cells with TNP really resulted in a modified exposure of H-2 class I allodeterminants. H-2 specific antibodies bound to various TNP-coated cells that remain negative without TNP treatment. Hence, although TNP treatment seems to

**Table 3.** Induction of modified exposure of allogeneic determinants on spleen cells after treatment with TNP

| McAb | Specificity | H-2 hyplotypes | | | | | | | |
|---|---|---|---|---|---|---|---|---|---|
| | | k | b | d | q | p | f | r | s |
| 36.7.5S | TNP− | 4[a] | 0 | 0 | 0 | 0 | 0 | 0 | 0 |
| | TNP+ | 4 | 3 | 0/1 | 2 | 1 | 1 | 0/1 | 3 |
| 28.14.8S | TNP− | 0 | 3 | 3 | 4 | 0 | 0 | 0 | 0 |
| | TNP+ | 2 | 3 | 3 | 4 | 1 | 2 | 2 | 2 |
| 27.11.13S | TNP− | 0 | 3 | 3 | 3 | 0 | 0 | 0 | 0 |
| | TNP+ | 2 | 4 | 3 | 4 | 0/1 | 2 | 2 | 2 |

[a] Figures indicate crude classification of fluorescence as measured by flow cytometry; 0 = background, 4 = highest intensity with 100% of positive cells

induce a wide range of allospecific determinants, antibodies recognizing those determinants were not induced by injections of syngeneic TNP-treated cells.

## Discussion

Cytotoxic T-lymphocytes directed against SV, TNP, and exogenous $\beta_2$m on syngeneic cells have combined specificity for the antigen and for products of the K and/or D region of the H-2 complex (Levy et al. 1980; Kast et al. 1984; Kurtz et al. 1985). Here we described three models in which we searched for antibodies having the same combined specificity in recognizing their antigens as the antigen receptors on CTLs. The existence of MHC-restricted antibodies was supported by a limited number of reports referring to MHC and influenza virus (Wylie et al. 1982), SV40 (Froscher and Klinman 1986), TNP (Levy et al. 1984), human H-Y (Van Leeuwen et al. 1979), and murine leukemia virus (Ivanyi et al. 1979) antigens.

Especially the results with SV40 showed that MHC-restricted antibodies can be induced in H-$2^b$ mice as a major part of the humoral immune response against injections with syngeneic virus-infected cells. We immunized H-$2^b$ mice with three types of syngeneic spleen cell, namely cells coated with SV, cells treated with TNP, and cells stimulated with ConA in culture medium containing FCS. It is known that the three cell types express cell surface structures that are able to induce antigen specific CTLs and that can function as target structures for CTLs. Two cell types, ConA blasts and TNP-treated spleen cells, have beside these T cell specific structures also unique structures on their membrane that can be recognized by anti-class I antibodies but are dependent on the presence of the foreign antigen (Levy et al. 1984; Kievits and Ivanyi 1987; Kahn-Perles et al. 1987; Jefferies and MacPherson 1987; Mierau et al. 1987). ConA blasts cultured in FCS will exchange endogenous $\beta_2$m for bovine $\beta_2$m from the medium. The association of H-$2^b$ class I heavy chains with bovine $\beta_2$m results in the induction of new allogeneic determinants that are serologically detectable (Kievits and Ivanyi 1987; Kahn-Perles et al. 1987; Jefferies and MacPherson 1987; Mierau et al. 1987). If H-$2^b$ lymphocytes are treated with TNP, similar allogeneic determinants are induced that are detectable with an H-$2K^k$ specific McAb (Levy et al. 1984). However, after immunization of mice with one of the two cell types, antibodies were obtained that were specific for bovine $\beta_2$m or TNP but not for determinants formed by H-$2^b$ plus the antigen.

Antibodies specific for bovine $\beta_2$m might be interpreted easily as H-2 specific because of their precipitation of H-2 class I molecules and their polymorphic reaction pattern on a panel of ConA$^{fcs}$ blast cells from various H-2 haplotypes. This polymorphism was due to an H-2 dependent quantitative difference in the capacity to exchange mouse $\beta_2$m for bovine $\beta_2$m.

Rather different results were obtained after syngeneic immunization with SV+ spleen cells. We observed the induction of lymphocytotoxic antibodies that were not absorbed out from the sera by the antigen alone. These antibodies were found to be H-2 class I specific and their reaction pattern was dependent on two factors: (a) the presence of SV on the membrane and (b) the H-2 haplotype of the target cell. By recognizing self-MHC+SV better than MHC alone, these antibodies mimic antiviral H-2 restricted antibodies. However, a detailed analysis of the specificity of the antibodies suggests that the reaction patterns of the sera are due to an assemply of various H-2 alloreactive antibodies rather than to antiviral MHC-restricted antibodies. This was supported by the finding that the increased binding to SV+ auto- and allogeneic cells was not SV specific but was also found on target cells coated with FLU. How viruses such as Sendai and influenza induce the presumed changes in the exposure of H-2 public determinants that result in increased antibody binding is unknown. Viral neuraminidase has been suggested to alter the binding affinity for H-2 specific antibodies after infection (Liberti et al. 1979). However, the molecular basis of these alterations remains unclear. Our results are not in agreement with the findings on syngeneic immunization with SV40 (Froscher and Klinman 1986) or FLU (Wylie et al. 1982) infected cells in which antiviral MHC-restricted antibodies were a normal part of the primary B cell repertoire.

One explanation can be found in the method of immunization. We analyzed serum antibodies after multiple injections with syngeneic cells and thus induced in a secondary immune response. Therefore in addition the primary response to injections with syngeneic SV+ spleen cells was studied by the technique of spleen fragment cultures (Wylie et al. 1982). However, the results in general were identical to those described after repeated injections with syngeneic SV+ lymphocytes (Kievits et al. 1987 a). Again H-2 specific antibodies were induced and no antiviral MHC-restricted antibodies (Kievits et al., 1988).

Another explanation can be found in the interpretation of the reaction pattern of the induced antibodies. When only a limited number of allogeneic target cells and one antigen were tested, the specificity of the here described H-2 specific antibodies would appear completely different (MHC-restricted or anti-MHC+X). Examination of a more extended panel of H-2 haplotypes showed that the induced lymphocytotoxic antibodies bound with high affinity to allogeneic (SV−) target cells and with low affinity to syngeneic target cells. These cross-reactions were strong on SV+ cells than on SV− cells but were not SV specific. Nevertheless, it remains an interesting fact that H-2 specific antibodies were induced after syngeneic immunization with SV+ cells. The absence of such antibodies after injections with syngeneic TNP+ or $ConA^{fcs}$ cells suggests a specific role for SV rather than random activation of B cells due to polyclonal activation after multiple injections. Selectivity in antibody inductions seems also to be evident from the observations that no class II specific antibodies were induced and some anti-class I specificities appeared more often than others. Furthermore, various inbred strains responded differently to injections with syngeneic SV+ cells. The immune response of the various inbred strains ranged between narrow alloreactive H-2 specific antibodies to widely allospecific antibodies with or without cross-reactions to syngeneic cells (Kievits et al., 1988). Experiments are now in progress to examine which part of the SV is responsible for the induction of the H-2 specific antibodies detected after injections of syngeneic SV+ spleen cells.

## Summary

In recent years it has become clear that the T cell receptor (TcR) recognizes its antigen as peptides presented by the appropriate MHC molecules. To answer the question of whether B cell receptors (immunoglobulins) are also able to recognize TcR structures, an effort has been made to induce MHC-restricted antibodies by injections of syngeneic lymphocytes expressing antigenic determinants that are known to be recognized by TcR. H-$2^b$ mice were injected with syngeneic spleen cells treated either with Sendai virus (SV), trinitrophenyl, or foreign $\beta_2$-microglobulin. The immune serum from each individual mouse was tested for antigen specific antibodies that were restricted in their binding by self-MHC molecules. The sera from mice immunized with syngeneic spleen cells treated with trinitrophenyl or foreign $\beta_2$-microglobulin contained antibodies that were specific for the antigen only. No MHC-restricted antibodies were present.

Syngeneic immunization with SV-treated (SV+) cells resulted in the induction of lymphocytotoxic polymorphic serum antibodies that mimic MHC-restricted antibodies by recognizing SV+ syngeneic target cells better than normal cells. However, a detailed examination of the reaction pattern of these antibodies showed that they were directed against allogeneic class I molecules. Our results do not support the existence of MHC-restricted antigen-specific antibodies.

*Acknowledgments.* This investigation was supported in part by the Foundation for Medical Research MEDIGON, which is subsidized by the Netherlands Organization for the Advancement of Pure Research (ZWO) (grant no. 509-087), and in part by the Deutsche Forschungsgemeinschaft (grant no. 5/1-2).

## References

Froscher BG, Klinman NR (1986) Immunization with SV40-transformed cells yields mainly MHC-restricted monoclonal antibodies. J Exp Med 164: 196–210

Gascoigne NRJ, Chien YH, Becher DM, Kavaler J, Davis MM (1984) Genomic organization and sequence of T-cell receptor $\beta$-chain constant and joining regions genes. Nature 310: 387

Gorczynski RM, Kennedy MJ, MacRae S, Steele EJ, Cunningham AJ (1980) Restriction of antigen recognition in mouse B lymphocytes by genes mapping within the major histocompatibility complex. J Immunol 124: 590–596

Hayday AC, Diamond DJ, Tanigawa G, Heilig JS, Folsom V, Saito H, Tonegawa S (1985) Unusual organization and diversity of T cell receptor $\alpha$ chain genes. Nature 316: 828

Honjo T (1983) Immunoglobulin genes. Annu Rev Immunol 1: 499

Hood LM, Kronenberg M, Hunkapiller T (1985) T cell antigen receptors and the immunoglobulin supergene family. Cell 40: 225

Hudson L, Hay FC (1980) Practical immunology. Blackwell, London

Ivanyi P, Melief CJM, van Mourik P, Vlug A, de Greeve P (1979) Lymphocytotoxic antibodies produced by H-2 allo-immunization distinguish between MuLV-positive and -negative substrains of the same H-2 haplotype. Nature 282: 843–845

Jefferies WA, MacPherson GG (1987) Expression of the W6/32 HLA epitope by cells of rat, mouse, human and other species: critical dependence on the interaction of specific MHC heavy chains with human or bovine $\beta_2$-microglobulin. Eur J Immunol 17: 1257

Kahn-Perles B, Boyer C, Arnold B, Sanderson AR, Ferrier P, Lemonnier FA (1987) Acquisition of HLA class-I W6/32-defined antigenic determinant by heavy chains from different species following association with bovine $\beta_2$-microglobulin. J Immunol 138: 2190

Kappler J, Kubo R, Haskins K, Hannum C, Marrack P, Pigeon M, McIntyre B, Allison J, Trowbridge I (1983) The MHC restricted antigen receptor on cells in mouse and men. Identification of constant and variable peptides. Cell 35: 295
Kast WM, de Waal LP, Melief CJM (1984) Thymus dicates major histocompatibility complex (MHC) specificity and immune response gene phenotype of class II MHC-restricted T cells but not class I MHC-restricted T cells. J Exp Med 160: 1752–1766
Katz DH, Skidmore BJ, Katz LR, Bogowitz CA (1978) Adaptive differentiation of murine lymphocyte. I. Both T and B lymphocytes differentiating in $F_1 \rightarrow$ parental chimeras manifest preferential cooperative activity for partner lymphocytes derived from the same parental type corresponding to the chimeric host. J Exp Med 148: 727–745
Kievits F, Ivanyi P (1987) Monomorphic anti-HLA monoclonal antibody (W6/32) recognizes polymorphic H-2 heavy chain determinants exposed by association with bovine or human but not murine $\beta$-microglobulin. Hum Immunol 20: 115–126
Kievits F, Rocca A, Opolski A, Limpens J, Leupers T, Kloosterman T, Boerenkamp WP, Pla M, Ivanyi P (1987a) Induction of H-2 specific antibodies by injections of syngeneic Sendai virus-coated cells. Eur J Immunol 17: 27–35
Kievits F, Ivanyi P, Krimpenfort P, Berns A, Ploegh H (1987b) HLA-restricted recognition of viral antigens in HLA transgenic mice. Nature 329: 447–449
Kievits F, Boerenkamp WJ, Ivanyi P (1988) Immunization with syngeneic sendai virus-infected cells induce no MHC-restricted antibodies but antibodies specific for H-2 class I determinants Immunogenetics, in press
Kurtz ME, Graff RJ, Adelman A, Morgan MM, Click RE (1985) CTL and serologically defined antigens of $\beta$2m, H-3 region. J Immunol 135: 2847
Levy RB, Gilheany P, Shearer GM (1980) Role of self and foreign antigenic determinants in allogeneic and self-restricted cytotoxic T cell recognition. J Exp Med 152: 405
Levy RB, Dower SK, Shearer GM, Segal DM (1984) Trinitrophenyl modification of H-2k and H-2b spleen cells result in serological detection of Kk like determinants. J Exp Med 159: 1464–1472
Liberti PA, Hackett CJ, Askonas BA (1979) Influenza virus infection of mouse lymphoblasts alters the binding affinity of anti-H-2 antibody: requirement for viral neuraminidase. Eur J Immunol 9: 751–757
Mierau R, Robinson PJ, Sanderson AR, Genth E, Cramer M (1987) Antigenic determinants shared between HLA-A, -B, -C antigens and H-2 class I molecules modified by bovine beta-2-microglobulin. Immunogenetics 26: 351–355
Nisbet-Brown E, Singh B, Diener E (1981) Antigen recognition. V. Requirement for histocompatibility between antigen-presenting cell and B-cell in the response to a thymus-dependent antigen, and lack of allogeneic restriction between T and B cells. J Exp Med 154: 676–687
Novotny J, Tonegawa A, Saito H, Kranz DM, Eisen HN (1986) Secondary, tertiary and quartenary structure of T-cell specific immunoglobulin like polypeptide chains. Proc Natl Acad Sci USA 83: 742
Opolski A, Kievits F, Ivanyi P (1986) Polymorphic and autoreactive H-2 specific monoclonal antibodies isolated after injections of syngeneic Sendai virus-coated lymphocytes. Immunogenetics 24: 402–408
Pestalozzi B, Stitz L, Zinkernagel RM (1987) Monoclonal antibodies against viral determinants are not restricted to the K/D end of the MHC. J Exp Med 166: 295–299
Schmidt W (1987) H-2 specific antibodies induced by injection of syngeneic leukemia cells. Immunogenetics 25: 215–221
Tamminen WL, Wraith D, Barber BH (1987) Searching for MHC-restricted anti-viral antibodies: antibodies recognizing the nucloprotein of influenza virus dominate the serological response of C57BL/6 mice to syngeneic influenza-infected cell. Eur J Immunol 17: 999–1006

Tonegawa S (1983) Somatic generation of antibody diversity. Nature 302: 575
Van Leeuwen AE, Goulmy A, van Rood JJ (1979) Major histocompatibility complex restricted antibody reactivity mainly, but not exclusively, directed against cells from male donors. J Exp Med 150: 1075–1083
Wylie DE, Sherman LA, Klinman NR (1982) Participation of the major histocompatibility complex in antibody recognition of viral antigens expressed on infected cells. J Exp Med 155: 403–414
Zinkernagel RM, Doherty PC (1979) MHC-restricted cytotoxic T cells: Studies on the biological role of polymorphic major transplantation antigens determining T cell restriction-specificity. Adv Immunol 27: 51–177

# Searching for MHC-Restricted Monoclonal Antibodies Recognizing the Determinants Seen by Anti-Influenza Virus Cytotoxic T-Lymphocyte Receptors

Wendy L. Tamminen and Brian H. Barber

Department of Immunology, Medical Sciences Building, University of Toronto, Ontario M5S 1A8, Canada

## Introduction

Cytotoxic T-lymphocyte (CTL)-mediated killing of virus-infected cells is known to be both virus specific and class I MHC restricted. This implies that the T cell receptor (TcR) on the virus specific CTL must recognize with specificity both self and nonself structures (Zinkernagel and Doherty 1979). Although recent studies have suggested that viral antigens may be processed and presented to CTLs through a system analogous to that described for class II-mediated presentation of antigen to helper T cells (Townsend et al. 1985, 1986; Staerz et al. 1987; Maryanski et al. 1986; Gotch et al. 1987; Bastin et al. 1987), it is still not clear where or how these T cell target structures are formed.

A direct approach to understanding the molecular nature of the CTL target structure would be to isolate and examine modified self-MHC structures from the surface of virus-infected cells. The ideal reagent for such an isolation would be purified TcR from a virus-specific CTL. Unfortunately, these molecules have not yet been obtained in a functional, soluble form. As an alternative, we have attempted to isolate a monoclonal antibody having the same specificity as an influenza-specific, class I-restricted TcR.

At the level of protein structure, it is know that there is significant similarity between the antigen-binding region of immunoglobulins (as reflected in their $V_H$ and $V_L$ domains), and the recognition units of the TcR (as defined by their $V\alpha$ and $V\beta$ domains) (Kronenberg et al. 1986; Collins et al. 1985). Thus, if a TcR can adopt a three-dimensional conformation which permits it to recognize a modified self (i.e., MHC+X) structure, then it seems reasonable to expect that an antibody could also be found which could react with the same (MHC+) structure.

Support for the idea of overlap between the B and T cell repertoires comes from studies which report that anibodies can be identified which are class I MHC restricted in a manner analogous to CTLs (Wylie et al. 1982; Van Leeuwen et al. 1979; Ivanyi et al. 1979; Levy et al. 1984; Froscher and Klinman 1986). In particular, Wylie et al. (1982) and Froscher and Klinman (1986) have suggested that syngeneic immunization with virus-modified cells can result in the induction of virus-specific antibodies which are mainly MHC restricted, reflecting the possibility of a sizeable overlap in the repertoire of structures which can be seen by antibodies and TcR. Although our initial, limited attempt to produce MHC-restricted, influenza-specific hybridoma antibodies after syngeneic immunization with influenza-infected cells did not yield any such antibodies (Tamminen et al. 1987), we felt that their potential utility warranted a continuation of this search. We have now extended this effort using two variations of our original approach to this problem.

Firstly, we have immunized mice with L cells transfected with both the H-2D$^b$ and the influenza nucleo-protein (NP) genes. Since these cells are known targets for NP-specific, D$^b$-restricted CTLs (Townsend et al. 1984), the aim was to maximize the expression of CTL-recognizable structures on the immunogen while minimizing the number of unrelated, viral serological determinants. Secondly, we have repeated the syngeneic immunization of mice with influenza-infected EL4 tumor cells and broadened the hybridoma screening panel to include, in addition to infected and uninfected EL4, spleen cell targets from mouse strains of 11 different MHC haplotypes. Because a percentage of virus-specific T cells are known to display alloreactivity (Matis et al. 1987 and references therein) it is possible that a B cell recognizing T cell determinants on an infected cell might also display alloreactivity. Such alloreactivity on the appropriate target might be more likely to result in a detectable assay signal than binding tothe actual CTL target structures on the infected cell, since the density of these target structures is unknown and may be very low. Here we report our initial results identifying a limited number of candidate MHC-restricted antibodies which may be useful in the further study of CTL target structures on influenza-modified cells.

## Material and Methods

### *Mice*

Mice of the following stranis were obtained from Jackson Laboratories (Bar Harbor, ME): C57BL/6J(H-2$^b$), (C3H/HeJ × C57BL/6J)$F_1$ (H-2$^k$ × H-2$^b$), Balb/cJ(H-2$^d$), A.CA/Sn(H-2$^f$), I/LnJ(H-2$^j$), CBA/J(H-2$^k$), P/J(H-2$^p$), DBA/1J(H-2$^q$), RIIIS/J(H-2$^r$), A.SW/Sn-(H-2$^s$), PL/J(H-2$^u$), and SM/J(H-2$^v$).

### *Virus*

The recombinant influenza virus strain X-31 was grown as described previously (15) from seed stock generously provided by J.J. Skehel (WHO Influenza Reference Centre, London, U.K.).

### *Cell Lines*

The following cell lines were obtained from the American Type Culture Collection (ATCC, Rockville, MD) and maintained in culture as described elsewhere (15): EL4 (H-2$^b$, T cell lymphoma), P815 (H-2$^d$, mastocytoma), BW5147.G.1.4 (H-2$^k$, T cell lymphoma), and Sp2/O (H-2$^d$, H chain-, L chain- myeloma). The L cell transfectant lines

L/B1.28 (Ltk-, transfected with the H-2D$^b$ gene) and L/B1.28/NP1 (Ltk-, transfected with H-2D$^b$ and influenza NP genes) were a generous gift from A.R.M. Townsend and are described in Townsend et al. (1984).

### *Influenza Infection of Murine Tumor Cells*

Tumor cells were infected as described previously (Tamminen et al. 1987), using 500 HAU of infectious allantoic fluid per $1 \times 10^7$ cells. An initial 1-h period of infection was followed by a 5-h incubation to allow viral protein expression before harvesting for use.

### *Immunization of Mice*

Cells to be used as immunogen were harvested, irradiated (5000 rads), washed 3 times in phosphate buffered saline (PBS), and resuspended at $5 \times 10^7$ cells/ml for i.p. injection of $1 \times 10^7$ cells. C57BL/6 mice were challenged once or twice with influenza-infected EL-4 cells. (C3H/He × C57BL/6)F$^1$ mice were challenged twice with the D$^b$/NP transfectant, L/B1.28/NP1 were trypsinized and then incubated in bacteriological petri dishes for 16–18 h to allow for adherence-independent recovery of their surface antigens before injection. Immune spleens were used for fusion 3–4 days after the last immunization.

### *Cell Fusion*

Fusion experiments were carried out as described in detail previously (Tamminen et al. 1987), using polyethylene glycol (1.0 ml per $1 \times 10^8$ total cells) to mediate fusion between equal numbers of immune spleen cells and Sp2/0 myeloma cells. After fusion, cells were plated in 24- or 96-well plates and fed medium containing hypoxanthine/aminopterin/thymidine for selection and establishment of hybrids in culture. Supernatants were collected for testing when wells were at least on-third confluent.

### *Preparation of Antigen Plates for Enzyme-Linked Immunosorbent Assay (ELISA)*

Infected and unifected tumor cell targets as well as whole virus (X-31) targets were prepared as described previously (Tamminen et al. 1987). For L-cell transfectant targets, the cells were seeded into flat-bottom, 96-well PVC plates at $1 \times 10^4$ cells/well in 100–200 μl and left in culture for 2–3 days. The cells were then glutaraldehyde fixed (0.05%) in the wells and stored at 4 °C. For spleen cell targets, spleen cell suspensions were prepared and depleted of surface Ig$^+$ and Fc-receptor$^+$ cells using a method based on that described by Wysocki and Sato (1978). These cells were then plated as described previously for tumor cells (Tamminen et al. 1987) using $2.5 \times 10^5$ cells/well.

*ELISA and Control Antibodies*

The ELISA assays performed on the fixed target plates were carried out as described previously (Tamminen et al. 1987). The control antibody-producing lines EX6 (anti-influenza HA) and EX8 (anti-influenza NP) were isolated previously in this laboratory (Tamminen et al. 1987) and HB28 (BBM.1, anti-human $\beta_2$-microglobulin) was obtained from ATCC. The corresponding monoclonal antibodies (McAbs) were utilized as culture supernatants. The mean and standard deviation of irrelevant antibody binding was established on each target in each assay. Any result greater than five standard deviations above the mean was judged to be positive. The irrelevant Ab controls were as follows: EX6 on the L cell transfectant and spleen cell targets; HB28 on the whole virus and infected and uninfected tumor cell targets.

## Results

*Identification of Hybridomas Reactive with the $D^b$/NP Transfectant*

Three separate fusions were performed with immune spleen cells from (C3H/HeJ × C57BL/6J)$F_1$ mice immunized twice with the $D^b$/NP transfectant cells. Supernatants of growing hybrids were screened for binding on the $D^b$/NP transfectant and lack of binding on the $D^b$-only transfectant as such a reactivity pattern would indicate specitic recognition of a determinant involving the influenza NP or fragments thereof. Of 344 growing hybrids, 43 showed positive binding on the $D^b$/NP transfectant (Table 1). In 14 of these cases, binding to the $D^b$-only target was just below the cut-off line set for a positive result (see Materials and Methods), while the positive binding to the $D^b$/NP target was only marginally higher. Thus, in terms of finding a good differential response, these hybrids were not as encouraging as suggested by their reactivity pattern designation (RP A, Table 1). The remaining 29 hybrids positive on both the $D^b$/NP and $D^b$-only targets were considered potentially interesting as they showed preferential binding on the double transfectant (RP B, Table 1). However, in further testing of 5 and 16 representative hybrids of reactivity patterns A and B respectively, none showed a virus-specific or MHC-restricted binding pattern on EL4, P815, or BW5147 tumor cells in either the infected or uninfected state.

**Table 1.** Hybridoma responses using LB1.28/NP1 as immunogen

| Reactivity pattern (RP) | $D^b$-only | $D^b$/NP | Total # observed |
|---|---|---|---|
| A | − | + | 14 |
| B | + | ++ | 29 |

*Identification of Hybridomas Raised in Response to Influenza-Infected EL4*

Four separate fusions were performed using immune spleen cells from mice immunized either once or twice with X-31-infected EL4 tumor cells. To identifiy hybrids which had reactivity against infected EL4 cells specifically, the supernatants from growing hybrids were screened for reactivity on uninfected and infected EL4 cells as well as on whole virus targets. In addition, the supernatants were simultaneously screened on a panel of spleen cell targets derived from inbred mouse strains of 11 different haplotypes in order to identify any possible allo-H-2 cross-reactivity. Of the 651 growing hybrids which were screened, 209 and 442 were derived from once- and twice-immunized mice, respectively. There were no differences clearly evident between the responses seen in either of these two cases. The initial screening results for all the hybrids are shown in Table 2. Reactivity patterns 1 and 2 represent hybrids which reacted with the infected cell, while reactivity patterns 3 and 4 represent hybrids which showed binding to the allogeneic spleen cell targets. There were no cases in which hybrids showed significant reactivity on both the infected cell and the allo-targets.

A high frequency of alloreactivity was observed in the response, with 108 of 651 hybrids showing positive reactivity on the allogeneic targets (RP 3 and 4, Table 2). Of the 108 which were positive on the allo-targets, 63 also showed binding to spleen cells of the syngeneic b haplotype (RP 3, Table 2) while 45 bound to the

**Table 2.** Hybridoma responses using X-31-infected EL4 cells as immunogens

| Reactivity pattern (RP) | EL4 +X-31 | EL4 | X-31 | Spleen cell targets | | Total no. observed |
|---|---|---|---|---|---|---|
| | | | | $H\text{-}2^b$ | Allogeneic | |
| 1 | + | – | + | – | – | 87 |
| 2 | + | – | – | – | – | 12 |
| 3 | – | – | – | + | + | 63 |
| 4 | – | – | – | – | + | 45 |

**Table 3.** Distribution of alloreactivity in response to X-31-infected EL4

| | Level[a] | Number of hybrids secreting reactive antibody | | | | | | | | | | |
|---|---|---|---|---|---|---|---|---|---|---|---|---|
| | | b | d | f | j | k | p | q | r | s | u | v |
| Reactivity pattern 3[b] | Low | 19 | 16 | 7 | 17 | 5 | 18 | 17 | 21 | 18 | 18 | 15 |
| | Medium | 23 | 10 | 14 | 9 | 22 | 7 | 9 | 10 | 12 | 14 | 6 |
| | High | 21 | 10 | 16 | 7 | 34 | 7 | 5 | 9 | 6 | 18 | 7 |
| Reactivity pattern 4[b] | Low | 0 | 12 | 24 | 3 | 17 | 5 | 5 | 12 | 7 | 9 | 6 |
| | Medium | 0 | 2 | 2 | 0 | 5 | 0 | 0 | 2 | 0 | 3 | 2 |
| | High | 0 | 0 | 0 | 0 | 5 | 0 | 0 | 0 | 0 | 0 | 0 |

[a] Level of assay signal defined as follows with respect to specific OD units: low $<0.5$; medium $=0.5$–$1.0$; high $>1.0$

[b] As shown in Table 2

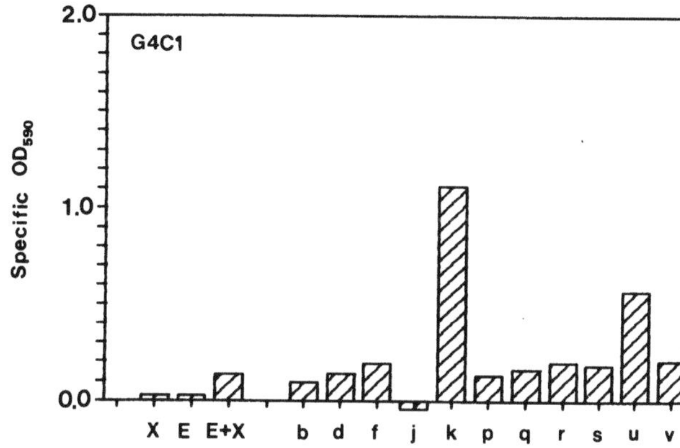


**Fig. 1.** Binding of hybrid G4C1 to initial screening targets. Specific ELISA binding to whole X-31 virus (X), EL4 (E), X-31-infected EL4 (E+X), and spleen cell targets of various haplotypes (Denoted by *lowercase letters*). All results are single point assays

allo-targets only (RP 4, Table 2). The majority of both the RP 3 and RP 4 allo-specific hybrids which showed medium or high level binding had broadly reactive binding profiles across the panel of allo-targets, with 67% and 90% respectively binding to more than three different haplotypes. The chart in Table 3 represents the binding distribution of the alloreactive hybrids with respect to hablotype and level of binding. This distribution shows that allo-targets of the k, f, and u haplotypes were the ones most frequently recognized in this initial screening assay, with binding to the k haplotype being the highest and most frequent. Based on this broadness of specificity, as well as the binding to the syngeneic target, the majority of hybrids showing alloreactivity were not considered likely to be cross-reactive with T-cell-like (MHC + flu) determinants.

The more narrow, CTL-like binding specificities we identified in the RP 4 (allo-only) subset are represented by the initial screening results of G4C1, shown in Fig. 1. The supernatant from this hybrid showed high level binding on one target only, namely the spleen cells of the k haplotype. In addition, it bound to the u haplotype target at an intermediate level and to the f and v haplotypes at low levels. After initial screening, this hybrid and several others with similar reactivity patterns appear to be the best candidates to be allo-cross-reactive, (MHC + flu)-specific hybrids.

The more standard part of the initial screening analysis involved infected and uninfected EL4 as well as whole influenza virus (X-31) as targets (Table 2, RP 1 and 2). Of the 99 hybrids which were reactive with X-31-infected EL4, but not uninfected EL4, 87 also bound to the whole virus target (RP 1), indicating that they are virus specific but lack any MHC restriction in their specificity. However, 12 of the 99 hybrids seemed to be specific for infected EL4 only, as they did not react with either uninfected EL4 or the intact virus antigen (RP 2). Having met the first criterion for a virus-specific, MHC-restricted antibody, these hybrids were further tested on a panel of infected and uninfected tumor cells [i.e., EL4 (H-$2^b$), P815 (H-$2^d$), and BW5147 (H-$2^k$)] to determine whether the binding was haplotype dependent. Of the ten hybrids which were maintained and tested in this secondary screening, seven

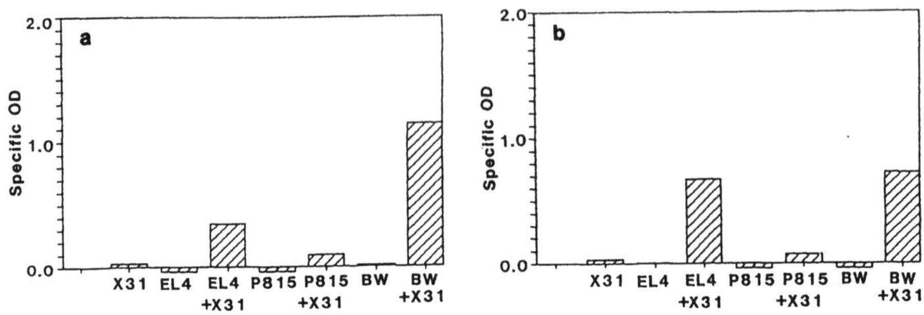


**Fig. 2 a, b.** Binding of hybrid antibodies to whole virus and infected and uninfected tumor cells. **a** hybrid F2C2, **b** hybrid F6A1. All values are the means of duplicates

showed a very low initial level of binding on the syngeneic, infected EL4 target, making it difficult to compare the binding on the different haplotypes quantitatively. Nevertheless, three of the hybrids clearly showed differential binding to the infected cells of different haplotypes. The binding of the two most reactive of these hybrids, F2C2 and F6A1, is shown in Fig. 2. These two antibodies appeared to react preferentially with infected cells of the b and k haplotypes, with much lower binding on infected cells of the d haplotype.

Recognizing the possibility that individual tumor cells might differ in their ability to express influenza genes, we sought to guard against mistakenly attributing such variance to a haplotype dependence in antibody reactivity. The results in Table 4 demonstrate that saturating concentrations of control antibodies EX6 and EX8, specific for influenza HA and NP respectively, do detect different amounts of these viral proteins on each target type, although the variation is much more significant in the case of the NP. Thus, to normalize for differential viral protein "availability," the binding results for F2C2 and F6A1 on each of the three infected targets were

**Table 4.** Reactivity of hybrids and controls with infected cells of different haplotype

| Hybrid | Specific OD[a] | | |
|---|---|---|---|
| | EL4+X-31 ($H\text{-}2^b$) | P815+X-31 ($H\text{-}2^d$) | BW5147+X-31 ($H\text{-}2^k$) |
| EX6 (Anti-HA) | 1.401 | 1.296 | 1.775 |
| EX8 (Anti-NP) | 0.970 | 0.500 | 1.723 |
| F2C2 | 0.344 | 0.092 | 1.134 |
| Ratio to HA | 0.245 | 0.007 | 0.639 |
| Ratio to NP | 0.354 | 0.184 | 0.658 |
| F6A1 | 0.673 | 0.071 | 0.728 |
| Ratio to HA | 0.480 | 0.005 | 0.410 |
| Ratio to NP | 0.694 | 0.142 | 0.423 |

[a] All values represent the mean of duplicates

divided by the control binding value obtained with both the anti-HA and anti-NP controls. The resulting ratios for F2C2 and F6A1 on each infected target, as well as the specific OD value prior to normalization, are shown in Table 4. For both hybrids, a differential reactivity still exists after normalization to either NP or HA expression, although a more dramatic differential remains after normalization to HA expression. Thus, these antibodies may represent examples of virus-specific antibodies which are partially MHC restricted.

## Discussion

In attempting to isolate virus-specific, MHC-restricted antibodies, the objective has been to obtain a reagent that would allow a direct approach to studying the molecular nature of what a CTL recognizes on the surface of an infected cell. In this study, we have tried to establish hybridomas of such specificity by immunizing syngeneic mice with influenza-modified cells of two different types. Firstly, mouse L cells cotransfected with both the influenza NP ant the H-$2D^b$ genes were used for syngeneic immunization of (C3H/HeJ × C5BL6)$F_1$ mice. It was hoped that the relative lack of viral serological determinants on this immunogen would permit an improved B cell response to the CTL-recognizable structure(s) on the transfectant. However, no NP-specific H-2 restricted McAbs were isolated using this approach. The positive reactivity seen with the $D^b$/NP and $D^b$-only targets may have been due to general anti-H-2 reactivity; however, this has yet to be tested.

We also used X-31-infected EL4 as the immunogen for C57BL/6 mice in an extension of our earlier study (Tamminen et al. 1987). In this case, a greater number of hybrids were screened and several antibodies were identified which react with infected cells, but not with virus alone. Although these antibodies could be recognizing either the viral M2 or nonstructural (NS) proteins, components which are produced in the infected cell but are not found in the virion (Ada and Janes 1986), this seems unlikely, because their reactivity pattern demonstrated a degree of haplotype-dependent binding which cannot easily be attributed to differing degrees of target cell infection. Immunoprecipitation studies are now in progress to determine whether class I MHC antigens are involved in the structures seen by these candidate MHC-restricted McAbs.

By testing on a panel of spleen cells from various haplotypes, it was found that many of the hybrids which were negative on the infected-cell screening target showed some degree of alloreactivity. In most cases, these hybrids had a broad range of alloreactivity across the haplotypes tested, often binding to syngeneic H-$2^b$ spleen cells. Fewer hybrids had the more restricted allorecognition pattern closer to that expected for an (MHC+X)-specific allo-cross-reactive CTL. It must now be determined whether these antibodies actually recognize class I MHC antigens and if so, whether they have been induced specifically by cross-reactivity with an (MHC+flu) determinant on the immunogen. Class I-specific alloreactivity after syngeneic immunization with virus-modified cells has been seen in other systems (Kievits et al. 1987 and other chapters in this book), and in the case of Sendai virus-coated cells as immunogen, the viral component of the immunogen was essential in eliciting the anti-H-2 antibodies (Kievits et al. 1987). In addition, exchange of class I MHC-associated murine $\beta_2$-microglobulin with bovine $\beta_2$-microglobulin from the fetal calf serum in culture could result in the mimicry of allo-H-2 determinants, transforming the immunogen from a syngeneic to an allogeneic form, thereby inducing anti-H-2 alloantibodies (see other chapters in this volume).

A number of fundamental problems have been brought to light in our search for MHC-restricted, virus-specific antibodies. For example, what level of reactivity is to be expected from the direct binding of an (MHC + influenza)-specific antibody to infected, syngeneic target cells at saturating concentrations of antibody? This cannot be answered with confidence because the number of modified class I structures present per cell is unknown. If the number is only a few hundred to a few thousand, which does not seem unreasonable, our ability to identify binding to these structures may be just at, or below, the limits of detection defined for the currently available screening procedures. This means that those hybridoma supernatants which give low assay signals cannot be dismissed as uninteresting, because this may reflect the inherent low level of target structures on infected cells. Screening on a panel of allo-targets for possible (MHC + X)-reactive clones which cross-reacted with higher density allo-MHC determinants was our attempt to circumvent this potential detection problem.

Another question arising from the search for MHC-restricted antibodies is the degree of stringency that one can reasonably expect in assessing their relative reactivities on infected cells of different haplotypes. Because the antibody binding assays are quite different from the chromium release assay used to assess CTL restriction, looking for an all-or-none antibody binding pattern may be setting the restriction standard unrealistically high. Thus, in terms of the present study, those infected-cell-specific and alloreactive antibodies found with broader reactivity patterns than expected by CTL standards may still merit further investigation.

There are reports in the literature indicating that for both influenza virus-infected (Wylie et al. 1982) and SV-40 transformed (Froscher and Klinman 1986) cells, MHC-restricted antibodies dominate the antiviral serological response in syngeneic mice. Nevertheless, we and others (Tamminen et al. 1987; Kievits et al. 1987; Pestalozzi et al. 1987) have been unable to find these frequencies reflected in different attempts to identify MHC-restricted hybridoma antibodies. The question of whether or not MHC-restricted antibodies exist is of fundamental importance, since a negative answer implies the existence of a "hole" in the B cell repertoire which prevents the generation of antibodies recognizing the determinants seen by T cells. This could reflect a lack of the appropriate V region Ig genes, or regulatory controls preventing the activation of these B cells. It is also possible that such a "hole" in the B cell repertoire might exist to prevent a blockade of T-cell mediated responses to infected cells and/or other antigen-presenting cells.

In summary, the syngeneic immunization of mice with both influenza-infected and NP-transfected cells has been used in an attempt to generate monoclonal, hybridoma-derived antibodies specific for the (MHC + X) determinants seen by influenza-specific CTLs. Although we have identified a number of candidate MHC-restricted antibodies, they appeared with a low frequency, and their reactivity patterns suggest a somewhat broader specificity than is usually seen for CTL receptors. Further work will be required to define the infected-cell structures seen by these antibodies and to determine whether or not they correspond to CTL determinants.

*Acknowledgments.* This work was supported by a Medical Research Council of Canada grant to B. H. B. W. L. T. is the recipient of an NSERC 1967 Science and Engineering Scholarship.

## References

Ada GL, Jones PD (1986) The immune response to influenza infection. Curr Top Microbiol Immunol 128: 1

Bastin J, Rothbard J, Davey J, Jones I, Townsend A (1987) Use of synthetic peptides of influenza nucleoprotein to define epitopes recognized by class I-restricted cytotoxic T lymphocytes. J Exp Med 165: 1508

Collins MKL, Owen MJ (1985) The T cell antigen receptor. Biochem J 230: 281

Froscher BG, Klinman NR (1986) Immunization with SV40-transformed cells yields mainly MHC-restricted monoclonal antibodies. J Exp Med 164: 196

Gotch F, Rothbard J, Howland K, Townsend A, McMichael A (1987) Cytotoxic T lymphocytes recognize a fragment of influenza virus matrix protein in association with HLA-A2. Nature 326: 881

Ivanyi P, Melief CJM, van Mourik P, Vlug A, de Greeve P (1979) Lymphocytotoxic antibodies produced by H-2 allo-immunization distinguish between MuLV-positive and -negative substrains of the same H-2 haplotype. Nature 282: 843

Kievits F, Rocca A, Opolski A, Limpens J, Leupers T, Kloosterman T, Boerenkamp WJ, Pla M, Ivanyi P (1987) Induction of H-2-specific antibodies by injections of syngeneic Senai virus-coated cells. Eur J Immunol 17: 27

Kronenberg M, Siu G, Hood LE, Shastri N (1986) The molecular genetics of the T-cell antigen receptor and T-cell antigen recognition. Annu Rev Immunol 4: 529

Levy RB, Dower SK, Shearer GM, Segal DM (1984) Trinitrophenyl modification H-$2^k$ and H-$2^b$ spleen cells result in enhanced serological detection of $K^k$-like determinants. J Exp Med 159: 1464

Maryanski JL, Pala P, Corradin G, Jordan BR, Cerottini J-C (1986) H-2-restricted cytolytic T cells specific for HLA can recognize a synthetic HLA peptide. Nature 324: 578

Matis LA, Sorger SB, McElligott DL, Fink PJ, Hedrick SM (1987) The molecular basis of alloreactivity in antigen-specific, major histocompatibility complex-restricted T cell clones. Cell 51: 59

Pestalozzi B, Stitz L, Zinkernagel RM (1987) Monoclonal antibodies against viral determinants are not restricted to the K/D end of the major histocompatibility complex. J Exp Med 166: 295

Staerz UD, Karasuyama H, Garner AM (1987) Cytotoxic T lymphocytes against a soluble protein. Nature 329: 449

Tamminen WL, Wraith D, Barber BH (1987) Searching for MHC-restricted anti-viral antibodies:antibodies recognizing the nucleoprotein of influenza virus dominate the response of C57BL/6 mice to syngeneic influenza-infected cells. Eur J Immunol. 17: 999

Townsend ARM, McMichael AJ, Carter NP, Huddleston JA, Brownlee GG (1984) Cytotoxic T cell recognition of the influenza nucleoprotein and hemagglutinin expressed in transfected mouse L cells. Cell 39: 13

Townsend ARM, Goth FM, Davey J (1985) Cytotoxic T cells recognize fragments of the influenza nucleoprotein. Cell 42: 457
Townsend ARM, Rothbard J, Gotch FM, Bahadur G, Wraith D, McMichael AJ (1986) The epitopes of influenza nucleoprotein recognized by cytotoxic T lymphocytes can be defined with short synthetic peptides. Cell 44: 959
Van Leeuwen A, Goulmy E, van Rood JJ (1979) Major histocompatibility complex-restricted antibody reactivity mainly, but not exclusively, directed against cells from male donors. Exp Med 150: 1075
Wylie DE, Sherman LA, Klinman NR (1982) Participation of the major histocompatibility complex in antibody recognition of viral antigens expressed on infected cells. J Exp Med 155: 403
Wysocki LJ, Sato VL (1978) "Panning" for lymphocytes: A method for cell selection. Proc Natl Acad Sci USA 75: 2844
Zinkernagel RM, Doherty PC (1979) MHC-restricted cytotoxic T cells. Adv Immunol 27: 51

# Immunization with Fibroblasts Transfected with a Cloned Retroviral DNA Induces H-2 Specific Antibodies in Syngeneic Recipients

MARIKA PLA, ADAM OPOLSKI, ANNA ROCCA, and LAURENT DEGOS

Laboratory of Mouse Immunogenetics, U93 INSERM Saint Louis Hospital, 2, place du Dr. A. Fournier, 75010 Paris, France

Alloimmune sera induced by immunization between mice from H-2 congeneic strains are a potent tool for the detection of an extensive polymorphism of H-2 class I and class II molecules. In recent years a new category of H-2 class I specific antibodies has been described. These antibodies have been observed in mice prior to any intentional alloimmunization and are considered to be naturally occurring class I specific antibodies (Ivanyi et al. 1982; Lynes et al. 1984; Černý-Provazník et al. 1985 a, b, 1986). In other experiments H-2 specific antibodies have been induced by immunization with syngeneic cells. Various approaches have been applied: injection of syngeneic cells (Ivanyi et al. 1980; Čapková et al. 1983; Takahashi et al. 1987), syngeneic concanavalin A-induced blasts (Mierau and Cramer 1984), syngeneic tumor cells (Schmidt and Alonzo 1983; Philipps et al. 1985), and syngeneic Sendai virus-coated cells (Opolski et al. 1986; Kievits et al. 1987 a, b). The specificities of these H-2 class I specific antibodies were variable and ranged from narrow to widely cross-reacting serotypes. Autoreactivity was occasionally present but the antibodies were never exclusively or preferentially directed against self-H-2 antigens. The mechanism responsible for their induction is unknown.

Several class I and viral genes have been isolated by molecular cloning and have been shown to be expressed on the surface of mouse cells after transfection. The main aim of our study was to develop an experimental system using cells transfected with the genes which code for viral and/or H-2 antigens. Further manipulation of these genes will allow us to define with more precision the antigenic determinants involved in the induction of the H-2 specific antibodies. In the present report we describe the analysis of humoral immune response against fibroblasts transfected with genes which code for the virus (Akv murine leukemia virus) and/or the H-2 molecule ($K^b$) in syngeneic animals. It has been shown that these transfected cells are recognized in vitro and in vivo by the virus-specific cytolytic T cells (Abastado et al. 1985; Plata et al. 1987).

Groups of (C3H × C57BL/10)$F_1$ mice (H-$2^{k/b}$) were injected i.p. weekly with $10^7$ (a) untransfected fibroblasts of C3H origin (L), (b) fibroblasts transfected with the $K^b$ gene (L + $K^b$), (c) Akv MuLV gene (L + V), or (d) fibroblasts transfected with both the $K^b$ and the MuLV genes (L + $K^b$ + V). After the fourth injection, the mice were bled from the orbital sinus and the individual sera were assayed in a complement-dependent cytotoxic test (Frelinger et al. 1974) against a panel of splenocytes from 11 H-2 congeneic strains.

Of the mice injected with L + $K^b$ + V or L + V fibroblasts, 100% (25/25) and 60% (6/10) respectively, produced lymphocytotoxic antibodies (Tables 1, 2). None of the sera from the mice (20–25 in each group) injected with L or L + $K^b$ fibroblasts contained lymphocytotoxic antibodies (data not shown). The specificity of the lymphocytotoxic sera was tested on syngeneic ($H\text{–}2^{k/b}$) cells and on a panel of splenocytes from 11 H-2 congeneic strains with B10 genetic background carrying the following H-2 haplotypes: b, d, f, j, k, p, q, r, s, u, and v. The polymorphic reactivity pattern of the sera (Tables 1, 2) suggests that they are H-2 specific. The H-2 specificity was also confirmed by further analysis using recombinant H-2 haplotypes (data not shown).

**Table 1.** Reaction pattern of sera from mice immunized with syngeneic fibroblasts transfected with the Akv and $K^b$ genes on a panel of splenocytes from 11 H-2 congeneic strains with B10 background

| No. | H-2 haplotype of target cells | | | | | | | | | | | Number of positive targets |
|---|---|---|---|---|---|---|---|---|---|---|---|---|
| | b | k | d | q | v | r | s | u | p | j | f | |
| 14 | –[a] | – | 8[b] | – | 8 | – | – | – | – | – | – | 2 |
| 6 | – | – | 16 | 8 | – | – | – | – | – | – | – | 2 |
| 17 | – | – | 8 | – | 8 | – | – | 16 | – | – | – | 3 |
| 1 | – | – | 16 | – | – | 16 | 8 | 4 | – | – | – | 4 |
| 4 | – | 16 | 32 | 32 | 16 | – | – | – | – | – | – | 4 |
| 11 | – | – | 4 | 2 | 4 | – | 4 | – | 4 | – | – | 5 |
| 15 | – | 64 | 16 | 64 | 8 | – | 16 | – | – | – | – | 5 |
| 24 | – | 16 | – | – | 8 | 4 | – | – | 4 | 8 | – | 5 |
| 25 | 128 | – | 128 | – | – | 64 | – | 16 | – | 128 | – | 5 |
| 13 | – | – | 16 | 32 | 32 | 8 | 4 | 16 | – | – | – | 6 |
| 22 | 4 | 8 | – | – | 16 | 8 | 32 | – | 16 | – | – | 6 |
| 2 | 8 | – | 16 | 16 | 8 | 32 | 8 | – | 16 | – | – | 7 |
| 3 | – | 8 | – | 4 | 4 | 4 | 8 | 4 | 4 | – | – | 7 |
| 5 | 64 | – | 32 | 64 | 8 | – | 16 | 8 | – | – | 8 | 7 |
| 9 | – | 16 | 4 | 16 | 8 | – | 4 | 4 | – | 8 | – | 7 |
| 18 | 32 | 32 | – | 32 | 16 | 16 | – | 8 | – | 64 | – | 7 |
| 19 | 4 | 16 | – | 8 | – | 16 | 16 | 64 | – | – | 32 | 7 |
| 21 | – | 32 | – | 32 | – | 8 | 8 | 8 | 4 | 16 | – | 7 |
| 23 | – | 16 | 32 | – | – | 16 | – | 32 | 16 | 16 | 32 | 7 |
| 7 | 32 | 32 | 32 | 32 | 32 | 64 | 8 | – | 16 | – | 8 | 9 |
| 12 | 8 | – | 64 | 8 | 16 | 16 | 16 | 32 | 8 | 32 | – | 9 |
| 20 | – | 128 | 4 | 64 | – | 256 | 64 | 8 | 16 | 256 | 64 | 9 |
| 8 | 64 | 32 | 128 | 32 | 32 | 32 | 16 | – | 16 | 64 | 16 | 10 |
| 10 | 128 | 128 | 64 | 16 | 32 | 256 | 32 | 8 | – | 256 | 128 | 10 |
| 16 | 128 | 128 | 16 | 64 | 64 | 128 | 64 | 16 | 32 | 256 | 128 | 11 |
| Total positive 25/25 | 11 | 15 | 19 | 18 | 18 | 17 | 17 | 15 | 12 | 11 | 8 | |
| 100% | 44 | 60 | 76 | 72 | 72 | 68 | 68 | 60 | 48 | 44 | 32 | |

[a] Negative in ½ dilution
[b] Numbers indicate reciprocals of lymphocytotoxic titer

**Table 2.** Reaction pattern of sera from the mice immunized with syngeneic fibroblasts transfected with the Akv gene on a panel of splenocytes from 11 H-2 congeneic strains with B10 background

| No. | H-2 haplotype of traget cells | | | | | | | | | | | Number of positive targets |
|---|---|---|---|---|---|---|---|---|---|---|---|---|
| | b | k | j | d | p | s | f | q | r | u | v | |
| 4 | –[a] | – | – | – | – | – | – | – | – | – | – | 0 |
| 5 | – | – | – | – | – | – | – | – | – | – | – | 0 |
| 7 | – | – | – | – | – | – | – | – | – | – | – | 0 |
| 10 | – | – | – | – | – | – | – | – | – | – | – | 0 |
| 8 | – | 4[b] | – | – | – | – | – | – | – | – | – | 1 |
| 2 | – | 4 | – | – | – | – | – | – | – | – | – | 1 |
| 3 | 4 | – | 8 | – | – | 4 | – | – | – | – | – | 3 |
| 9 | 8 | – | – | – | – | 8 | 16 | – | – | – | – | 3 |
| 6 | 16 | 16 | 16 | 8 | 8 | – | – | – | – | – | – | 5 |
| 1 | 8 | 8 | 4 | 16 | 2 | – | – | – | – | – | – | 5 |
| Total positive 6/10 | 4 | 4 | 3 | 2 | 2 | 2 | 1 | 0 | 0 | 0 | 0 | |
| 60% | 40 | 40 | 30 | 20 | 20 | 20 | 10 | 0 | 0 | 0 | 0 | |

[a] Negative in ½ dilution
[b] Numbers indicate reciprocals of lymphocytotoxic titer

The fact that the identity in the H-2I region was not sufficient to ensure a positive reaction and that 100% of the spleen cells were lysed, strongly indicates that the lymphocytotoxic antibodies are class I specific.

The reaction pattern of the lymphocytotoxic sera of the mice immunized with doubly transfected fibroblasts (L + $K^b$ + V) was rarely narrow (Table 1). None of the sera reacted with only one haplotype and about 70% of the sera recognized more than five haplotypes. The H-$2^d$ cells were the most frequently recognized targed cells (76% of sera). Reactivity on syngeneic cells (H-$2^{k/b}$) had the same frequency. However, the serological analysis on H-$2^b$ and H-$2^k$ target cells showed that 32% of the mice immunized with L + $K^b$ + V fibroblasts produced antibodies cytotoxic for H-$2^k$ but not for H-$2^b$ cells and 16% of the mice produced antibodies positive only on H-$2^b$ cells. Twenty-eight percent of sera contained both anti-h-$2^b$ and anti-H-$2^k$ antibodies. Autoreactivity was always accompanied by lymphocytotoxic reactions on other haplotypes.

The reaction pattern of the six lymphocytotoxic sera from mice immunized with fibroblasts transfected only with the Akv gene (Table 2) was rather narrow. Two sera were reactive with three haplotypes (nos. 3 and 9) and two sera recognized five haplotypes (nos. 1 and 6). All four sera contained antibodies positive on H-$2^b$ target cells and two sera (nos. 2 and 8) were positive only on H-$2^k$ splenocytes.

In the present experiments, H-2 specific antibodies were induced in 60%–100% of mice immunized with syngeneic fibroblasts transfected with a cloned retroviral DNA. The induction of lymphocytotoxic antibodies was dependent on the presence of both viral and H-2 antigens on the surface of the injected fibroblasts since the immunization of H-$2^{k/b}$ mice with the fibroblasts not expressing viral antigens did not induce any lymphocytotoxic antibodies and the H-$2^k$ fibroblasts expressing viral and $K^b$ molecules resulted in more efficient antibody induction in comparison with those without the $K^b$ molecules. The reaction pattern of the sera of mice injected with the doubly transfected fibroblasts ($L+K^b+V$) often broadly cross-reacted while the sera of mice injected with the fibroblasts transfected only with the Akv gene ($L+V$) had rather narrow serotypes. These results suggest that the reaction pattern of the sera is dependent on the presence of the H-2 molecules on the surface of the injected cells. Reactivity on the syngeneic splenocytes (H-$2^{k/b}$) was frequent (77% of lymphocytotoxic sera). However, with a very few exceptions (2 out 26 sera) it was not exclusive and was accompanied by cytotoxic reactions on other haplotypes. The serological analysis of the sera from the mice immunized with $L+K^b+V$ fibroblasts on H-$2^b$ and H-$2^k$ target cells showed that antibodies cytotoxic for H-$2^k$ occurred more frequently than those cytotoxic for H-$2^b$ cells (32% and 16% of the sera, respectively).

Taken together, we have shown that the immunization with fibroblasts transfected with a cloned retroviral DNA induced H-2 specific antibodies. Double transfection of mammalian cells with cloned genes which encode transplantation antigens and viral proteins is thus a useful approach for the analysis of the induction of these antibodies. It can now be extended to a more detailed analysis of the viral and H-2 genes in order to define the regions in these genes that code for the determinants which are recognized by the B cell receptor.

*Acknowledgments.* We thank Drs. J.-P. Abastado and F. Plata for the kind gift of the transfected fibroblasts and Dr. P. Ivanyi for a critical reading of the typescript.

## References

Abastado JP, Plata F, Morello D, Daniel-Vedele F, Kourilsky P (1985) H-2-Restricted cytolysis of L cells doubly transfected with a cloned H-$2K^b$ gene and cloned retroviral DNA. J Immunol 135: 3512–3519

Čapková J, Hašek M, Kousalová M, Holáň V, Boubelík M (1983) Alloreactive cytotoxic anti-H-2 antibodies in sera of the syngeneically immunized C57BL/10ScSnPh, BALB/cJPh, and B10.A/SnPh mice. Folia Biol (Praha) 29: 200–212

Černý-Provazník R, Radl J, Leupers T, van Mourik P, Brodijk R, de Greeve P, Ivanyi P (1985a) Anti-MHC immunity detected prior to intentional alloimmunization. I. Naturally occurring H-2-specific antibodies in C57BL/KaLwRij (H-$2^b$) mice. Nat Immun Cell Growth Regul 4: 138–159

Černý-Provazník R, van Mourik P, Radl J, Leupers T, Limpens JC, Ivanyi P (1985b) Anti-MHC immunity detected prior to intentional alloimmunization. II. Monoclonal H-2-specific antibodies obtained from a normal C57BL/KaLwRij (H-$2^b$) mouse. Nat Immun Cell Growth Regul 4: 160–168

Černý-Provazník R, Kloosterman T, Verkerk D, van Mourik P, Reboul M, Frangoulis B, Pla M, Ivanyi P (1986) Anti-MHC immunity detected prior to intentional alloimmunization. IV. Natural monoclonal H-2-specific antibodies. J Immunogenet 13: 287–297

Freilinger JA, Niederhuber JE, David CS, Shreffler DC (1974) Evidence for the expression of Ia (H-2-associated) antigens on thymus derived lymphocytes. J Exp Med 140: 1273–1284

Ivanyi P, van Mourik P, Breuning M, Melief CJM (1980) Anti-H-2 antibodies induced by syngeneic immunization. Immunogenetics 10: 319–332
Ivanyi P, van Mourik P, Breuning M, Kruisbeek AM, Kröse CJM (1982) Natural H-2-specific antibodies in sera of aged mice. Immunogenetics 15: 95–102
Kievits F, Rocca A, Opolski A, Limpens J, Leupers T, Kloosterman T, Boerenkamp WJ, Pla M, Ivanyi P (1987a) Induction of H-2-specific antibodies by injections of syngeneic Sendai virus-coated cells. Eur J Immunol 17: 27–35
Kievits F, Opolski A, Boerenkamp WJ, Pla M, Ivanyi P (1987b) Monoclonal H-2 class I specific antibodies insolated after immunization of C57BL/6 mice with syngeneic Sendai-virus-coated cells. In: David CS (ed) H-2 antigens: genes, molecules, function. Plenum, New York, pp 669–680
Lynes MA, Flaherty L, Michaelson J, Collins JJ, Rinchik EM (1984) Allogeneic reactivity in normal mouse serum. J Immunogenet 11: 189–196
Mierau R, Cramer M (1984) Serological and immunochemical analysis of H-2 class I molecules encoded by the $D^b$ region. Immunogenetics 20: 341–345
Opolski A, Kievits F, Ivanyi P (1986) Polymorphic and autoreactive H-2-specific monoclonal antibody isolated after injections of syngeneic Sendai virus-coated lymphocytes. Immunogenetics 24: 402–408
Philipps C, McMillan M, Flood PM, Murphy DB, Forman J, Lancki D, Womack JE, Goodenow RE, Schreiber H (1985) Identification of a unique tumor-specific antigen as a novel class I major histocompatibility molecule. Proc Natl Acad Sci USA 82: 5140–5144
Plata F, Langlade-Demoyen P, Abastado JP, Berbar T, Kourilsky P (1987) Retrovirus antigens recognized by cytolytic T lymphocytes activate tumor rejection in vivo. Cell 48: 231–240
Schmidt W, Alonzo A (1983) Biologic significance of autoantibodies against H-2 antigens in syngeneic antitumor sera. Transplant Proc 15: 2118–2120
Takahashi T, Matsudaira Y, Obata Y, Moriwaki K (1987) An autoreactive H-2-specific monoclonal antibody with allospecificity. Immunogenetics 26: 105–106

# Sendai Virus Infection of Tumor Cells Increases the Production of Autoreactive H-2 Specific Antibodies in Syngeneic Recipients

Adam Opolski, Laurent Degos, and Marika Pla

Laboratory of Mouse Immunogenetics, U93 INSERM, Saint Louis Hospital, 2, place du Dr. A. Fournier, 75010 Paris, France

The major histocompatibility complex (MHC) phenotype of neoplastic cells reflects both the cells tumorigenic history and its internal state of differentiation and metabolic activity. In many cases MHC expression probably also affects the outcome of the disease. An increasing body of evidence indicates that various agents are able to increase the immunogenicity of malignant cells. An increase in the immunogenicity of tumor cells was observed following in vitro treatment with various drugs (Koyama and Ishii 1969; Tsukagoshi and Hashimoto 1973; Frost et al. 1983; Carlow et al. 1985; Gorelik et al. 1985; Flood et al. 1987), UV light treatment (Peppoloni et al. 1985), and virus infection (Wakamiya et al. 1986; Williams et al. 1986) of tumor cells. Recently, we have shown (Opolski et al. 1987) that the median survival time of mice inoculated with Sendai virus-infected EL4 tumor cells was significantly longer (in some animals a complete rejection was observed) than in those inoculated with untreated EL4 cells. Moreover, most of the mice which rejected Sendai virus-infected EL4 cells were found to be resistant to subsequent inoculation of untreated EL4 tumor cells (Opolski et al. 1987). What remains unknown is which of the changes on the cell surface are responsible for the conversion of the weakly or nonimmunogenic tumor cells into a highly immunogenic form. Numerous experimental data indicate that cell surface antigens are recognized by T-lymphocytes in association with MHC products (Zinkernagel and Doherty 1975). One might postulate that conversion of nonimmunogenic tumor cells into a highly imunogenic form would be paralleled by changes in expression of MHC products on the surface of tumor cells (Goodenow et al. 1985).

Injections of Sendai virus-coated syngeneic cells can trigger the production of auto- and alloreactive H-2 class I specific lymphocytotoxic antibodies (Opolski et al. 1986; Kievits et al. 1987 a, b). Some data suggest that viral infection might have consequences for the allorepertoire (Zinkernagel and Doherty 1979; Finberg et al. 1978; Braciale et al. 1981; Askonas and Lin 1982; Ashwell et al. 1986). Up until now, it has been presumed that viral antigens can induce MHC-restricted antiviral cytotoxic T cells which also recognize some allodeterminants (Finberg et al.1978; Abramson-Leeman and Cantor 1983). However, recently Yang and Welsh (1986) reported that in vivo viral infection can also induce, in parallel with antiviral MHC-restricted cytotoxic T cells, alloreactive cytotoxic T cells that are not inhibited by syngeneic virus-infected cells.

How a virus can induce H-2 specific antibodies or cytotoxic T cells remains unclear. Whatever the mechanisms responsible for the induction, they may play a role in the immune surveillance of the tumor's host organism. In an attempt to examine the possible biological consequences of their presence we have investigated the antibody response to injections of untreated or Sendai virus-infected EL4 tumor cells.

C57BL/10 (H-$2^b$) mice were injected i.p. weekly with either irradiated (2500 rads) $10^7$ untreated (EL4−) or Sendai virus-infected (EL4+) EL4 cells. EL4+ cells were obtained by coculturing $5 \times 10^7$ EL4 cells with $5 \times 10^{10}$ particles of Sendai virus in 25 ml culture medium for 18 h at 37 °C. After the fourth injection, the mice were bled from the orbital sinus and the individual sera were assayed in a complement-dependent cytotoxic test (Frelinger et al. 1974) against a panel of various target cells (EL4−, EL4+, untreated and Sendai virus-infected splenocytes, and Con A blasts from inbred mouse strains with various H-2 haplotypes and various genetic backgrounds).

The summary of the reaction pattern of these sera is shown in Table 1. Nine of the 24 mice (37%) injected with EL4− cells and 12 of the 24 mice (50%) injected with EL4+ cells produced lymphocytotoxic antibodies. None of the tested sera were positive only on EL4 target cells (untreated and Sendai virus-infected) or on untreated syngeneic (H-$2^b$) splenocytes or Con A blasts. Two of the 24 mice (8%) immunized with EL4+ cells produced antibodies which reacted with both EL4− and EL4+ cells and, when tested on a panel of splenocytes and Con A blasts from eight H-2 congeneic strains, recognized only Sendai virus-infected H-$2^b$ target cells (e.g., serum no. 7 in Table 2). All other lymphocytotoxic sera (10/24) from the mice immunized with EL4+ cells were reactive with both syngeneic and allogeneic target cells. On the other hand, only 8% of mice (2/24) immunized with EL4− cells produced antibodies, which were positive on both syngeneic and allogeneic target cells. Most of the lymphocytotoxic sera (7/9) were cytotoxic only for allogeneic cells. Thus our results show that Sendai virus infection of EL4 cells significantly increases the induction of autoreactive antibodies.

**Table 1.** Summary of the reaction pattern of sera from C57BL/10 mice immunized with untreated (EL4−) or Sendai virus-infected (EL4+) EL4 cells on untreated (−) and Sendai virus-infected (+) target cells: EL4 cells and a panel of syngeneic (syn) and allogeneic (allo) splenocytes and Con A blasts from eight congeneic strains with B10 genetic background

| Target recognized | | | | | | | | | | Number (%) of lymphocytotoxic sera from the mice immunized with | |
|---|---|---|---|---|---|---|---|---|---|---|---|
| EL4 | | Splenocytes | | | | Con A blasts | | | | | |
| EL4− | EL4+ | syn− | syn+ | allo− | allo+ | syn− | syn+ | allo− | allo+ | EL4− | EL4+ |
| − | − | − | − | − | − | − | − | − | − | 15/24 (63%) | 12/24 (50%) |
| + | + | − | − | − | − | − | − | − | − | 0 | 0 |
| + | − | + | − | − | − | + | − | − | − | 0 | 0 |
| + | + | − | + | − | − | − | + | − | − | 0 | 2/24 (8%) |
| + | + | + | + | + | + | + | + | + | + | 2/24 (8%) | 10/24 (42%) |
| + | + | − | − | + | + | − | − | + | + | 7/24 (29%) | 0 |

**Table 2.** Reaction pattern of sera from the mice immunized with untreated (EL4−: nos. 5 and 11) or Sendai virus-infected (EL4+: nos. 3, 7, and 13) EL4 cells on a panel of untreated and Sendai virus-infected splenocytes from H-2 congeneic mouse strains with various H-2 haplotypes and on various genetic backgrounds

| Target | H-2 genotype | | | Sendai virus infection | Mice immunized with | | | | |
|---|---|---|---|---|---|---|---|---|---|
| | | | | | EL4− | | EL4+ | | |
| | K | I | D | | No. 5 | 11 | 3 | 7 | 13 |
| EL4 | b | b | b | − | −[a] | 128[b] | 32 | 64 | 256 |
| | | | | + | − | 128 | 256 | 256 | 2048 |
| B10 | b | b | b | − | − | 64 | − | − | 128 |
| | | | | + | − | 64 | 4 | 16 | 512 |
| A.BY | b | b | b | − | − | 32 | − | − | 32 |
| | | | | + | − | 32 | 4 | 8 | 128 |
| C3H.B10 | b | b | b | − | − | 64 | − | − | 64 |
| | | | | + | − | 64 | 8 | 16 | 256 |
| B10.D2 | d | d | d | − | − | − | − | − | − |
| | | | | + | − | − | − | − | − |
| B10.M | f | f | f | − | − | 16 | − | − | 4 |
| | | | | + | − | 16 | 8 | − | 8 |
| B10.BR | k | k | k | − | 16 | 128 | − | − | 16 |
| | | | | + | 16 | 128 | 4 | − | 32 |
| B10.P | p | p | p | − | − | 64 | − | − | 64 |
| | | | | + | − | 128 | − | − | 256 |
| B10.G | q | q | q | − | − | 16 | − | − | 64 |
| | | | | + | − | 32 | − | − | 256 |
| B10.RIII | r | r | r | − | − | 16 | − | − | 32 |
| | | | | + | − | 16 | − | − | 64 |
| B10.S | s | s | s | − | − | − | − | − | − |
| | | | | + | − | − | − | − | − |
| A.TL | s | k | d | + | NT[c] | − | − | NT | − |
| B10.A (3R) | b | b/k | d | − | NT | 32 | − | − | 64 |
| | | | | + | NT | 32 | 4 | 4 | 128 |
| HTG | d | d | b | − | NT | 32 | − | − | 64 |
| | | | | + | NT | 32 | 8 | 4 | 64 |

[a] −, negative in the dilution ½
[b] Numbers indicate reciprocals of lymphocytotoxic titers
[c] NT, not tested

The reaction pattern of the lymphocytotoxic sera from individual mice, as tested on a panel of various H-2 congeneic strains (Table 2), was dependent on the H-2 haplotype, and ranged from almost private (serum no. 5) to widely cross-reactive (sera nos. 11 and 13) serotypes. The fact that 100% of spleen cells were lysed and that the identity in the H-2I region was not sufficient to ensure the positive reactions (Table 2, B10.BR and ATL target cells) strongly indicates that the lymphocytotoxic antibodies are class I specific. The results concerning recombinant H-2 haplotypes (Table 2) show that the sera contained autoreactive lymphocytotoxic antibodies which recognized both the $K^b$ [B10.A(3R)] and $D^b$ (HTG) region products. Two sera (nos. 3 and 7) from the mice immunized with EL4+ cells were positive on $K^b$ and/or $D^b$ splenocytes (Table 2) and Con A blasts (data not shown) only when they were infected with the Sendai virus. These sera were cytotoxic for both the EL4− and the EL4+ cells. The presence of antibody types such as these has been observed so far only in the sera of mice immunized with EL4+ cells. An attempt to generate monoclonal antibodies which recognize the same structure on virus-infected cells as those seen by cytotoxic T cell receptors has been recently reported using syngeneic virus-infected cells (Tamminen et al. 1987; Pestalozzi et al. 1987). Such antibodies, if they mimicked the T cell receptor specificity, would be expected to be both virus specific and restricted in their property to be bound by the MHC antigens. None of the ten infected-cell-specific monoclonal antibodies generated in mice after multiple injections of syngeneic, influenza-infected EL4 cells were restricted by the MHC in the recognition of the viral antigens (Tamminen et al. 1987). Three monoclonal antibodies have been isolated from mice immunized with syngeneic Sendai virus-coated cells (Opolski et al. 1986; Kievits et al. 1987b), which, whilst not recognizing either viral antigen or self-MHC+X composite determinants, nevertheless recognize allogeneic H-2 determinants. However, the possibility that MHC-restricted antibodies could be generated is supported by a number of reports in the literature on the subject (Van Leeuwen et al. 1979; Wylie et al. 1982; Froscher and Klinman 1986; Schmidt 1987). In the present report it has been suggested that the induced cytotoxic antibodies (serum no. 7 in Table 2) imitated MHC-restricted antibodies. However, a more detailed study, especially on the monoclonal level, is required to confirm their reaction pattern and to avoid any misleading interpretation.

Taken together, our data show the presence of H-2 class I specific antibodies in the sera of mice immunized with EL4 tumor cells. The infection of tumor cells with the Sendai virus significantly increases the production of H-2 class I specific autoreactive antibodies. Recently, the induction of H-2 specific antibodies without any intentional alloimmunization has been described in other experimental models. Various approaches have been applied: injection of syngeneic cells (Ivanyi et al. 1980; Čapková et al. 1983; Takahashi et al. 1987), Con A-induced blasts (Mierau and Cramer 1984), syngeneic tumor cells (Schmidt and Alonzo 1983; Philipps et al. 1985), and syngeneic virus (Sendai)-coated cells (Opolski et al. 1986; Kievits et al. 1987a, b). The mechanisms responsible for the induction of such antibodies and their biological consequences are unknown at present. In our experiments (Opolski et al. 1987) there was no correlation between the presence of autoreactive H-2 specific antibodies and the resistance of the mice which rejected EL4+ cells to the subsequent EL4 challenge (only 20% of resistant mice had detectable levels of autoreactive antibodies).

More experiments are needed in order to make possible a description of some of the relevant molecular interactions and to allow the dissection of the bases causing this intriguing phenomenon.

## References

Abramson-Leeman SR, Cantor H (1983) Specificity of T cell clones for antigen and autologous major histocompatibility complex products determines specificity for foreign major histocompatibility complex products. J Exp Med 158: 428–437

Ashwell JD, Chen C, Schwartz RH (1986) High frequency and nonrandom distribution of alloreactivity in T cell clones selected for recognition of foreign antigen in association with self class II molecules. J Immunol 136: 389–395

Askonas BA, Lin YL (1982) An influenza specific T-killer clone is restricted to H-2 $L^d$ and cross-reacts with $D^k$ region. Immunogenetics 16: 83–87

Braciale TJ, Andrew ME, Braciale VL (1981) Heterogeneity and specificity of cloned lines of influenza-virus-specific cytotoxic T lymphocytes. J Exp Med 153: 910–923

Čapková J, Hašek M, Kousalová M, Holáň V, Boubelík M (1983) Alloreactive cytotoxic anti-H-2 antibodies in sera of the syngeneically immunized C57BL/10ScSnPh, BALB/cJPh, and B10. A/SnPh mice. Folia Biol (Praha) 29: 200–212

Carlow D, Kerbel R, Feltis T, Elliott B (1985) Enhanced expression of class I major histocompatibility complex gene ($D^k$) products on immunogenic variants of a spontaneous murine carcinoma. JNCI 75: 291–302

Finberg R, Burakoff SJ, Cantor H, Benacerraf B (1978) Biological significance of alloreactivity: T cells stimulated by Sendai virus-coated syngeneic cells specifically lyse allogeneic target cells. Proc Natl Acad Sci USA 75: 5145–5149

Flood PM, Schreiber H, Ron Y (1987) Protective immunity to progressive tumors can be induced by antigen presented on regressor tumors. J Immunol 138: 3573–3579

Frelinger JA, Niederhuber JE, David CS, Shreffler DC (1974) Evidence for the expression of Ia (H-2-associated) antigens on thymus derived lymphocytes. J Exp Med 140: 1273–1284

Froscher BG, Klinman NR (1986) Immunization with SV40-transformed cells yields mainly MHC-restricted monoclonal antibodies. J Exp Med 164: 196–210

Frost P, Kerbel R, Bauer E, Tartamella-Biondo R, Cefalu W (1983) Mutagen treatment as a means for selecting immunogenic variants from otherwise poorly immunogenic malignant murine tumors. Cancer Res 43: 125–132

Goodenow RS, Vogel JM, Linsk RL (1985) Histocompatibility antigens on murine tumors. Science 230: 777–783

Gorelik E, Peppoloni S, Overton R, Herberman RB (1985) Increase in H-2 antigen expression and immunogenicity of BL6 melanoma cells treated with *N*-methyl-nitronitrosoguanine. Cancer Res 45: 5341–5347

Ivanyi P, van Mourik P, Breuning M, Melief CJM (1980) Anti-H-2 antibodies induced by syngeneic immunization. Immunogenetics 10: 319–332

Kievits F, Rocca A, Opolski A, Limpens J, Leupers T, Kloosterman T, Boerenkamp WJ, Pla M, Ivanyi P (1987 a) Induction of H-2-specific antibodies by injections of syngeneic Sendai virus-coated cells. Eur J Immunol 17: 27–35

Kievits F, Opolski A, Boerenkamp WJ, Pla M, Ivanyi P (1987 b) Monoclonal H-2 class I specific antibodies isolated after immunization of C57BL/6 mice with syngeneic Sendai virus-coated cells. In: David CS (ed) H-2 antigens: genes, molecules, function. Plenum, New York, p 669–680

Koyama K, Ishii K (1969) Induction of non-transplantable mutant clones from an ascites tumor. Gann 60: 367–374

Mierau R, Cramer M (1984) Serological and immunochemical analysis of H-2 class I molecules encoded by the $D^b$ region. Immunogenetics 20: 341–345

Opolski A, Kievits F, Ivanyi P (1986) Polymorphic and autoreactive H-2-specific monoclonal antibody isolated after injections of syngeneic Sendai virus-coated lymphocytes. Immunogenetics 24: 402–408

Opolski A, Degos L, Pla M (1988) Rejection of Sendai virus-infected EL4 thymoma cells by mice. Cancer Immunol Immunother (submitted)

Peppoloni S, Herberman RB, Gorelik E (1985) Induction of highly immunogenic variants of Lewis lung carcinoma tumor by ultraviolet irradiation. Cancer Res 45: 2560–2566

Pestalozzi B, Stitz L, Zinkernagel RM (1987) Monoclonal antibodies against viral determinants are not restricted to the K/D end of the major histocompatibility complex. J Exp Med 166: 295–299

Philipps C, McMillan M, Flood PM, Murphy DB, Forman J, Lancki D, Womack JE, Goodenow RE, Schreiber H (1985) Identification of a unique tumor-specific antigen as a novel class I major histocompatibility molecule. Proc Natl Acad Sci USA 82: 5140–5144

Schmidt W (1987) H-2-specific antibodies induced by injection of syngeneic leukemia cells. Immunogenetics 25: 215–221

Schmidt W, Alonzo A (1983) Biologic significance of autoantibodies against H-2 antigens in syngeneic antitumor sera. Transplant Proc 15: 2118–2120

Takahashi T, Matsudaira Y, Obata Y, Moriwaki K (1987) An autoreactive H-2-specific monoclonal antibody with allospecificity. Immunogenetics 26: 105–106

Tamminen WL, Wraith D, Barber DH (1987) Searching for MHC-restricted anti-viral antibodies: Antibodies recognizing the nucleoprotein of influenza virus dominate the serological response of C57BL/6 mice to syngeneic influenza-infected cells. Eur J Immunol 17: 999–1006

Tsukagoshi S, Hashimoto Y (1973) Increased immunosensitivity in nitrogen mustard-resistant Yoshida sarcoma. Cancer Res 33: 1038–1042

Van Leeuven A, Goulmy E, van Rood JJ (1979) Major histocompatibility complex-restricted antibody reactivity mainly, but not exclusively, directed against cells from male donors. J Exp Med 150: 1075–1083

Wakamiya N, Wang Y-L, Imai H, Gu H-X, Ueda S, Kato S (1986) Feasibility of UV-inactivated vaccinia virus in the modification of tumor cells for augmentation of their immunogenicity. Cancer Immunol Immunother 23: 125–129

Williams ME, Cox DC, Stevenson JR (1986) Rejection of reovirus-treated L1210 leukemia cells by mice. Cancer Immunol Immunother 23: 87–92

Wylie DE, Sherman LA, Klinman NR (1982) Participation of the major histocompatibility complex in antibody recognition of viral antigens expressed on infected cells. J Exp Med 155: 403–414

Yang H, Welsh RM (1986) Induction of alloreactive cytotoxic T cells by acute virus infection of mice. J Immunol 136: 1186–1193

Zinkernagel RM, Doherty PC (1975) H-2 compatibility requirement for T cell mediated lysis of target cells infected with lymphocytic choriomeningitis virus. Different cytotoxic T-cell specificities are associated with structures coded for in H-2K or H-2D. J Exp Med 141: 1427–1436

Zinkernagel RM, Doherty PC (1979) MHC-restricted cytotoxic T cells: studies on the biological role of polymorphic major transplantation antigens determining T cell restriction specificities, function, and responsiveness. Adv Immunol 27: 51–177

# Immunization with Fibroblasts Expressing Human $\beta_2$-Microglobulin Induces H-2 Specific Antibodies in Syngeneic Recipients

ANNA ROCCA, LAURENT DEGOS, and MARIKA PLA

Laboratory of Mouse Immunogenetics, U93 INSERM, Saint Louis Hospital, 2, place du Dr. A. Fournier, 75010 Paris, France

It is now generally accepted that the expression of some antigenic determinants of major histocompatibility complex (MHC) class I molecules results from a complex interaction between a heavy chain, encoded by highly polymorphic loci within the MHC, and the $\beta_2$-microglobulin ($\beta_2$-m) encoded by a rather monomorphic non-MHC locus. In an attempt to evaluate the structural consequences of the association of murine class I heavy chains with xenogeneic $\beta_2$-m, we investigated the antibody response of the mice injected with syngeneic fibroblasts expressing human $\beta_2$-m.

C3H mice (H-$2^k$) were injected i.p. weekly with $10^7$ fibroblasts (L) of C3H origin or fibroblasts transfected with the gene coding for the human $\beta_2$-m (L+$\beta_2$-m). After the 12th injection, the mice were bled from the orbital sinus and the individual sera were assayed in a complement-dependent microcytotoxicity test (Frelinger et al. 1974) against a panel of splenocytes from various H-2 congeneic strains. The lymphocytotoxic tests were performed in the absence of any heterologous sera.

The reaction pattern of sera coming from two independent experiments in which mice were immunized with L+$\beta_2$-m are presented in Table 1. Of the 20 mice immunized with fibroblasts expressing the human $\beta_2$-m, 11 (55%) produced lymphocytotoxic antibodies. The reaction pattern of these sera on the panel of splenocytes from eight H-2 congeneic strains suggested that the lymphocytotoxic antibodies are H-2 specific. In both experiments the reactivity of lymphocytotoxic sera was variable and ranged from narrow to widely cross-reacting serotypes. It was also observed that most of the positive sera were reactive with more than one haplotype and that H-$2^p$ and H-$2^r$ cells were the most frequently recognized target cells (40% of sera), whereas H-$2^s$ target cells were recognized with the lowest frequency (10% of the sera). Interestingly, 20% (4/20) of the mice immunized with L+$\beta_2$-m fibroblasts produced antibodies which were cytotoxic for syngeneic lymphocytes. This autoreactivity was always accompanied by cytotoxic reactions on other haplotypes.

Additionally, all the 20 sera were positive on all human cells (peripheral blood lymphocytes from 19 unrelated individuals) and cell lines (9 EBV cell lines, established from unrelated individuals). There was no correlation between the reactivity on human cells and the lymphocytotoxic reactivity on murine splenocytes. The absence of reactivity in the case of Daudi and K562 cell lines, which are known not to express the heterodimer HLA class I heavy chain and human $\beta_2$-m on their surface, suggested that the sera contain antibodies recognizing the human $\beta_2$-m.

Similar serological studies on the panel of H-2 congeneic strains were made with

**Table 1.** Reaction pattern of sera from H-$2^k$ mice immunized with syngeneic fibroblasts expressing human $\beta_2$-microglobulin tested on a panel of splenocytes from eight H-2 congeneic strains with B10 genetic background

| Serum no. | H-2 haplotype of target cells | | | | | | | |
|---|---|---|---|---|---|---|---|---|
| | s | k | b | q | f | d | p | r |
| 5 | –[a] | – | – | – | – | – | – | – |
| 7 | – | – | – | – | – | – | – | – |
| 8 | – | – | – | – | – | – | – | – |
| 10 | – | – | – | – | – | – | – | – |
| 12 | – | – | – | – | – | – | – | – |
| 14 | – | – | – | – | – | – | – | – |
| 15 | – | – | – | – | – | – | – | – |
| 17 | – | – | – | – | – | – | – | – |
| 19 | – | – | – | – | – | – | – | – |
| 3 | – | – | – | – | – | – | – | 8[b] |
| 4 | – | – | – | – | – | – | – | 16 |
| 13 | – | – | – | – | – | – | 16 | 4 |
| 2 | – | – | – | 4 | – | – | 8 | 16 |
| 9 | – | – | – | – | – | 4 | 16 | – |
| 11 | – | – | 4 | – | 64 | 64 | 32 | – |
| 16 | 8 | – | 32 | – | 8 | 64 | 64 | – |
| 6 | – | 32 | 128 | 64 | 16 | 8 | 128 | 16 |
| 18 | – | 32 | 64 | 128 | 64 | 8 | 64 | 32 |
| 20 | – | 128 | – | 128 | 16 | 128 | – | 64 |
| 1 | 64 | 128 | 128 | 64 | 64 | 128 | 128 | 128 |
| Total positive | | | | | | | | |
| 11/20 | 2 | 4 | 5 | 5 | 6 | 7 | 8 | 8 |
| 55% | 10 | 20 | 25 | 25 | 30 | 35 | 40 | 40 |

[a] –, negative in ½ dilution
[b] Numbers indicate reciprocals of lymphocytotoxic titer

**Table 2.** Reactivity of the serum no. 6 (Table 1) before and after absorption with various cells ($5 \times 10^7$) on untransfected fibroblasts (L), on human $\beta_2$-m expressing fibroblasts (L+$\beta_2$-m), and on H-$2^k$ and H-$2^p$ splenocytes

| Absorbed on | Reciprocal of LCT titer on target cells | | | |
|---|---|---|---|---|
| | L | L+$\beta_2$m | H-$2^k$ | H-$2^p$ |
| Unabsorbed | 128 | 1024 | 32 | 128 |
| L | – | 320 | – | – |
| L+$\beta_2$-m | – | – | – | – |
| H-$2^p$ splenocytes | 80 | 640 | – | – |

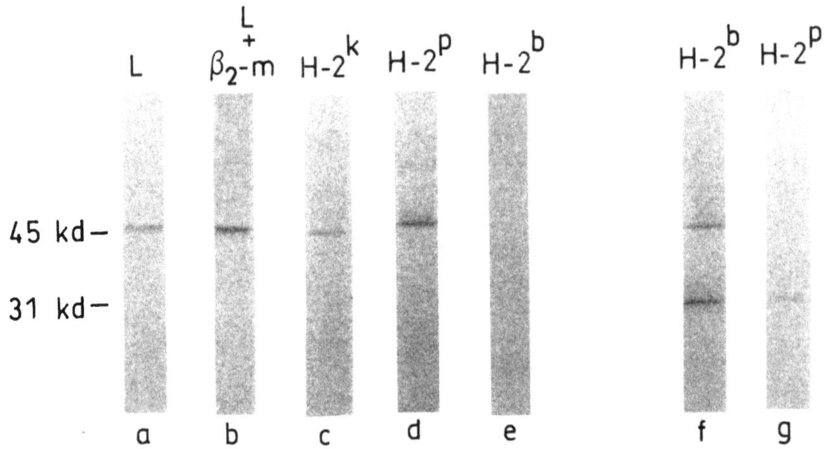


**Fig. 1.** SDS-PAGE analysis of immunoprecipitates prepared from Nonidet P-40 lysates of internally labeled L and L+$\beta_2$-m fibroblasts, and H-2$^k$, H-2$^p$, and H-2$^b$ splenocytes using sera from H-2$^k$ mice (pool of sera nos. 2, 9, and 20) immunized with syngeneic fibroblasts expressing human $\beta_2$-m (L+$\beta_2$-m) *(lanes a–e)* or H-2$^p$ anti-H-2$^b$ serum *(lanes f and g)*

the sera coming from the mice immunized with untransfected fibroblasts (L), and none of the sera was found to be lymphocytotoxic (data not shown).

In an attempt to characterize further the structures recognized by the antibodies present in the sera coming from the mice immunized with L+$\beta_2$-m, absorption studies were performed. The results presented in Table 2 show that the absorption on L cells does not remove all reactivity against L+$\beta_2$-m fibroblasts. The ability of H-2$^p$ splenocytes to absorb the reactivity on H-2$^k$ cells but not on L and L+$\beta_2$-m fibroblasts strongly suggests that in parallel with the H-2 specific antibodies, the sera of mice immunized with L+$\beta_2$-m contain antibodies specific for fibroblasts.

Immunoprecipitation studies were performed using a pool of the lymphocytotoxic sera. Figure 1 shows that sera coming from the mice immunized with L+$\beta_2$-m fibroblasts (lanes a–e) contained antibodies which were able to precipitate class I murine heavy chain on L and L+$\beta_2$-m fibroblasts, but also on syngeneic (H-2$^k$) and allogeneic (H-2$^p$) spleen cells. No precipitation was obtained when anti-L+$\beta_2$-m serum pool was assayed on the H-2$^b$ spleen cells.

Taken together, our data show that the target structures recognized by lymphocytotoxic antibodies present in the sera of mice immunized with syngeneic fibroblasts expressing human $\beta_2$-m are H-2 class I molecules. The induction of H-2 specific antibodies without any intentional alloimmunization has been described in other experimental models: by injection of syngeneic cells (Ivanyi et al. 1980; Čapková et al. 1983; Takahashi et al. 1987), syngeneic Con A-induced blasts (Mierau and Cramer 1984), syngeneic tumor cells (Schmidt and Alonzo 1983; Philipps et al. 1985), and syngeneic Sendai virus-coated cells (Opolski et al. 1986; Kievits et al. 1987 a, b). The important question has arisen as to what is the trigger structure responsible for the induction of these H-2 specific antibodies. Kievits et al. (1987 a) proposed that the coating of lymphocytes with Sendai virus might cause changes of cell surface antigens (by a distant signal, without forming a stable complex), resulting in a changed exposure of some cross-reacting determinants of H-2 class I molecules. It has been reported (Bernabeau et al. 1984; Kubota 1984; Ferrier et al. 1985; Kahn-Perles et al. 1987; Kievits and Ivanyi 1987) that some xenogeneic $\beta_2$-m binds to the H-2 class I heavy chains on the cell surface during the culture of the cells in a medium supplemented with xenogeneic serum. The newly formed "hybrid" molecules exhibit alterations of their serological reactivities that suggest the occurrence of structural modifications of both light (Ferrier et al. 1985) and heavy chains (Ferrier et al. 1985; Kahn-Perles et al. 1987; Kievits and Ivanyi 1987). Thus we presume that a new configuration generated by the association of murine class I heavy chain and human $\beta_2$-m is responsible for the induction of the H-2 class I specific antibodies in the sera of mice immunized with syngeneic fibroblasts expressing human $\beta_2$-m.

## References

Bernabeau C, van de Rijn N, Lerch PG, Terhorst P (1984) $\beta_2$-microglobulin from serum associates with MHC class I antigens on the surface of cultured cells. Nature 308: 642–645

Čapková J, Hašek M, Kousalová M, Holáň V, Boubelík M (1983) Alloreactive cytotoxic anti-H-2 antibodies in sera of the syngeneically immunized C57BL/10ScSnPh, BALB/cJPh, and B10.A/SnPh mice. Folia Biol (Praha) 29: 200–212

Ferrier P, Layet C, Caillol DH, Jordan BR, Lemonnier FA (1985) The association between murine $\beta_2$-microglobulin and HLA class I heavy chains results in serologically detectable conformational changes of both chains. J Immunol 135: 1281–1287

Frelinger JA, Niederhuber JE, David CS, Shreffler DC (1974) Evidence for the expression of Ia (H-2-associated) antigens on thymus derived lymphocytes. J Exp Med 140: 1273–1284

Ivanyi P, van Mourik P, Breuning M, Melief CJM (1980) Anti-H-2 antibodies induced by syngeneic immunization. Immunogenetics 10: 319–332

Kahn-Perles B, Boyer C, Arnold B, Sanderson AR, Ferrier P, Lemonnier FA (1987) Acquisition of HLA class I W6/32 defined antigenic determinant by heavy chains from different species following association with bovine $\beta_2$-microglobulin. J Immunol 138: 2190–2196

Kievits F, Ivanyi P (1987) Monomorphic anti-HLA monoclonal antibody (W6/32) recognizes polymorphic H-2 heavy chain determinants exposed by association with bovine or human but not murine $\beta_2$-microglobulin. Hum Immunol 20: 115–126

Kievits F, Rocca A, Opolski A, Limpens J, Leupers T, Kloosterman T, Boerenkamp WJ, Pla M, Ivanyi P (1987a) Induction of H-2-specific antibodies by injections of syngeneic Sendai virus-coated cells. Eur J Immunol 17: 27–35

Kievits F, Opolski A, Boerenkamp WJ, Pla M, Ivanyi P (1987b) Monoclonal H-2 class I specific antibodies isolated after immunization of C57BL/6 mice with syngeneic Sendai-virus-coated cells. In: David CS (ed) H-2 antigens: genes, molecules, function. Plenum, New York, pp 669–680

Kubota K (1984) Association of serum $\beta_2$-microglobulin with H-2 class I heavy chains on the surface of mouse cells in culture. J Immunol 133: 3203–3210

Mierau R, Cramer M (1984) Serological and immunochemical analysis of H-2 class I molecules encoded by the $D^b$ region. Immunogenetics 20: 341–345

Opolski A, Kievits F, Ivanyi P (1986) Polymorphic and autoreactive H-2-specific monoclonal antibody isolated after injections of syngeneic Sendai virus-coated lymphocytes. Immunogenetics 24: 402–408

Philipps C, McMillan M, Flood PM, Murphy DB, Forman J, Lancki D, Womack JE, Goodenow

RE, Schreiber H (1985) Identification of a unique tumor-specific antigen as a novel class I major histocompatibility molecule. Proc Natl Acad Sci USA 82: 5140–5144
Schmidt W, Alonzo A (1983) Biologic significance of autoantibodies against H-2 antigens in syngeneic antitumor sera. Transplant Proc 15: 2118–2120
Takahashi T, Matsudaira Y, Obata Y, Moriwaki K (1987) An autoreactive H-2-specific monoclonal antibody with allospecificity. Immunogenetics 26: 105–106

# MHC-Specific Monoclonal Antibodies Induced by Injection of Syngeneic Leukemia Cells

WILHELM SCHMIDT

Institut für Zellbiologie (Tumorforschung), Universitätsklinikum Essen (GH), Hufelandstrasse 55, 4300 Essen 1, Federal Republik of Germany

## Introduction

The immune system is activated by foreign antigenic determinants in different ways. B cells directly recognize epitopes of antigenic molecules via their surface Ig-receptors whereas T cells only recognize foreign antigens in the context of major histocompatibility complex (MHC) molecules on the surface of somatic cells. Current experiments with MHC class II antigens require that antigenic fragments are bound to MHC molecules and thus presented to T cell receptors (Allen et al. 1987; Buus et al. 1987). Neither the molecular details of the interaction nor the determinants of the bound antigenic fragments which are recognized by T cell receptors are known. The present models of MHC restriction of T cells imply that T cells may recognize novel conformations, induced in the antigenic fragment upon association with certain determinants of the MHC molecules (associative recognition model), or conformational alterations induced in MHC molecules upon association with antigen (altered self model). Additional molecules (L3T4/Ly2, $T_3$) then are involved in stabilizing the $T_c$ receptor mediated interaction and in transduction of the activation signals across the cell membrane (Walker et al. 1984; Alcover et al. 1987).

## H-2 Restricted Antibodies

Novel conformation determinants on MHC bound antigenic fragments and/or the MHC molecules themselves may not only be recognized by T cells but could also lead to stimulation of appropriate receptors on B cells, and thus to the production of antibodies. These would be specific for an MHC molecule and a foreign antigen and therefore could be considered as MHC restricted.

In order to find out if such antibodies indeed exist, I performed immunizations of AKR mice with syngeneic leukemia cells. The rejection of the leukemia cells 369 grown in tissue culture is accompanied by a strong cytotoxic T-lymphocyte (CTL) response. The CTLs are $H\text{-}2K^k$ restricted and probably recognize gag-related murine leukemia virus (MuLV) glycoproteins as antigens (Schäfer and Schmidt 1986). They do not lyse L cells or cultured sarcoma cells of $H\text{-}2^k$ haplotype.

Sera taken from AKR mice immunized with 369 leukemia cells indeed contained antibody which precipitated H-2K and $\beta_2$m and also viral gp 70 from leukemia cell lysates. Antibodies against H-2 antigens only appeared after repeated immunizations, but then constituted the major portion of cytotoxic and precipitating antibodies in the sera. The antisera also precipitated H-2K antigens from other H-2 positive AKR leukemia cells but not from normal AKR lymphocytes (Schmidt 1987 a).

I sought to eliminate the possibility of coprecipitation of H-2 molecules by antiviral antibodies and to ascertain whether H-2 specific antibodies were reacting with mutated H-2K antigens of leukemia cells or whether they recognized conformational determinants induced by association with foreign antigen. Therefore hybridoma cells were produced by fusion of lymphocytes from immune AKR mice with nonproducer myeloma cells.

From three fusions, 21 hybridoma secreting monoclonal antibodies (McAbs) were obtained, which showed distinctly different patterns of reactivity. All McAbs were reacting with the immunizing AKR leukemia cells in a cellular radiobinding assay. Seven of the McAbs precipitated H-2K antigens from 369 tumor cell lysates, but the majority of McAbs did not (Fig. 1). These latter McAbs were noncytotoxic and reacted in cellular binding assays with many different types of tumor cells and with normal lymphocytes of AKR and various other strains of mice; they will not be considered further in this context.

Antibodies 3–7, 4–3, 4–5, and 4–15 were obtained from the first fusion. They reacted in cytotoxic and cellular binding assays with H-2 positive but not with H-2 negative leukemias or with normal AKR lymphocytes. Thus their reactivities depended both on the presence of H-2K$^k$ and on additional molecules present on

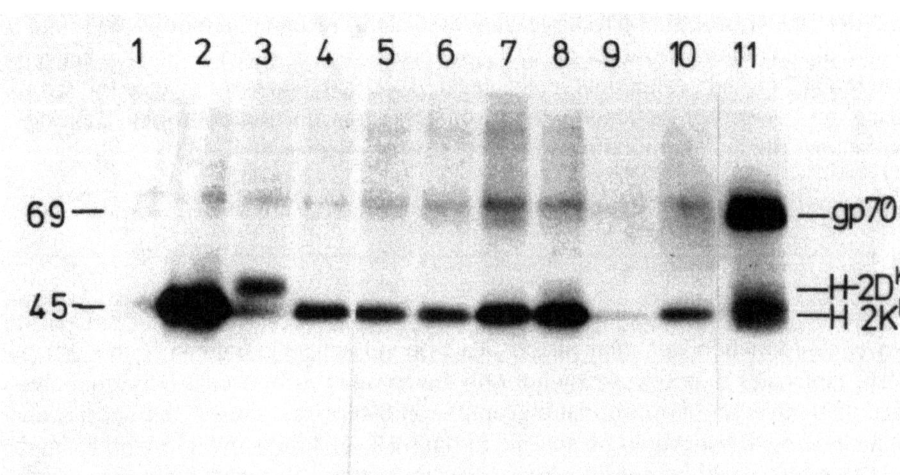


**Fig. 1.** Precipitation of H-2K$^k$ antigens by syngeneic antileukemia McAbs. Antigens: lysates from $^{125}$I-labeled 369 leukemia cells. Antibodies: *1,* control without McAb; *2,* 100/5 (H-2K$^k$ specific); *3,* H-2D$^k$ specific alloantiserum; *4,* McAb 3–7; *5,* McAb 4–3; *6,* McAb 4–5; *7,* McAb 4–15; *8,* McAb 7–5; *9,* McAb 6–18; *10,* McAb 6–22; *11,* Goat anti-MuLV gp 70 serum. The separation of the immunoprecipitates was by 7%–15% gradient SDS-PAGE

leukemia cells but not on lymphocytes. All four McAbs precipitated H-2K molecules from 369 and other AKR leukemia cells, but not from normal AKR lymphocytes (Fig. 1). However, they did react with H-$2^k$ lymphocytes after mitogen stimulation and culture in vitro, suggesting that they were not directed against mutated H-2K antigens on leukemia cells. This notion was also supported by the finding that the McAbs did not react with 369 leukemia cells or AKR lymphoblasts when they were grown in medium supplemented with 1% normal mouse serum (NMS) (of BALB/c origin) instead of fetal calf serum (FCS). However, it also indicated that some FCS components might be recognized in association with H-$2K^k$ antigens.

It is well known that H-2 antigens are heterodimeric molecules, composed of a polymorphic cell membrane bound heavy chain encoded in the MHC and a non-covalently associated $\beta_2$-microglobulin ($\beta_2$m); $\beta_2$m is thought to facilitate the cell membrane integration of H-2 molecules and to stabilize their tertiary structure. However, it can readily be exchanged for exogenous $\beta_2$m when present in the medium (Schmidt et al. 1981; Kubota 1984). Since it had been shown that FCS contains free bovine $\beta_2$m (Bernabeu et al. 1984), I sought to establish whether bovine $\beta_2$m was relevant for the activity of the antileukemia McAbs. Therefore, 369 tumor cells were grown in medium supplemented with 1% NMS for at least five cell doubling times. Under these conditions all cell surface H-2 molecules should be associated with endogenous, murine $\beta_2$m. Then aliquots of the cells were incubated with purified bovine or human $\beta_2$m to facilitate exchange of endogenous $\beta_2$m on the cell surface for their foreign analogues. The exchange reaction was monitored by the use of radiolabeled $\beta_2$m and by precipitation of foreign $\beta_2$m with anti-H-2 alloantisera from cell lysates (Schmidt et al. 1981). Cells with substituted $\beta_2$m and control cells were reacted in cellular binding assay with the McAbs and also radio-labeled and lysed for immunoprecipitation.

All four McAbs reacted with and precipitated H-$2K^k$ molecules from cells after incubation with bovine $\beta_2$m, i.e., when the H-2K heavy chains were associated with bovine $\beta_2$m, but not when associated with murine or human $\beta_2$m (Fig. 2). The McAbs did not react with free bovine $\beta_2$m in a radioimmunoassay, nor were they inhibited by free bovine $\beta_2$m when binding to 369 tumor cells, indicating that they were not directed against determinants of free bovine $\beta_2$m. To find out whether they were recognizing determinants of $\beta_2$m induced upon association with H-2 molecules, they were tested on cells of other H-2 genotypes. Therefore, EL4 (H-$2^b$), MethA (H-$2^d$), and also IC9 (H-$2^{bxk}$) cells, which only express H-$2D^b$ molecules but which had previously been transfected with cloned H-$2K^k$ (IC9$K^k$) or H-$2K^d$ (IC9$K^d$) (Arnold et al. 1985), were grown in medium supplemented with 1% NMS. Aliquots of these cells were incubated with bovine $\beta_2$m and assayed with the McAbs in a cellular binding assay, together with control cells without $\beta_2$m exchange. The McAbs only reacted with IC9$K^k$ cells after incubation with bovine $\beta_2$m, but not with IC9$K^d$, EL4, or MethA cells, even when their H-2 antigens were associated with bovine $\beta_2$m (Schmidt 1987b; and Fig. 2).

These results show that the McAbs specifically reacted with complexes of H-$2K^k$ and bovine $\beta_2$m. They confirm that they are not directed against conformational determinants of bovine $\beta_2$m induced upon association with murine H-2 antigens and suggest that the McAbs may recognize conformational alterations of H-$2K^k$ molecules induced by association with foreign (bovine) $\beta_2$m. An alternative pos-

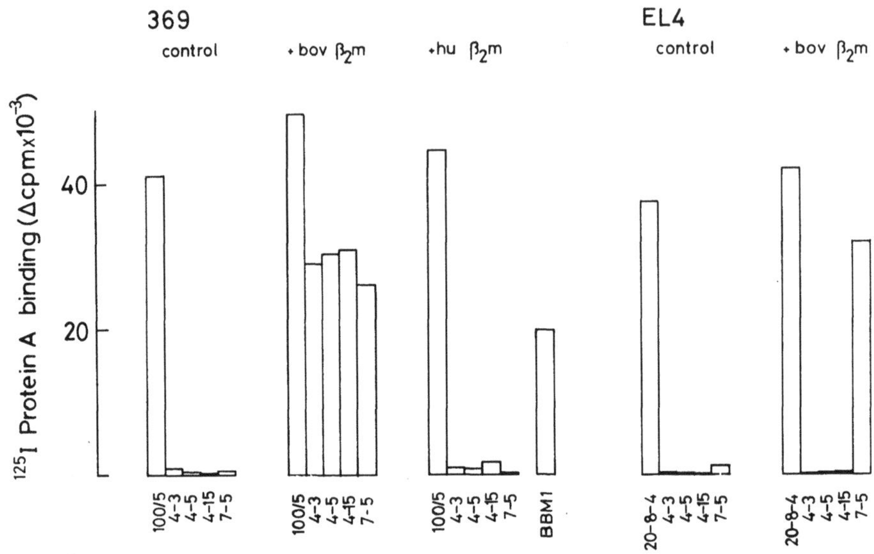


**Fig. 2.** Antigen and MHC specificity of syngeneic antileukemia McAbs. Reactivity in a cellular radiobinding assay to 369 (H-$2^k$) and EL4 (H-$2^b$) leukemia cells is shown. Cells were grown in medium supplemented with 1% NMS (control). Aliquots of cells were preincubated with bovine *(+bov $\beta_2m$)* or human $\beta_2$m *(+hu $\beta_2m$)* before the assay. Reactivities of antileukemia McAbs 4–3, 4–5, 4–15, and 7–5 are shown; McAbs 100/5 (H-$2K^k$ specific), 20-8-4 (H-$2K^b$ specific), and BBMI (human $\beta_2$m-specific) served as controls

sibility is that the McAbs recognize interaction determinants in the complex formed by H-$2K^k$ heavy chain and bovine $\beta_2$m towards which both components contribute.

## Antibodies Specific for Monomorphic Determinants

From fusion #3 McAb 7-5 was obtained which reacted with and precipitated H-2 antigens from lysates of 369 leukemia cells when grown in the presence of FCS (Fig. 1) but not in NMS. When 369 cells grown in the presence of NMS were preincubated with bovine $\beta_2$m, McAb 7-5 again reacted and precipitated H-$2K^k$ and H-$2D^k$ antigens, thus differing from the McAb described above (Schmidt 1987b). Furthermore, it also reacted with and precipitated H-2 antigens from lysates of EL4 and MethA cells when their H-2 antigens were associated with bovine $\beta_2$m (Fig. 2) whereas the first group of McAbs did not. However, like the McAb described before, it did not react with free bovine $\beta_2$m nor was it blocked by it in the cellular binding assay. These results indicated that McAb 7-5 recognized a determinant of bovine $\beta_2$m exposed upon association with murine H-2 antigens, or alternatively a monomorphic determinant of H-2 molecules when associated with bovine $\beta_2$m.

## Other MHC Class I Specific Antibodies

Two antibodies, 6–18 and 6–22, were obtained from fusion # 2 which also precipitated H-2K$^k$ antigens from 369 leukemia cells (Fig. 1). In contrast to the McAb described above, McAbs 6–18 and 6–22 also reacted with 369 cells when grown in the presence of NMS. However, no H-2K$^k$ antigens were precipitated from lysates of such cells, but a class I molecule with a 40-kd heavy chain and $\beta_2$m. Furthermore, the McAbs also reacted with H-2 negative leukemia cells K36 and 422 and with embryonal carcinoma cells F9 and PCC4 (Table 1). The antigens recognized by the McAbs on these cells have not been characterized yet.

The results obtained so far suggest that McAbs 6–18 and 6–22 recognize H-2K$^k$ molecules when associated with bovine $\beta_2$m but also additional MHC class I molecules, independent of bovine $\beta_2$m, possibly also present on H-2 negative leukemia and embryonal carcinoma cells. One antibody, 7–8, was obtained from fusion # 3 which showed a reaction pattern similar to 6–18 and 6–22, but which did not precipitate H-2K$^k$ associated with bovine $\beta_2$m. It did, however, also precipitate a 40-kd polypeptide together with $\beta_2$m from 369 cells.

Interestingly anti-MHC class I antibodies similar to those described here were obtained without intentional immunization. While screening 40 sera from preleukemic AKR mice, three sera were found which precipitated 45-kd and 11-kd proteins from lysates of 369 leukemia cells. The precipitation was not inhibited by prior removal of H-2K$^k$ antigens by affinity chromatography (Fig. 3). This suggested that these sera contained antibodies which were not directed against H-2K (45-kd) or H-2D (48-kd) but against additional MHC class I molecules.

## Biological Significance

Although $\beta_2$m is the natural subunit of H-2 molecules, it can also be considered a prototype foreign antigen. Exogenously added, allogeneic or xenogeneic $\beta_2$m binds to H-2 antigens by displacing the endogeneous $\beta_2$m, and can induce an H-2

**Table 1.** Reactivity of antileukemia sera with leukemia and embryonal carcinoma cells

| Cell line | Origin | $^{125}$I protein A binding ($\Delta$ cpm)[a] | | | | |
|---|---|---|---|---|---|---|
| | | H-2K[b] | 3–7 | 6–18[c] | 6–22[c] | 7–8[c] |
| 369 | AKR | 154677 | 38509 | 23306 | 34142 | 19943 |
| 422 | AKR | 435 | 323 | 53399 | 29350 | 38949 |
| EL4 | B6 | 79420 | 165 | 6650 | 47842 | 86990 |
| F9 | 129 | 0 | 0 | 358 | 23016 | 56675 |
| PCC4 | 129 | 137 | 32 | 8723 | 20741 | ND |

[a] Reactivity of the McAbs was tested in a cellular radiobinding assay
[b] Anti-H-2K McAbs 100/5 and 20-8-4 were used as controls
[c] A second antibody (sheep anti-mouse IgI) was used to enhance protein A binding

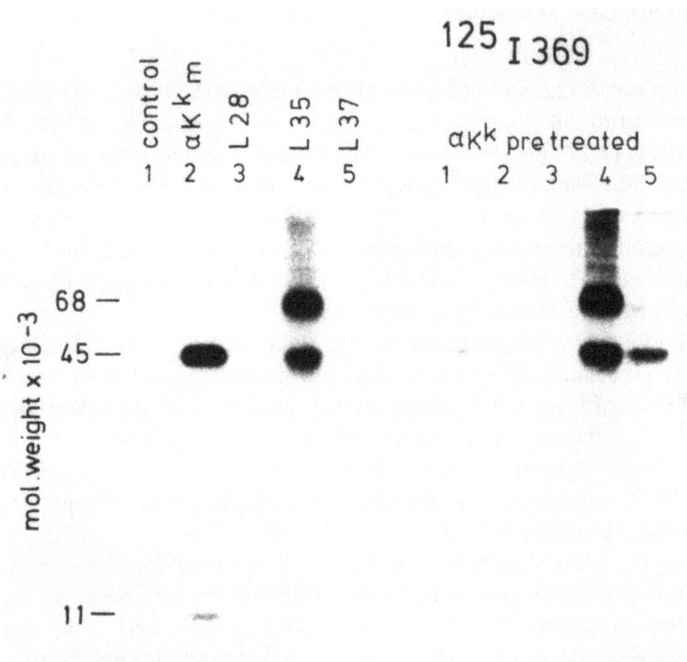


**Fig. 3.** Presence of MHC class I specific antibodies in sera from preleukemic AKR mice. Antigens: lysates from $^{125}$I-labeled 369 cells either untreated *(left side)* or after passage through an affinity column (McAb 100/5 coupled to Sepharose) to remove H-2K$^k$ antigens (*right side:* $\alpha$K$^k$ treated). Antibodies: *1,* control without antibody; *2,* McAb 100/5; *3–5,* sera from preleukemic AKR mice 28, 35, and 37, respectively

restricted CTL response (Rammensee et al. 1986). The production of McAbs specific for H-2K$^k$ as well as for bound bovine $\beta_2$m demonstrates that not only T cells but also B cells can be MHC restricted. MHC-restricted antibodies may recognize either specific structural alterations of H-2 molecules induced by association with a foreign antigen, or alternatively conformational determinants of a complex towards which the bound antigen as well as the MHC molecule contributes. The existence of MHC-restricted antibodies therefore demonstrates that the information necessary for "associative recognition" can be inherent in one particular conformation of the interaction complex. Since the antigen binding sites of T cell receptors probably are of similar size to those of immunoglobulins, it is likely that T cells also recognize these or related determinants. Thus the McAbs will provide a tool for elucidating their nature.

MHC specific antisera have already been found in other systems (Van Leeuwen et al. 1979; Wylie et al. 1982; Froscher and Klinman 1986). In some cases only alloreactive antibodies were found, suggesting a modified exposure of MHC determinants by virus infection (Opolski et al. 1986). However, alloreactivity also is a feature of certain virus specific CTLs. It is not known, however, whether MHC-restricted antibodies are formed normally during the course of a cellular immune response. They may only be present in small amounts and have a regulatory role within the idiotype network. Their presence in some cases is associated with pathological conditions in which immune responses are involved (Klöting and Stark 1985; Manson 1987). If present in higher amounts, they could directly interfere with or augment the function of effector cells. In this context it will be important to demonstrate whether CTLs specific for bovine $\beta_2$m are present in our immune mice. Interestingly, no McAbs specific for MuLV structural proteins were obtained in the experiments described here.

The demonstration of antibodies against MHC class I molecules other than H-2 antigens implies the presence of such molecules on the surface of the immunizing AKR leukemia cells. These antibodies cannot be directed against Qa 2, 3 or the classical TL antigens because they are not expressed on AKR (Qa $2^b$, $Tla^b$) leukemia cells. They do, however, express Qa-$1^b$ antigens. Several MHC class I genes of the Qa/Tla region with very low polymorphism have been reported to show a widespread tissue expression (Palmer and Frelinger 1987; Transy et al. 1987). Our McAbs should enable us to characterize the additional MHC class I antigens and to assign the encoding genes.

*Acknowledgments. I am grateful to G. Kaltenmaier for technical assistance and to the Deutsche Forschungsgemeinschaft for financial support.*

## References

Alcover A, Ramarli D, Richardson NE, Chang HC, Reinherz EL (1987) Functional and molecular aspects of human T lymphocyte activation via $T_3$–T and T11. Immunol Rev 95: 5–36

Allen PM, Matsueda GR, Evans RJ, Dunbar JB, Marshall GR, Unanue ER (1987) Identification of the T-cell and Ia contact residues of a T-cell antigenic epitope. Nature 327: 713–715

Arnold B, Horstmann Ü, Kuon W, Burgert HG, Hämmerling GJ, Kvist S (1985) Alloreactive cytolytic T cell clones preferentially recognize conformational determinants on histocompatibility antigens: analysis with genetically engineered hybrid antigens. Proc Natl Acad Sci USA 82: 7030–7034

Bernabeu C, Rijn M, Lerch PG, Terhorst CP (1984) $\beta_2$-microglobulin from serum associates with MHC class I antigens on the surface of cultured cells. Nature 308: 642–645

Buus S, Sette A, Colon SM, Miles C, Grey HM (1987) The relation between major histocompatibility complex (MHC) restriction and the capacity of Ia to bind immunogenic peptides. Science 235: 1353–1358

Froscher BG, Klinman NR (1986) Immunisation with SV40-transformed cells yields mainly MHC-restricted monoclonal antibodies. J Exp Med 164: 196–210

Klöting I, Stark O (1985) Occurrence of antibodies against gene products of the major histocompatibility complex in normal sera of diabetes-prone BB rats. Folia Biol (Praha) 31: 34–37

Kubota K (1984) Association of serum $\beta_2$-microglobulin with H-2 class I heavy chains on the surface of mouse cells in culture. J Immunol 133: 3203–3210

Manson LA (1987) Progressive tumor growth and host immune response (Abstr). Proc Am Assoc Cancer Res 28: 348

Opolski A, Kievits F, Ivanyi P (1986) Polymorphic and autoreactive H-2 specific monoclonal antibody isolated after injection of syngeneic Sendai virus-coated lymphocytes. Immunogenetics 24: 402–408

Palmer MJ, Frelinger JA (1987) Widespread transcription of a Qa region gene in adult mice. J Exp Med 166: 95–108

Rammensee HG, Robinson PJ, Crisanti A, Bevan MJ (1986) Restricted recognition of $\beta_2$-microglobulin by cytotoxic T lymphocytes. Nature 319: 502–504
Schäfer A, Schmidt W (1986) Resistance to cellular immune response in AKR leukemias. Eur J Immunol 16: 753–759
Schmidt W (1987a) H-2 specific antibodies induced by injection of syngeneic leukemia cells. Immunogenetics 25: 215–221
Schmidt W (1988) Monoclonal antibodies recognize interaction determinants of MHC plus foreign $\beta_2$-microglobulin. (Submitted) Immunogenetics, in press
Schmidt W, Festenstein H, Ward PJ, Sanderson AR (1981) Interspecies exchange of $\beta_2$-microglobulin and associated MHC and differentiation antigens. Immunogenetics 13: 483–491
Transy C, Nash SR, David-Watine B, Cochet M, Hunt SW, Hood LE, Kourilsky P (1987) A low polymorphic mouse H-2 class I gene from the Tla complex is expressed in a broad variety of cell types. J Exp Med 166: 341–361
Van Leeuwen A, Goulmy E, van Rood JJ (1979) Major histocompatibility complex-restricted antibody reactivity mainly, but not exclusively, directed against cells from male donors. J Exp Med 150: 1075–1083
Walker ZD, Hogarth PM, Murray BJ, Lovering KE, Classon BJ, Chambers GW, McKenzie IFC (1984) Ly antigens associated with T cell recognition and effector function. Immunol Rev 82: 47–77
Wylie DE, Sherman LA, Klinman WR (1982) Participation of the major histocompatibility complex in antibody recognition of viral antigens expressed on infected cells. J Exp Med 155: 403–414

# Monoclonal Antibodies to Murine Tumors as Probes for Genetic Analysis of Individually Specific Tumor Antigens

Patricia L. Ward and Hans Schreiber

Department of Pathology, University of Chicago, 5841 S. Maryland Ave., P.O. Box 414 Chicago, IL 60637, USA

## Introduction

The existence of tumor-specific antigens was first demonstrated in chemically induced tumors in mice (Gross 1943; Foley 1953). These tumors displayed an enormous antigenic diversity as immunization with tumor tissue but not normal donor tissue led to strong resistance to a challenge from the same tumor but not to reproducible resistance against any other independently induced tumor (Prehn and Main 1957).

The seemingly infinite variety of tumor-specific antigens has been postulated to arise from a family of related but highly polymorphic genes such as those of the major histocompatibility complex (MHC) or the immunoglobulin genes (Burnet 1970). While some evidence for these possibilities exist (Invernizzi and Parmiani 1975; Callahan et al. 1983; Pravtcheva et al. 1981), the difficulty in obtaining serological probes with unique tumor specificity as well as the limited number of conclusive studies makes it impossible to determine whether these examples are the exception or the rule for the origin of other unique tumor antigens.

Previous studies in our laboratory generated monoclonal antibodies against the UV-induced fibrosarcoma 1591 to try to elucidate the biochemical and molecular nature of a tumor-specific transplantation antigen. These antibodies were found to recognize novel class I MHC molecules, one of which is a target for tumor rejection in the syngeneic host (Philipps et al. 1985; Stauss et al. 1986a, b).

In order to determine conclusively whether these tumor-specific antigens are the direct result of the carcinogenic effects of ultraviolet light causing somatic mutations, it is critical to examine on the genetic level normal tissues from the host of tumor origin. We have recently generated a bank of UV-induced and spontaneous tumors from which we have isolated concurrently normal tissues from each mouse which gave rise to a tumor. We have recently raised two additional antibodies against these syngeneic tumors in the effort to identify the genetic origin(s) of tumor-specific antigens.

## Materials and Methods

### *Mice*

C3H/HeN ($MTV^-$) mice were obtained from the National Cancer Institute – Frederick Cancer Research Facility. This colony is monitored regularly for genetic stability by skin transplantation and analysis of isoenzyme patterns.

### *Tumors*

All tumors were induced in Frederick C3H/HeN ($MTV^-$) mice. One group was derived by Kripke (1981) and includes the 1591 tumor. The other was derived recently by the same protocol as Kripke in our laboratory (unpublished results). The derivation, antigenic phenotype, and culture of the parental 1591 tumor and its variants have been described in detail (Wortzel et al. 1983; Urban et al. 1982a, b; Urban and Schreiber 1983).

### *Immunological, Biochemical, and Genetic Analysis*

Derivation of the CP28 and CP3F4 antibodies and properties of the anti-class I MHC (H-2) ($L^d$, $D^d$, $K^k$, $D^k$) monoclonal antibodies, MHC-gene transfected L cells, and the anti-$L^d$ cytotoxic T-cell line L3 have been described (Philipps et al. 1985). Microcytotoxicity assays, chromium release assays, two-dimensional gel electrophoresis, and isoenzyme analysis were done as previously described (Philipps et al. 1985).

### *Isolation of Novel Class I Genes*

Construction and screening of the 1591 genomic library, transfection of cloned class I genes, and analysis by Southern blot have been described in detail (Stauss et al. 1986a). Briefly, genomic DNA from 1591 tumor cells was partially digested with the restriction enzyme MboI, and the fragments inserted into the lambda vector EMBL3 (Frischauf et al. 1983) and filter hybridized with a class I specific probe pH-2IIa (Steinmetz et al. 1981). Phage clones containing complete class I genes were transfected into C3H L cells and analyzed with the FACS IVB for class I antigen expression using the CP28 and CP3F4 antibodies specific for the 1591 tumor as well as the antibodies 30-5-7 and 28-14-8 which were raised against the H-2 $L^d$ antigen (Ozato et al. 1980) and cross-react with the 1591 tumor (Philipps et al. 1985). Conditions of Southern blot analysis and origin of the MHC class I specific probe 149.6-6 have been described (Stauss et al. 1986a).

## Results and Discussion

The monoclonal antibody CP28 raised against the 1591 tumor was tested for specificity using the cell sorter. Figure 1 shows that a total of 37 other independently

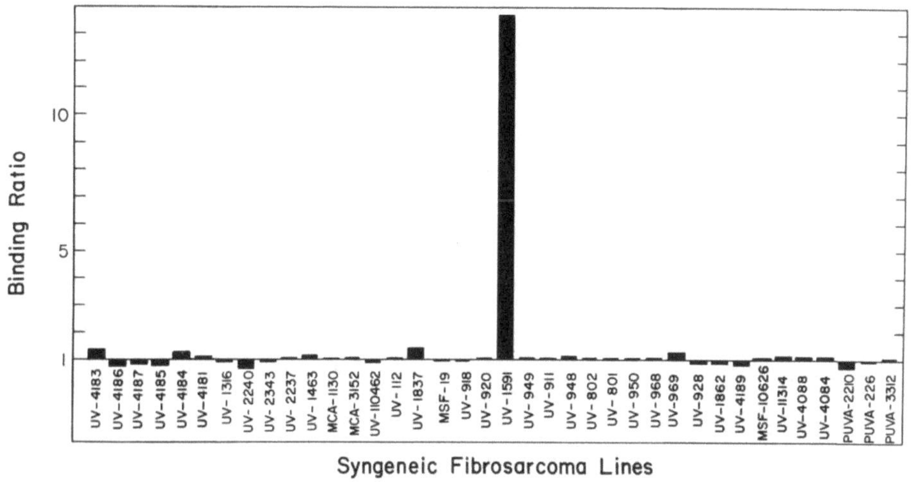


**Fig. 1.** Unique tumor specificity of the syngeneic monoclonal antibody CP28 when tested on a battery of 37 other syngeneic fibrosarcoma lines. Cells from each line were incubated first with CP28 antibody and then with fluorescein-coupled goat anti-mouse immunoglobulin antibodies. Cells were analyzed with the FACS IV and the amount of fluorescence was recorded in logarithmic channel numbers. The binding ratio is the ratio of the amount of fluorescence after staining with both antibodies divided by the amount of fluorescence after staining with the second antibody alone, for each cell line

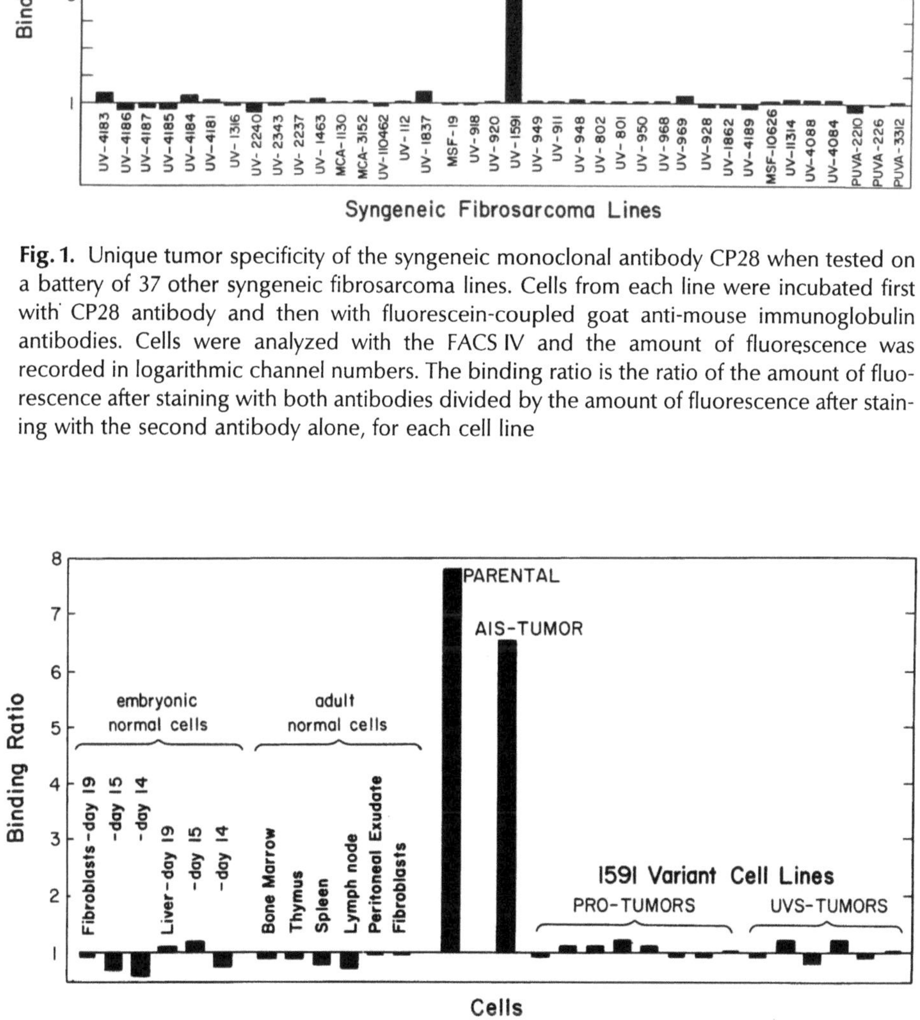


**Fig. 2.** Absence of the CP28 marker on 1591 tumor variants that were reisolated from normal hosts (PRO-tumors) or from UV-irradiated hosts (UVS-tumors), both of which are capable of a specific cytotoxic T-lymphocyte (CTL) response to 1591. The marker is not lost from tumors that grow progressively in idiotypically suppressed mice (AIS-tumors) incapable of generating a tumor-specific CTL response. Antibody CP28 does not bind to the various normal syngeneic adult and embryonic cells tested. The binding ratios were determined as described in the legend of Fig. 1 (From Philipps et al. 1985)

derived syngeneic fibrosarcomas were analyzed and found not to react with CP28. This antibody also did not react with normal syngeneic fibroblasts, embryonic cells, or any 1591 variants which have been isolated (Fig. 2). The antibody CP3F4 derived in an identical manner against 1591 was also specific as far as has been tested (data not shown).

Lymphocytes from other mouse strains were screened next for reactivity with the CP28 monoclonal antibody as this approach can sometimes give a clue about the type of molecule with which an antibody reacts. This is based on the assumption that a mutational change in a normal surface molecule may lead to new epitopes that are more likely to cross-react with molecules of similar origin from other strains because the antigenic polymorphism in the other strain may closely simulate the mutation. It was found that both CP28 and CP3F4 lysed more than 95% of BlO.D2 (H-$2^d$) spleen cells in the presence of complement. Further mapping studies using H-2 recombinant strains localized the determinant detected by CP28 in H-$2^d$ strains to the D end of the H-2 complex and showed that CP3F4 cross-reacts strongly with the K end and weakly with a D end controlled determinant. FACS analysis of C3H L cells transfected with H-$2^d$ class I genes showed that CP28 reacted with $D^d$ transfectants and CP3F4 with $K^d$ transfectants. These data along with the finding that CP28 and CP3F4 each precipitated a different 45-kd molecule that was associated with a 12-kd molecule, suggested the possibility that the 1591 tumor expresses novel MHC class I genes (Philipps et al. 1985). While previous studies have suggested the existence of altered MHC class I molecules on malignant cells (Invernizzi and Parmiani 1975; Callahan et al. 1983), these findings were inconclusive particularly with regard to the question of whether altered MHC class I molecules are related to unique tumor-specific antigens.

A genomic library from 1591 tumor DNA and isolated clones which contained class I sequences was constructed (Stauss et al. 1986a). Clones which contained complete class I genes were transfected into mouse L cells and analyzed for reactivity with the monoclonal antibodies which recognize the novel class I products on 1591. These include the syngeneic tumor-specific CP28 and CP3F4 antibodies and the H-2 $L^d$ specific antibodies 30-5-7 and 28-14-8. Three 1591 genes were identified and designated A216, A166, and A149 which together account for all the novel class I determinants expressed on this tumor (Fig. 3). The CP28 antibody recognized the A216 gene product and the A166 gene product is recognized by both CP3F4 and 30-5-7. The 1591 tumor also expresses the normal complement of $K^k$ and $D^k$ molecules in that all known monoclonal antibodies reactive with normal k haplotype MHC class I antigens also recognize the 1591 tumor.

In order to determine whether the novel class I antigens on the 1591 tumor express any epitopes found on H-$2^k$ class I antigens, we transfected Balb/c 3T3 cells with the cloned 1591 genes. A total of 216 transfected cells were recognized by antibodies directed against $K^k$ and $D^k$ while 166 and 149 transfectants showed no cross-reactivity with anti-$K^k$ or anti-$D^k$ monoclonal antibodies (Stauss et al. 1986a). Thus, we concluded that the 216 gene product bears normal H-$2^k$ determinants as well as unique epitopes recognized by CP28. Southern blot analysis of the 1591 tumor DNA (Stauss et al. 1986a) and the transfectants shows that these novel epitopes can generate restriction fragment polymorphisms when compared to normal C3H DNA (Fig. 4). A variant of 1591 which was selected with the cell sorter for

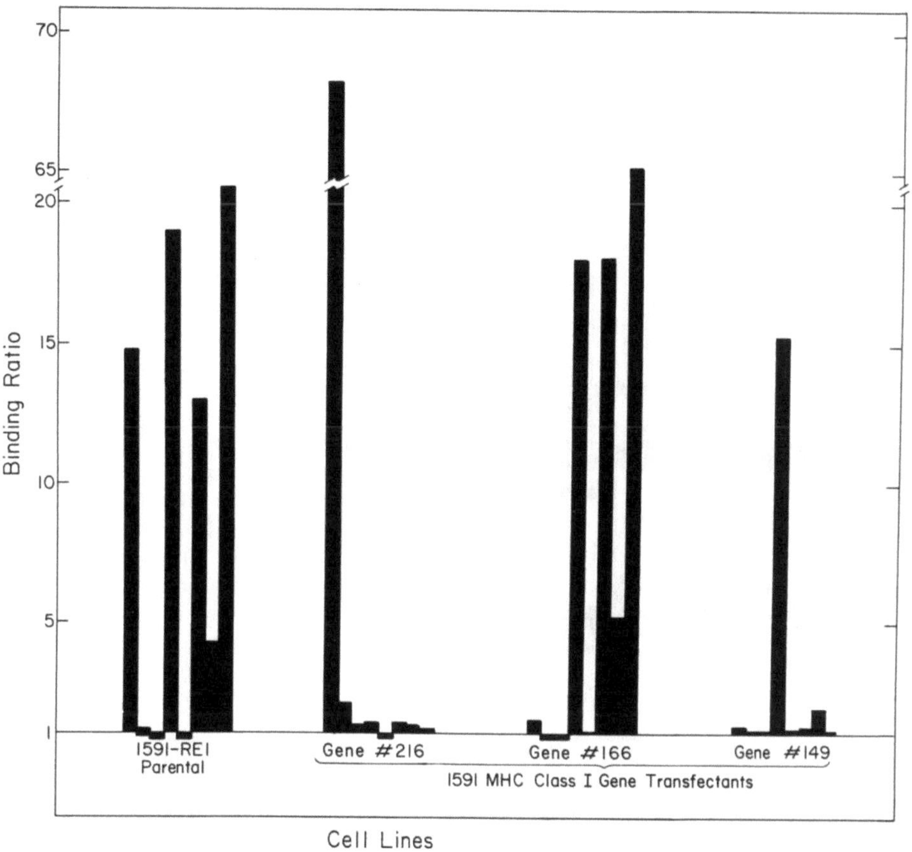


**Fig. 3.** The 1591 class I genes 216, 166, and 149 account for all the novel MHC class I antigens that have previously been detected on 1591-RE using MHC class I specific antibodies. The novel class I gene 216 selectively encodes for the antigen reactive with the syngeneic tumor-specific monoclonal antibody CP28. The fluorescence-activated cell sorter, FACS IVB, was used to determine indirect fluorescence listed as binding ratio as described in Fig. 1. The *bars* (left to right) indicate the binding of CP28, 34-2-12 ($D^d$), 34-5-8 ($D^d$), 28-14-8 ($L^dDD^{b,q}$), 34-4-20 ($L^dD^d$), 30-5-7 ($L^dD^qL^qL^b$), 23-5-21 ($D^bD^dD^{s,q,p}$), and CP3F4 to the parental 1591 regressor tumor or the gene transfected L cells (From Stauss et al. 1986b)

loss of the CP28 defined antigen also loses the polymorphic bands seen in the parental 1591 DNA. The presence of the 1-kb band in the 166 transfectant and the 950-bp and of 216 were confirmed by restriction enzyme mapping. So far we do not know the significance of the other two higher molecular weight polymorphic bands seen on the blot. Cells transfected with the genes containing these fragments did not react with any antibodies used (Stauss et al. 1986).

In order to determine whether any of the novel MHC class I genes isolated from the 1591 tumor were responsible for tumor rejection in vivo, the transfectants were analyzed for reactivity with cytolytic T cell lines specific for the 1591 tumor. The 216-transfected cell line was lysed by the anti-A cytolytic T cell line which had been

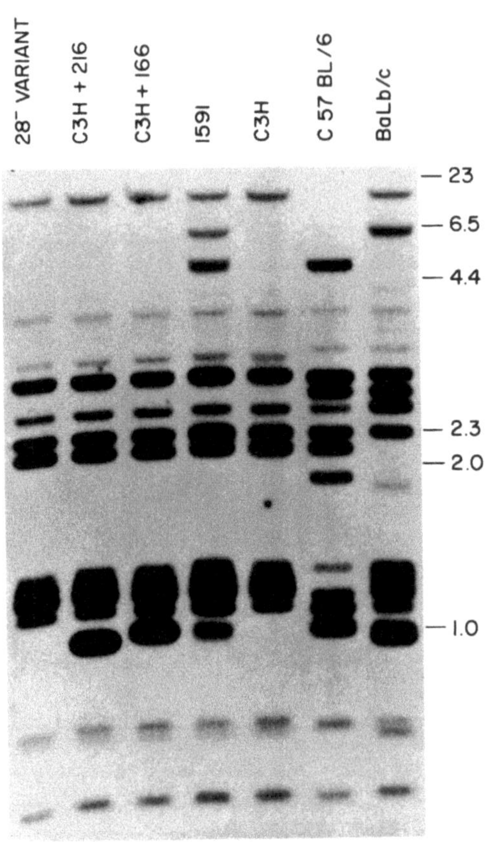


**Fig. 4.** Southern blot analysis of tumor DNA compared to normal C3H DNA. C3H spleen DNA *(lane 5)* was compared to 1591 tumor DNA *(lane 4)*. *Lanes 2 and 3* contain C3H DNA plus approximately 3 ng of gene 216 and 166 DNA, respectively. *Lane 1* contains DNA from a tumor variant that lost expression of the novel class I antigens. DNA of the inbred strains C57BL/6 *(lane 6)* and BALB/c *(lane 7)* is shown for comparison. All DNAs were digested with the restriction enzyme Pst 1 and the blots were probed with the 3′ class I probe 149.6-6. The position of the size markers is indicated (From Stauss et al. 1986 a)

used to select for progressor variants in vitro (Wortzel et al. 1984) and transfection of the 216 gene into a progressively growing $A^-$ variant of 1591 resulted in rejection of this tumor by normal mice (Stauss et al. 1986 b). Thus, while it is clear that expression of the novel class I genes of the 1591 tumor is responsible for mediating rejection, we would also like to explore the possible role of these antigens in tumorigenesis and what effects the expression of these genes might have on the immunobiology of the tumor and its malignant behavior. It also will be important to determine whether class I molecules are commonly a source of unique tumor antigens. To this end, we would like to show csonclusively that these antigens arise as a direct result of the carcinogenic effects of UV; that is, that they are a result of somatic mutations rather than preexistent in the germline genome.

Thus, we have recently generated a bank of UV-induced and spontaneous tumors where we have concurrently isolated normal cells from each mouse which gave rise to a tumor. While following the same approach as for the 1591 tumor, we will have the additional advantage of being able to compare on the genetic level, tumor cells with normal cells from the mouse which gave rise to the tumor (i. e., autochthonous normal cells). We are currently raising syngeneic monoclonal antibodies against these tumors in order to determine whether MHC genes commonly give rise to tumor-specific antigens. We have generated two additional antibodies whose molecular nature is thus far unknown. One of these is directed against a newly derived UV-induced regressor tumor and does not react with 13 other UV-induced tumors nor with autochthonous normal cells; the other is directed to a spontaneously occurring fibrosarcoma called Ag104A. This antibody does not react with normal fibroblasts from the animal which gave rise to the tumor nor with any other spontaneous or UV-induced tumor which we have tested. We are currently investigating the molecular nature of the antigens recognized by these monoclonal antibodies.

While the generation of specific monoclonal antibodies against UV-induced tumors has proven to be exceedingly difficult, the high immunogenicity of these tumors makes them ideal candidates for this type of approach. The molecular nature and genetic origin of tumor-specific antigens has remained obscure for decades. Generation of monoclonal antibody probes with unique tumor specificity together with autochthonous control cells would prove invaluable for molecular and genetic studies aimed toward resolving this fundamental question on the existence of tumor-specific antigens.

## Summary

Ultraviolet light induced tumors in mice are often highly immunogenic and exhibit unique (individually specific) antigens which cause tumor rejection in normal mice. Like tumor-specific transplantation antigens on chemically induced tumors, unique tumor-specific transplantation antigens on UV-induced tumors appear to display an enormous polymorphism in that each independently induced tumor appears to have its own individually specific antigen. Previous studies have elucidated the molecular nature and genetic origin(s) of tumor-specific antigens on a UV-induced fibrosarcoma designated 1591. Syngeneic monoclonal antibodies were raised against the 1591 tumor which had unique tumor specificity and did not react with normal syngeneic cells. The tumor-specific monoclonal antibodies cross-reacted with allogeneic MHC class I molecules, precipitated 45-kd and 15-kd molecules that after cloning and transfection of their encoding genes could be identified as unique tumor-specific transplantation (rejection) antigens. We are now raising syngeneic monoclonal antibodies against newly derived UV-induced and spontaneous tumors in order to determine whether altered MHC genes commonly encode tumor-specific antigens and whether such antigens are absent from autochthonous normal control cells. Thus far, we have generated two additional antibodies against these syngeneic tumors which may prove useful in identifying the genetic origin(s) of the unique antigens. The availability of autochthonous normal control cells will allow us to assess the possible contribution of germline mutations or residual heterozygosity to the appearance of these antigens.

## References

Burnet FM (1970) A certain symmetry: Histocompatibility antigens compared with immune receptors. Nature 226: 124–126

Callahan GN, Pardi D, Giedlin MA, Allison JP, Morizot DM, Martin WJ (1983) Biochemical evidence for expression of a semi-allogeneic, H-2 antigen by a murine adenocarcinoma. J Immunol 130: 471–480

Foley EJ (1953) Antigenic properties of methylcholanthrene-induced tumors in mice of the strain of origin. Cancer Res 13: 835–837

Frischauf A, Lehrach H, Poustka A, Murray N (1983) Lambda replacement vectors carrying polylinker sequences. J Mol Biol 170: 827–831

Gross L (1943) Intradermal immunization of C3H mice against a sarcoma that originated in an animal of the same line. Cancer Res 3: 326–333

Invernizzi G, Parmiani G (1975) Tumor-associated transplantation antigens of chemically-induced sarcomata cross reactivity with allogeneic histocompatibility antigens. Nature 254: 713–715

Kripke ML (1981) Immunologic mechanism in UV-radiation carcinogenesis. Adv Cancer Res 34: 69- 106

Ozato K, Hansen TH, Sachs D (1980) Monoclonal antibodies to mouse MHC antigens. J Immunol 125: 2473–2483

Philipps C, McMillan M, Flood P, Murphy D, Forman J, Lancki D, Womack J, Goodenow R, Schreiber H (1985) Identification of a unique tumor-specific antigen as a novel class I major histocompatibility molecule. Proc Natl Acad Sci USA 82: 5140–5144

Pravtcheva DD, DeLeo AB, Ruddle FH, Old LJ (1981) Chromosome assignment of the tumor-specific antigen of a 3-methylcholanthrene-induced mouse sarcoma. J Exp Med 154: 964–977

Prehn RT, Main JM (1957) Immunity to methylcholanthrene-induced sarcomas. JNCI 18: 769–778

Stauss HJ, Linsk R, Fischer A, Watts S, Banasiak D, Haberman A, Clark I, Forman J, McMillan M, Schreiber H, Goodenow R (1986a) Isolation of the MHC genes encoding the tumor-specific class I antigens expressed on a murine fibrosarcoma. J Immunogenet 13: 101–111

Stauss HJ, van Waes C, Fink MA, Starr B, Schreiber H (1986b) Identification of a unique tomor antigen as rejection antigen by molecular cloning and gene transfer. J Exp Med 164: 1516–1530

Steinmetz M, Frelinger JG, Fisher D, Hunkapiller T, Pereira D, Weissman SW, Uehara H, Nathenson S, Hood L (1981) Three cDNA clones encoding mouse transplantation antigens: homology to immunoglobulin genes. Cell 24: 125–134

Urban JL, Schreiber H (1983) Selection of macrophage resistant progressor tumor variants by the normal host: requirement for concomitant T cell-mediated immunity. J Exp Med 157: 642–656

Urban JL, Burton RC, Holland JM, Kripke ML, Schreiber H (1982a) Mechanisms of syngeneic tumor rejection: susceptibility of host-selected progressor variants to various immunological effector cells. J Exp Med 155: 557–573

Urban JL, Holland JM, Kripke ML, Schreiber H (1982b) Immunoselection of tumor cell variants by mice suppressed with ultraviolet radiation. J Exp Med 156: 1025–1041

Wortzel RD, Philipps C, Schreiber H (1983) Multiple tumor-specific antigens expressed on a single tumor cell. Nature 304: 165–167

Wortzel RD, Urban JL, Schreiber H (1984) Malignant growth in the normal host after variant selection in vitro with cytolytic T cell lines. Proc Natl Acad Sci USA 81: 2186–2190

# Syngeneic Monoclonal Antibodies Against Chemically Induced Tumors

Antonio Garrido, Ignacio Algarra, Concepción Delgado, Abelardo Caballero, Matias Pérez, and Federico Garrido

Servicio Análisis Clínicos, Hospital Virgen de las Nieves, Granada 18014, Spain

It has been demonstrated that humoral as well as cellular immune responses take place against chemically induced tumors (Lennox et al. 1981; Ogata et al. 1986). Indeed, syngeneic monoclonal antibodies (McAbs) and antitumor cytotoxic T-lymphocyte responses can be used to define cell surface structures that may be involved in tumor rejection (Simrell and Klein 1979; Fernández Cruz et al. 1987; Wortzel et al. 1983).

It is well established that tumor-associated transplantation antigens of chemically induced tumors do not cross-react. Immunization with solid tumors or tumor cells in culture is able to provide protection against a challenge by the immunizing tumor but there is no or only weak cross-reaction with other tumors induced with the same or a similar carcinogen (Prehn and Main 1957; Klein et al. 1960; Old et al. 1962). However, the chemical nature of the cell surface structure(s) involved in the rejection phenomena has not been defined in the majority of the chemically induced tumors.

## The GR9 Tumor System

The GR9 tumor is a methylcholanthrene-induced fibrosarcoma composed of different clones with different class I antigen expression (Table 1). There are clones with high, others with moderate, and a third group with low class I expression (A. Garrido et al. 1986; M$^{a}$L. Garrido et al. 1986). In addition we have correlated these phenotypes with local tumor growth; clones that grow faster are those with low class I expression; similarly most immunogenic clones are class I positive. We have not found class I gene rearrangement or deletion that can account for the phenotypic differences observed between class I positive and class I negative clones (Bonal et al. 1986). Furthermore, recombinant $\gamma$-interferon and not tetradecanoylphorbol-13-acetate (TPA) or retinoic acid can switch on class I expression to a similar or higher level as is observed with class I positive tumor cell clones (Garrido 1987).

**Table 1.** Heterogeneity of class I expression on GR9 tumor cell clones: influence on local tumor growth

| Clons | Class I[a] | | | Class II[a] | Local tumor growth[b] |
|---|---|---|---|---|---|
| | $K^d$ | $D^d$ | $L^d$ | | |
| G2 | +++ | +++ | ++ | − | − |
| D8 | +++ | +++ | ++ | − | − |
| A7 | ++ | +++ | + | − | − |
| D6 | ++ | ++ | ± | − | ± |
| B11 | + | + | ± | − | ± |
| B7 | + | ++ | ± | − | ± |
| B3 | + | + | ± | − | ± |
| C5 | ± | ± | ± | − | +++ |
| B10 | ± | ± | ± | − | +++ |
| B9 | ± | ± | ± | − | +++ |

[a] Results obtained using several anti $H\text{-}2^d$, $K^d$, $D^d$, $L^d$, and Ia McAbs. Details of the binding assay used were reported elsewhere (A. Garrido et al. 1986)
[b] Local tumor growth injecting $2.5 \times 10^5$ tumor cells subcutaneously

## Generation of Syngeneic Monoclonal Antibodies

We decided to analyze the presence of tumor-associated antigens on GR9 tumor by immunizing syngeneic BALB/c mice with A7 GR9 clone ($K^d$, $D^d$, $L^d$ positive). A7 clone is highly immunogenic: more than $10^6$ cells are required to obtain a locally growing tumor mass.

The following immunizing protocol was used – day 1: $0.5 \times 10^6$ cells; day 21: $10^6$ cells; day 31: $1.5 \times 10^6$ cells; day 41: $4 \times 10^6$ cells; day 51: $6 \times 10^6$ cells; day 61: $10^7$ cells; day 71: $6 \times 10^6$ cells; day 81: $10^7$ cells; day 91: $6 \times 10^6$ cells. Ninety-five days after the first inoculation with A7 cells, splenocytes were obtained and fused with NS1 myeloma according to the Khöler and Milstein technique (for details see Gomez-Morales et al. 1985).

## Pattern of Reactivity of Anti-GR9-A7 Monoclonal Antibodies

Several McAbs were obtained that reacted with the immunizing cell by indirect immunofluorescence (A7 cells). A7.1 McAb reacted with most normal tissues, including BALB/c; so it behaves as an autoantibody. A7.2 and A7.6 McAbs only reacted with tumor cells and not with normal cells obtained from different congeneic and H-2 recombinant strains of mice (see Table 2). A7.2 is an IgG1 and A7.6, IgG2a.

We also tested A7.2 and A7.6 McAbs with several mouse tumor cell lines, including chemically induced ones [MethA ($H\text{-}2^d$); MC6A ($H\text{-}2^b$); MCG3 ($H\text{-}2^b$); EL4 ($H\text{-}2^b$); TLC5 ($H\text{-}2^k$); Gardner ($H\text{-}2^k$); L1220 ($H\text{-}2^d$)] and virus-induced ones [Gi1 IV ($H\text{-}2^b$),

**Table 2.** Tissue distribution of the antigen recognized by the syngeneic antitumor McAbs

| | McAbs[a] A7.2/A7.6 |
|---|---|
| Tumor cell lines | |
| GR9 clones: A7, B9, D8, G2, D6, B11, C11, C5, B10 | + + + |
| GR9 clones: B7 | – |
| Tumor cell lines: GRA51, GRIR5, MCG11, GRB7, 1Esb MNNG | + + + |
| Tumor cell lines: MC6A, MCG3, EL4, TLC5, SL2, Gadner, L1210, MethA, Gi14, EqK1, EAKR, TLX9, L5178YE, YC8, LSTRA, Esb | – |
| Normal cells | |
| BALB/c tissues: spleen, lymph node, thymus, NIH 3T3 | – |
| Other strains of mice[b]: DBA/2 CBA/H, B10BR, B10, B10A(2R), B10A(4R) | – |

[a] A7.2 and A7.6 syngeneic McAbs were present in most GR9 tumor clones (with the exception of B3 and B7) and also in other tumors induced with methylcholanthrene and growing in monolayer (GR5.1, MCG11, GRIG5, and GRB7.1). The in vitro mutagenized Esb was also positive. The rest of the tumor cells tested were negative (including chemically, virally, and radiation induced leukemia/lymphomas). A binding assay was performed using an iodine-labeled sheep anti-mouse Ig (Fab2). The method has been previously described (A. Garrido et al. 1986). + + +, positive reactivity with A7.2 and A7.6 McAbs; –, no reactivity with A7.2 and A7.6 McAbs

[b] A7.2 and A7.6 McAbs were tested on a large panel of normal cells by P. Ivanyi with negative results

EqK1 (Gross virus, H-$2^b$); EAKR (Gross virus H-$2^b$); RBL-5 (Rauscher virus H-$2^b$); YC8 (Moloney virus H-$2^d$)]. In addition we included in the cell panel methylcholanthrene-induced BALB/c tumors produced in our laboratory that grow as monolayers (MCG11; GRA5.1; GRIR5; GRB7.1). The results are summarized in Table 2.

A7.2 and A7.6 McAbs reacted positively with most GR9 tumor cell lines; only two GR9 clones were negative. We also found a positive reaction to other methylcholanthrene tumors induced in our laboratory, as well as the in vitro MNNG-mutagenized Esb line (kindly tested by Dr. P. Altevogt, DKFZ, Heidelberg). The rest of the tumor cell lines used were negative. These cell lines included various tumors induced with Gross, Moloney, and Rauscher viruses that are known to express gp70 structures (Schirrmacher and Garrido 1977; Plata et al. 1983).

## Immunoprecipitation Studies

We analyzed the molecules recognized by A7.2 and A7.6 syngeneic antitumor McAbs. Immunoprecipitation and SDS-PAGE analysis were performed on iodine-labeled GR9-A7 tumor cells (Fig. 1). A single 70-kd peptide was precipitated. In a second experiment we used a goat anti-Rauscher antiserum and immunoprecipitated iodine-labeled GR9-A7 cells. High molecular weight bands were obtained

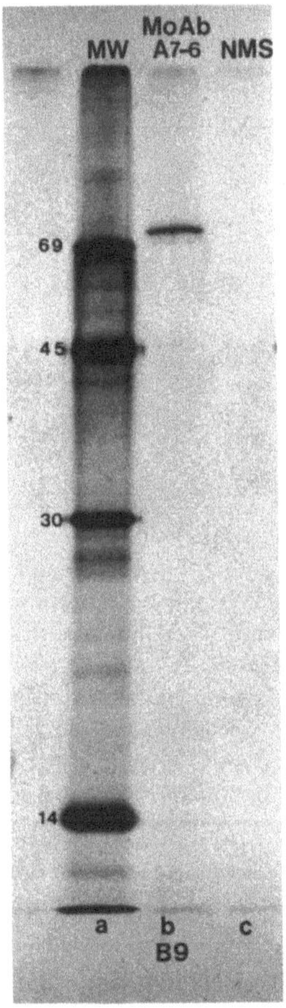


**Fig. 1.** Immunoprecipitation of the syngeneic antitumor McAb. The GR9-B9 clone was labeled with iodine-125, solubilized, and precipitated with the A7.6 McAb. The molecular weight was determined by SDS-PAGE analyses. A 70-kd peptide was precipitated

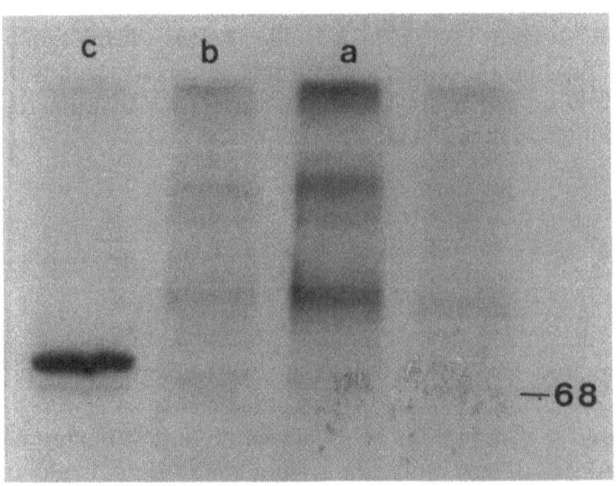


**Fig. 2.** Sequential immunoprecipitation. *Lanes a and b:* First and second immunoprecipitations using a goat anti-Rauscher antiserum. *Lane c:* Third step using the syngeneic antitumor McAb. There is no coprecipitation of the two antigens

(Fig. 2a, b). In this same lysate a sequential immunoprecipitation was carried out with our A7.2 McAbs (Fig. 2c). A clear 70-kd band could again be obtained, indicating that this molecule is not related to the one recognized by the goat anti-Rauscher serum.

## Discussion

From our results it seems that we have obtained monoclonal antibodies (A7.2 and A7.6) that probably react with a gp70-related structure. This peptide may be coded by an activated endogenous virus integrated in the GR9 genome. These syngeneic mouse McAbs react with type specific gp70 determinants, since these antibodies were negative on tumor cell lines known to express MuLV surface antigen (see Table 2). Similar results have been reported with antisera obtained by immunizing BALB/c mice with methylcholanthrene-induced tumors (Brown et al. 1978).

We have not been able to find a correlation between the presence of gp70 antigen defined by A7.2 and A7.6 McAbs and the specificity of the rejection antigen(s) present on GR9 tumor. Indeed, immunization of BALB/c mice with GR9-A7 cells did protect against a challenge of A7 and other GR9 clones. However, it did not protect when other methylcholanthrene tumors were used. These tumors included those obtained in our laboratory that were positive for gp70 antigens like GRB7.1 and GRA5.1 or tumor negative for gp70, like MethA (see Table 3). These results suggest that the tumor-associated transplantation antigens of GR9 tumor do not cross-react or cross-react weakly with other methylcholanthrene-induced tumors and that the gp70 peptide defined by our McAbs is not strongly involved in the rejection antigen. Similar results were reported by Lennox et al. (1981). Nevertheless, the presence of polymorphic gp70-like structures on methylcholanthrene-induced tumors cannot completely exclude its participation in the tumor rejection antigens.

We have not obtained evidence of alloreactivity in our syngeneic McAbs. A7.2 and A7.6 were negative on normal cells obtained from mice of different H-2 haplotypes. Previous publications by ourselves (Garrido et al. 1979) and by other labora-

**Table 3.** Tumor-associated transplantation antigens of GR9-A7 tumor clone

| Immunizing cells[a] | Challenge | Protection |
|---|---|---|
| GR9-A7 | GR9-A7 | Yes |
| GR9-A7 | GR9 tumor system: GR9, G2, B9, B10, B11 | Yes |
| GR9-A7 | Other methylcholanthrene tumors: GRB7, GRA5.1, MethA | No |

[a] The immunizing protocol was carried out as follows: Groups of five mice were injected with – day 1: $10^6$ mitomycin-treated GR9 A7 cells; day 10: $10^6$ GR9 A7 cells; day 20: $10^6$ GR9 A7 cells. Twenty days after the last injection each group of mice was challenged with $10^6$ cells of one tumor (for B9 and B10 GR9 clones $2.5 \times 10^5$ cells were used). As control, untreated mice were inoculated with the challenger tumor at the same time as the immunized ones

tories (Meschini and Parmiani 1978) indicated that anti-H-2 alloantibodies could be induced by syngeneic immunizations with tumor cells. Those tumors (MCG4 and C-1 respectively) were demonstrated to carry alien H-2 determinants that were responsible for the observed alloreactivity. On the GR9 tumor system, we have no evidence of alien H-2 molecules nor do we obtain syngeneic antibodies with alloreactive activity.

Finally, little is known about the biological role of these gp70 molecules. Further studies are required to analyze their function in the immune response against tumor development.

*Acknowledgments.* We thank Pavol Ivanyi for studies of A7.2 and A7.6 McAbs on normal cells from mice of different H-2 haplotypes, L. Eisenbach (Weizmann Institute) for supplying the goat anti-Rauscher antiserum, and Rosa Fernández for typing the manuscript.

This work was partially supported by the FISSS (Fondo de Investigaciones Sanitarias de la Seguridad Social) and CAICYT (Spain).

## References

Bonal FJ, Pareja E, Martin J, Romero C, Garrido F (1986) Repression of class I H-2K, H-2D antigens on GR9 methylcholanthrene-induced tumour cell clones is related to the level of DNA methylation. J Immunogenet 13: 179–186

Brown JP, Klitzman JM, Hellström I, Nowinski RC, Hellström KE (1978) Antibody response of mice to chemically induced tumors. Proc Natl Acad Sci USA 75: 955–958

Fernández Cruz F, Pérez M, Zabay JM, Garrido F (1987) In vivo generation, enhancement and recruitment of host cellular immune responses against tumor: Influence of MHC antigen expression on clones of tumor cell variants. In: Truitt RL, Gale RB, Bortin MM (eds) Cellular immunotherapy of cancer. Liss, New York: 115–125

Garrido A, Pérez M, Delgado C, Garrido MªL, Rojano J, Algarra I, Garrido F (1986) The influence of class I H-2 gene expression on local tumor growth. Description of a model obtained from clones derived from a solid BALB/c tumor. Clin Exp Immunogenet 3: 99–110

Garrido F (1987) The biological implications of the abnormal expression of histocompatibility antigens on murine and human tumors. In: David CS (ed) H-2 gene, molecules, functions. Plenum, Bar Harbor ME USA Vol 144: 623–639

Garrido F, Pérez M, Torres MD, Garcia Olivares E, Ivanyi P, Schirrmacher V (1979) A syngeneic anti tumor serum recognizing a complex H-2 alloantigen. Immunobiology 156: 110–120

Garrido MªL, Pérez M, Delgado C, Rojano J, Algarra I, Garrido A, Garrido F (1986) Immunogenicity of H-2 positive and H-2 negative clones of a murine tumour GR9. J Immunogenet 13: 159–167

Gomez-Morales M, Ruiz-Cabello F, López Nevot MA, Collado A, Cabrera T, San Miguel J, Garrido F (1985) A monoclonal antibody GR2110 reactive with a P24 antigen present in a subgroup of acute lymphoid leukemias. Hybridoma 4: 369–378

Klein G, Sjogren MO, Klein E, Hellström KE (1960) Demonstration of resistance against methylcholanthrene induced sarcomas in the primary autochthonous host. Cancer Res 20: 1561–1572

Lennox ES, Lowe AD, Cohn J, Evan G (1981) Specific antigens on methylcholanthrene-induced tumors of mice. Transplant Proc 13: 1759–1761

Meschini A, Parmiani G (1978) Anti H-2 alloantibodies elicited by syngeneic immunizations with a chemically induced fibrosarcoma. Immunogenetics 6: 117–123

Ogata M, Shimizu J, Kosaka H, Maekawa R, Shimizu K, Fujiwara H, Hamaoka T (1986) Expression of H-2 antigens and inductibility of antitumor immune responses on various tumor cell clones established from methylcholanthrene-induced fibrosarcomas. Jpn J Cancer Res 77: 1134–1141
Old LJ, Boyse EA, Clarke DA, Garwell E (1962) Antigenic properties of chemically induced tumors. Ann NY Acad Sci 101: 80–106
Plata F, Kalil J, Zilber MT, Fellous M, Levy D (1983) Identification of a viral antigen recognized by H-2 restricted cytolytic T lymphocytes on a murine leukemia virus induced tumor. J Immunol 131: 2551–2556
Prehn RT, Main JM (1957) Immunity to methylcholanthrene-induced sarcomas. JNCI 18: 769–778
Schirrmacher V, Garrido F (1977) Expression of H-2 specificities and of MuLV envelope antigens on murine tumors. In: Eijsvoogel VP (ed) Regulatory mechanisms in lymphocyte activation. Proceedings of the 11th Leukocytes Culture Conference. Academic, New York, pp 379–382
Simrell CR, Klein PA (1979) Antibody responses of tumor-bearing mice to their own tumors captured and perpetuated as hybridomas. J Immunol 123: 2386–2394
Wortzel RD, Philipps C, Schreiber H (1983) Multiple tumour-specific antigens expressed on a single tumour cell. Nature 304: 165–167

# Effect of Anti-HLA Class I Monoclonal Antibodies on the Proliferation of T Cells Induced by PHA-P. Comparison with the Effect on T Cell Activation via the CD2 and CD3 Pathways

S. Ferrone[1], M. De Felice[2], M. C. Turco[2], L. Corbo[2], and S. Venuta[2, 3]

[1] Department of Microbiology and Immunology, New York Medical College, Valhalla, N.Y., USA

[2] Istituto di Scienze Biochimiche, II Facolta di Medicina e Chirurgia, Napoli, Italy

[3] Istituto di Oncologia Clinica e Sperimentale, Facolta di Medicina e Chirurgia di Catanzaro, Universita di Reggio Calabria, Catanzaro, Italy

T-lymphocyte proliferation is known to occur through two pathways, one involving the CD3 molecule and the other one the CD2 molecule (for review, see Alcover et al. 1987). PHA-P has been suggested to induce T cell proliferation via the CD2 pathway, since rise in calcium mobilization is blocked by anti-CD2 monoclonal antibodies (McAbs) (O'Flynn et al. 1985). Furthermore PHA-P does not stimulate the proliferation of cells lacking the CD2 molecule (O'Flynn et al. 1986).

It has been recently shown that McAbs to monomorphic and polymorphic determinants of HLA class I antigens may enhance the proliferation of T cells stimulated with the anti-CD2 McAb 9 − 1 + 9.6 (Turco et al., 1988) and may inhibit that of T cells stimulated with the anti-CD3 McAb OKT3 (Akiyama et al. 1985; Bonagura et al. 1985; Chouaib et al. 1985; Turco et al. 1985; Taylor et al. 1986; Dasgupta et al. 1987 a, b; De Felice et al. 1987; Huet et al. 1987). These results suggest that HLA class I antigens play a regulatory role in the proliferation of T cells and that this activity is influenced by the type of mitogenic stimulus utilized.

To define the involvement of the CD2 and CD3 pathway in PHA-P induced T cell proliferation, we have investigated the regulatory role of HLA class I antigens in this phenomenon. Specifically, we have compared the effect of anti-HLA class I McAbs on the proliferation of T cells stimulated with PHA-P with that on the proliferation of T cells stimulated via the CD2 or the CD3 pathway with the McAb 9-1 + 9.6 and with the McAb OKT3, respectively.

The McAb W6/32 to a monomorphic determinant of HLA class I antigens (Barnstable et al. 1978) inhibits the proliferation of peripheral blood mononuclear cells (PBMCs) stimulated with PHA-P or with the McAb OKT3, but does not affect McAb 9-1 + 9.6 induced T cell proliferation. On the other hand, the anti-HLA-A2,A28 McAb CR11-351 (Russo et al. 1983) and the McAb Q6/64 recognizing a determinant restricted to the gene products of the HLA-B locus (Quaranta et al. 1981) inhibit T cell proliferation induced by PHA-P or by McAb OKT3, but enhance that induced by McAb 9-1 + 9.6. The anti-human $\beta_2$-microglobulin ($\beta_2$-m) McAb NAMB-1 (Pellegrino et al. 1982) does not affect the proliferation of PBMCs induced by the three mitogenic stimuli (Table 1).

**Table 1.** Effect of anti-HLA class I McAbs on the proliferation of PBMCs induced by PHA-P, McAb 9-1 + 9.6, and McAb OKT3[a]

| McAb | HLA specificity | PHA-P | | McAb 9-1 + 9.6 | | McAb OKT3 | |
|---|---|---|---|---|---|---|---|
| | | cpm ($\times 10^{-3}$) | Variation (%) | cpm ($\times 10^{-3}$) | Variation (%) | cpm ($\times 10^{-3}$) | Variation (%) |
| – | – | 100.6 ± 8.1 | | 13.8 ± 1.2 | | 95.5 ± 17.4 | |
| CR11-351 | A2, A28 | 18.1 ± 3.6 | −83 | 42.1 ± 3.1 | +205 | 17.5 ± 2.8 | −82 |
| Q6/64 | B locus restricted | 25.3 ± 1.6 | −76 | 70.5 ± 5.9 | +411 | 0.6 ± 0.1 | −99 |
| W6/32 | Framework | 1.5 ± 0.5 | −99 | 15.5 ± 1.7 | + 12 | 0.2 ± 0.1 | −99 |
| NAMB-1 | $\beta_2$-m | 96.8 ± 5.0 | − 3 | 14.4 ± 0.9 | + 1 | 86.1 ± 5.3 | −10 |

[a] PBMCs ($5 \times 10^5$/ml) were incubated for 3 days at 37 °C with PHA-P (0.3 µg/ml), McAb 9-1 + 9.6 (2 µg/ml), and McAb OKT3 (5 ng/ml) in the presence of anti-HLA class I McAbs (10 µg/ml). Then tritiated thymidine ($^3$H-dT) (1 µCi) was added to each well; after an additional 15 h incubation, cells were harvested on glass fiber strips and $^3$H-dT incorporation was measured in a beta-counter

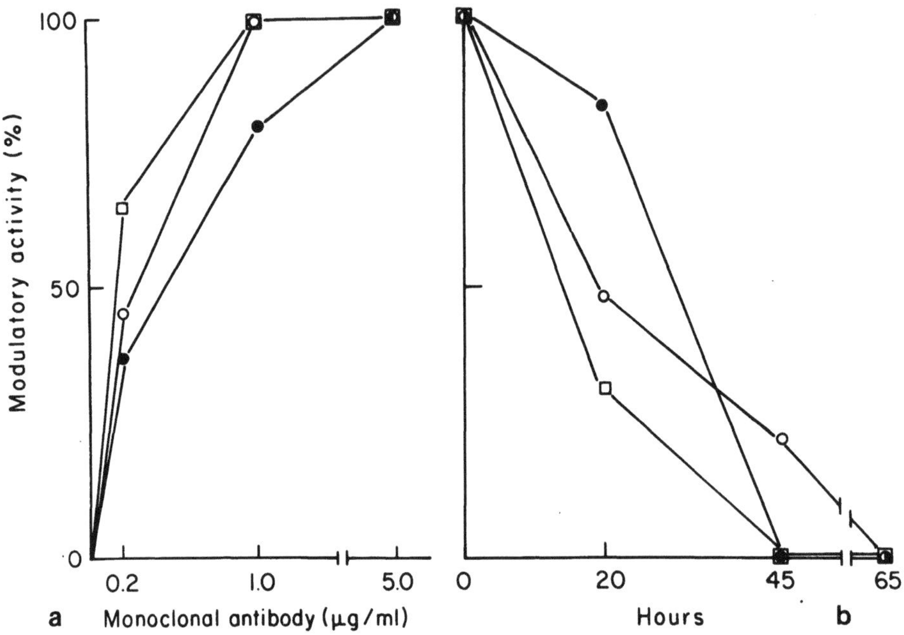


**Fig. 1.a** Dose-dependent modulation by anti-HLA-A2,A28 McAb CR11-351 and **b** kinetics of the modulation by anti-HLA class I McAb Q6/64 of the proliferation of HLA-A2 positive T cells stimulated with PHA-P (●—●), with the anti-CD2 McAb 9-1 + 9.6 (□—□), and with the anti-CD3 McAb OKT3 (○—○). In **a,** PBMCs were incubated at 37 °C with PHA-P (0.3 μg/ml), with McAb 9-1 + 9.6 (2 μg/ml), or with McAb OKT3 (5 ng/ml) in the presence of McAb CR11-351. In **b,** PBMCs were incubated with PHA-P (0.3 μg/ml), with McAb 9-1 + 9.6 (2 μg/ml), or with McAb OKT3 (5 ng/ml); at the indicated times the McAb Q6/64 (5 μg/ml) was added. In both experiments following a total 72-h incubation at 37 °C $^{3}$H-thymidine incorporation was measured. Results are expressed as % of maximal effect

The modulatory effect of anti-HLA class I McAbs is specific and reflects their interaction with the corresponding antigenic determinant, since (a) PBMCs obtained from HLA-A2,A28 negative individuals and stimulated with PHA-P, McAb 9-1 + 9.6, or McAb OKT3 are not affected by McAb CR11-351 and (b) $F(ab')_2$ fragments of anti-HLA class I McAbs have the same effect as whole IgG on PBMCs activated with PHA-P, McAb 9-1 + 9.6, or McAb OKT3 (not shown).

In the three systems investigated the effect of anti-HLA class I McAbs is dose dependent: the McAb CR11-351 displays more than 75% of its maximal modulatory activity when it is used at the final concentration of 1 μg/ml. Lower concentrations exert lower modulatory effects (Fig. 1). It should be noted that the modulatory effect of anti-HLA class I McAbs may be influenced by the concentration of PHA-P used, but is not affected by that of McAb 9-1 + 9.6 and McAb OKT3. Furthermore, in the three systems there is individual variability in the susceptibility to modulation by anti-HLA class I McAbs.

The kinetics of the modulatory effect of anti-HLA class I McAbs on the proliferation of T cells stimulated with PHA-P is slightly different from that on T cells stimu-

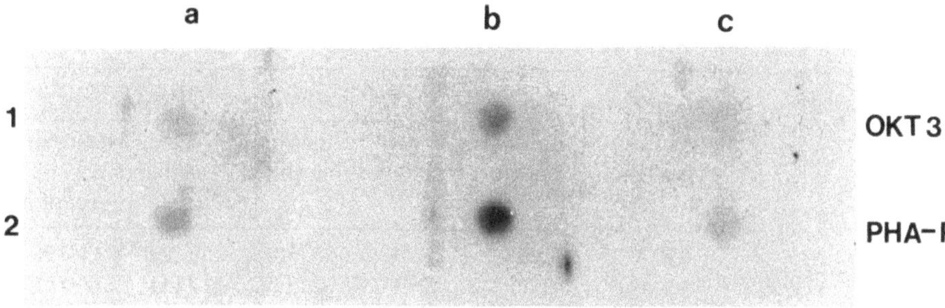


**Fig. 2.** Inhibition by anti-HLA class I McAbs of IL-2 receptor gene transcription in PHA-P and McAb OKT3 stimulated PBMCs. PBMCs from an HLA-A2 positive donor were incubated for 20 h at 37 °C with PHA-P (0.3 µg/ml) in the presence of anti-HLA class I McAb Q6/64 (5 µg/ml) *(lane c)* or with McAb OKT3 (5 ng/ml) in the presence of anti-HLA-A2,A28 McAb CR11-351 (5 µg/ml) *(lane c)*. Nuclei were then isolated and incubated at 26 °C for 20 min with $^{32}$P-$\alpha$-GTP (200 µCi) (Vannice et al. 1984). RNA was extracted (Smith et al. 1978), hybridized at 37 °C for 4 days to nitrocullulose filters containing 3 µg of denatured linear IL-2 receptor cDNA clone p IL-2R2, washed with SSPE-SDS, and autoradiographed for 4 days at −70 °C. *Lane a* indicates control PBMCs and *lane b* PBMCs stimulated with the mitogen in the absence of anti-HLA class I McAbs

lated with McAb 9-1 + 9.6 and McAb OKT3: in the latter two systems more than 50% of the modulatory effect is lost when anti-HLA class I McAb Q6/64 is added after 20 h of incubation of PBMCs with the mitogenic stimuli. On the other hand, about 80% of the maximal effect is still present when PBMCs are stimulated with PHA-P. In the three systems no marked effect is detected when anti-HLA class I McAbs are added to the cultures following 45 h of incubation with the mitogenic stimuli.

Additional experiments were then performed to compare the mechanism(s) underlying the modulatory effect of anti-HLA class I McAbs on the proliferation of T cells stimulated with PHA-P, McAb 9-1 + 9.6, or McAb OKT3. The anti-HLA class I McAbs W6/32, Q6/64, and CR11-351 inhibit the expression of Tac antigen and transferrin receptor by T cells stimulated with PHA-P and with McAb OKT3. In both systems the reduced expression of Tac antigen is associated with a decreased level of mRNA which hybridizes with the IL-2 receptor cDNA clone pIL-2R2 (Fig. 2). On the other hand, the three anti-HLA class I McAbs differ in their effect on the expression of Tac antigen and transferrin receptor by T cells stimulated with McAb 9-1 + 9.6. While the McAb W6/32 does not affect the percentage of T cells expressing the two receptors, the McAbs Q6/64 and CR11-351 enhance it. The increase induced by the latter is slightly lower than that induced by the former (Table 2).

To determine the cell targets of the modulatory activity of anti-HLA class I McAbs in the three systems analyzed, the anti-HLA-A2,A28 McAb CR11-351 was tested for its effect on the proliferation of HLA-A2 positive and negative T cells mixed with HLA-A2 negative and positive accessory cells and stimulated with PHA-P, McAb 9-1 + 9.6, or McAb OKT3. In the three systems the anti-HLA-A2,A28 McAb CR11-351 modulates the proliferation of HLA-A2 positive T cells when they are mixed with HLA-A2 positive or negative accessory cells. On the other hand, no effect is detected when HLA-A2,A28 negative T cells are mixed with HLA-A2 posi-

**Table 2.** Effect of anti-HLA class I McAbs on Tac antigen and transferrin receptor expression by T cells activated with PHA-P, McAb 9-1+9.6, and McAb OKT3[a]

| McAb | HLA specificity | PHA-P | | McAb 9-1+9.6 | | McAb OKT3 | |
|---|---|---|---|---|---|---|---|
| | | Tac antigen | Transferrin receptor | Tac antigen | Transferrin receptor | Tac antigen | Transferrin receptor |
| – | – | 62[b](112)[c] | 35 (85) | 35 (121) | 26 (90) | 59 (98) | 33 (79) |
| CR11-351 | A2,A28 | 13 (97) | 8 (49) | 52 (142) | 32 (92) | 12 (44) | 17 (75) |
| Q6/64 | B locus restricted | 11 (85) | 13 (63) | 67 (111) | 42 (89) | 9 (51) | 7 (47) |
| W6/32 | Framework | 8 (91) | 6 (62) | 28 (90) | 23 (79) | 7 (63) | 6 (51) |

[a] PBMCs ($5 \times 10^5$/ml) were incubated for 3 days at 37 °C with PHA-P (0.3 μg/ml), McAb 9-1+9.6 (2 μg/ml), and McAb OKT3 (5 ng/ml) in the presence of anti-HLA class I McAbs (10 μg/ml). Then cells were incubated with fluorescein isothiocyanate-conjugated McAb anti-Tac and anti-transferrin receptor for 30 min at 4 °C, washed Ortho Spectrum III

[b] % of positive cells

[c] Mean fluorescence intensity

tive accessory cells and stimulated with the three mitogenic stimuli (Table 3). These results indicate that the modulatory effect of anti-HLA class I McAbs on the proliferation induced by PHA-P, McAb 9-1+9.6, and McAb OKT3 reflects an interaction with T cells and does not appear to involve accessory cells.

Our results indicate that the effect of anti-HLA class I McAbs on T cell proliferation induced by PHA-P resembles more closely that on the proliferation of T cells activated via the CD3 pathway than via the CD2 pathway. This conclusion is surprising in view of the reported role of the CD2 pathway in PHA-P induced T cell proliferation. Several possibilities can be envisioned to account for our findings: First, under our experimental conditions PHA-P induces proliferation of T cells via the CD3 pathway. Alternatively, the different characteristics of T cell proliferation induced by distinct anti-CD2 McAbs (Huet et al. 1986) suggest that although PHA-P and McAb 9-1+9.6 both act through the CD2 pathway, they differ in some activation steps which display differential susceptibility to the modulatory activity of anti-HLA class I McAbs. Lastly, since activation of T cells may occur through a pathway involving both the CD2 and CD3 molecules (Yang et al. 1986), inhibition by anti-HLA class I McAbs of PHA-P induced T cell proliferation may reflect the interaction of CD2 and CD3 pathways in the activation process.

The mechanism(s) underlying the regulatory role of HLA class I antigens in T cell proliferation and the differential effect of anti-HLA class I McAbs on the phenomenon is only speculative at this time. The results we have presented suggest that HLA class I antigens modulate the level of molecules which play a crucial role in T cell activation. The differential effect of an anti-HLA class I McAb on the CD2 and CD3 pathways and of McAbs to distinct determinants of HLA class I antigens on a pathway may reflect changes in conformation of HLA class I antigens which may affect their interactions with molecules of the T cell activation pathways.

Whether the regulatory role of HLA class I antigens is restricted to the proliferation of T cells and whether the in vitro phenomena we have described have an in

**Table 3.** Cell target in the modulation by anti-HLA-A2,A28 McAb CR11-351 of the proliferation of T cells stimulated with PHA-P, McAb 9-1+9.6, and McAb OKT3[a]

| T cells HLA phenotype | Monocytes | Mitogen | McAb CR11-351 | $^3$H-dT incorporation cpm $\times 10^{-3}$ | % variation |
|---|---|---|---|---|---|
| A2,A26 | A24,A24 | PHA | − | 62.4 ± 5.5 | |
| A2,A26 | A24,A24 | PHA | + | 10.4 ± 1.0 | −83 |
| A24,A24 | A2,A26 | PHA | − | 45.6 ± 3.9 | |
| A24,A24 | A2,A26 | PHA | + | 42.4 ± 4.5 | −7 |
| A2,A26 | A24,A24 | McAb 9-1+9.6 | − | 22.8 ± 3.7 | |
| A2,A26 | A24,A24 | McAb 9-1+9.6 | + | 53.6 ± 3.2 | +135 |
| A24,A24 | A2,A26 | McAb 9-1+9.6 | − | 31.6 ± 2.8 | |
| A24,A24 | A2,A26 | McAb 9-1+9.6 | + | 35.6 ± 1.2 | +13 |
| A2,A26 | A24,A24 | McAb OKT3 | − | 35.7 ± 1.6 | |
| A2,A26 | A24,A24 | McAb OKT3 | + | 6.8 ± 0.9 | −81 |
| A24,A24 | A2,A26 | McAb OKT3 | − | 31.0 ± 2.0 | |
| A24,A24 | A2,A26 | McAb OKT3 | + | 28.5 ± 1.8 | −8 |

[a] T cells were purified from PBMCs by sequential centrifugation on a Percoll density gradient and incubation with anti-HLA class II McAb OKIa1 and rabbit complement. Monocytes were purified from PBMCs by centrifugation through a 46% Percoll density gradient. T cells ($5 \times 10^5$/ml) were mixed with monocytes ($5 \times 10^4$/ml) and incubated for 3 days at 37 °C with McAb 9.1+9-6 (2 µg/ml) or McAb OKT3 (5 ng/ml) in the presence of anti-HLA class I McAbs (10 µg/ml). Then $^3$H-dT incorporation was measured

vivo counterpart is not known at present. Our results suggest a mechanism to account for the immunological abnormalities in heroin addicts and in patients with hydatidosis since antibodies to monomorphic determinants of HLA class I antigens have been detected in both conditions (Ameglio et al. 1987, 1988). The antibodies may affect the immune system through their effect on T cell proliferation.

*Acknowledgments.* The authors wish to acknowledge the excellent secretarial assistance of Mrs. Edwina L. Jones and Miss Donna D. Jones.

This work was supported in part by Consiglio Nazionale delle Ricerche projects "Oncologia" and "Ingegneria genetica e basi molecolari delle malattie ereditarie," by grants from Associazione Italiana per la Ricerca sul Cancro, by NIH grants AI21384 and CA39559, and by a grant from the International Juvenile Diabetes Foundation.

## References

Akiyama Y, et al. (1985) Effect of monoclonal antibodies (MoAb) to class I and class II HLA antigens on lectin- and MoAb OKT3-induced lymphocyte proliferation. Cell Immunol 91: 477–481

Alcover A, et al. (1987) Functional and molecular aspects of human T lymphocyte activation via T3-Ti and T11 pathways. Immunol Rev 95: 5–159

Ameglio F, et al. (1987) Antibody reactivity to HLA classes I and II in sera from patients with hydatidosis. J Infect Dis 156: 673–676

Ameglio F, et al. (1988) A high proportion of sera of heroin addicts possesses anti-HLA class I and class II reactivity. Clin Immunol Immunopathol 46: 328–334

Barnstable CJ, et al. (1978) Production of monoclonal antibodies to group A erythrocytes, HLA and other human cell surface antigens – new tools for genetic analysis. Cell 14: 9–20

Bonagura VR, Nordli DR, Pernis B (1985) Antibodies against monomorphic determinants of the α chain of class I human leukocyte antigens inhibit the reaction of human lymphocytes to mitogen and tetanus toxoid. Immunol Invest 14: 233–243

Chouaib S, Welte K, Dupont B (1985) Differential effect of anti-$\beta_2$-microglobulin on IL 2 production and IL 2 receptor expression in the primary mixed lymphocyte culture reaction. J Immunol 134: 940–948

Dasgupta JD, Yunis EJ (1987 a) Receptor-like role of HLA-class I antigens: Regulation of T cell activation. J Immunol 139: 672–677

Dasgupta JD, et al. (1987 b) The role of class I histocompatibility antigens in the regulation of T-cell activation. Proc Natl Acad Sci USA 84: 1094–1098

De Felice M, et al. (1987) Differential regulatory role of monomorphic and polymorphic determinants of histocompatibility leukocyte antigen class I antigens in monoclonal antibody OKT3-induced T cell proliferation. J Immunol 139: 2683–2689

Huet S, et al. (1986) T cell activation via CD2 [T, gp50]: The role of accessory cells in activating resting T cells via CD2. J Immunol 137: 1420–1428

Huet S, et al. (1987) Role in T-cell activation for HLA class I molecules from accessory cells: Further distinction between activation signals delivered to T cells via CD2 and CD3 molecules. Proc Natl Acad Sci USA 84: 7222–7226

O'Flynn K, et al. (1985) Phytohaemagglutinin activation of T cells through the sheep red blood cell receptor. Nature 313: 686–687

O'Flynn K, et al. (1986) Different pathways of human T-cell activation revealed by PHA-P and PHA-M. Immunology 57: 55–60

Pellegrino MA, Ng AK, Russo C, Ferrone S (1982) Heterogeneous distribution of determinants defined by monoclonal antibodies on HLA-A and B antigens bearing molecules. Transplantation 34: 18–23

Quaranta V, Pellegrino MA, Ferrone S (1981) The monoclonal xenoantibody Q6/64 recognizes a determinant expressed by certain gene products of the A and B loci of the HLA region. Immunogenetics 14: 403–413

Russo C, Ng A-K, Pellegrino MA, Ferrone S (1983) The monoclonal antibody CR11-351 discriminates HLA-A2 variants identified by T cells. Immunogenetics 18: 23–35

Smith MM, Reeve AE, Huang RCC (1978) Analysis of RNA initiated in isolated mouse myeloma nuclei using purine nucleoside 5'[$\gamma$-S]triphosphates as affinity probes. Cell 15: 615–626

Taylor DS, Nowell PC, Kornbluth J (1986) Functional role of HLA class I cell-surface molecules in human T-lymphocyte activation and proliferation. Proc Natl Acad Sci USA 83: 4446–4450

Turco MC, et al. (1985) Regulatory role of a monomorphic determinant of HLA class I antigens in T cell proliferation. J Immunol 135: 2268–2273

Turco MC, et al. (1988) Enhancing effect of anti-HLA class I monoclonal antibodies on T cell proliferation induced via CD2 molecule. J Immunol 144: 2275–2281

Vannice JL, Taylor JM, Ringold GM (1984) Glucocorticoid-mediated induction of $\alpha_1$-acid glycoprotein: Evidence for hormone-regulated RNA processing. Proc Natl Acad Sci USA 81: 4241–4245

Yang SY, et al. (1986) A common pathway for T lymphocyte activation involving both the CD3-Ti complex and CD2 sheep erythrocyte receptor determinants. J Immunol 137: 1097–1100

# A Monoclonal Antibody Induced by H-2 Syngeneic ConA Blasts: Its Reactivity Pattern with Mouse and Human MHC Class I Antigens

MATTHIAS CRAMER[1], RUDOLF MIERAU[2], WOLFGANG KUON[3], ELISABETH WEISS[3], and PETER J. ROBINSON[4]

[1] Institute for Genetics, University of Cologne, Weyertal 121, 5000 Köln 41, Federal Republic of Germany

[2] Rheumaforschungsinstitut, Rheumaklinik Aachen, 5100 Aachen, Federal Republic of Germany

[3] Institute for Immunology, 8000 München 2, Federal Republic of Germany

[4] Institute for Immunology and Genetics, Deutsches Krebsforschungszentrum, 6900 Heidelberg, Federal Republic of Germany

## Introduction

Class I antigens encoded by the major histocompatibility complex (MHC) are notable for their high level of interspecies homology. The monoclonal antibody COB6-3 highlights some aspects of this feature. The antibody's production, characterization, and specificity are reviewed and summarized in this article.

## Induction Against H-2$^d$ Syngeneic ConA Blasts and Reactivity Towards H-2K$^d$ and H-2D$^b$

The reactivity pattern of COB6-3 was unexpected: COB6-3 was induced by immunizing a CAL-20 mouse with concanavalin A (ConA) blasts prepared from spleen cells of BALB/c mice, i.e., in an H-2 syngeneic strain combination. Nevertheless, in complement-dependent cytotoxicity assays and according to electrophoretic analysis of immunoprecipitates (Mierau and Cramer 1984), the antibody proved to be specific for certain H-2 class I antigens. Using cells from a panel of mouse inbred strains, we found that COB6-3 mainly reacted with molecules encoded by the D$^b$ region, and, to a lesser extent, with K$^d$ encoded antigens. The COB6-3 epitope was absent from the K$^b$, D$^d$, L$^d$, K$^k$, and D$^k$ antigens (Mierau and Cramer 1984).

## Analysis of Cell Lines Transfected with the Isolated H-2D$^b$ Gene

The antigen detected by COB6-3 on H-2D$^b$ positive cells was not identical with the classical H-2D$^b$ molecule as detected by the monoclonal antibody H141-31 (Lemke et al. 1979). This was shown by sequential precipitation experiments where preclearing a radiolabeled cell lysate by immunoprecipitation with COB6-3 did not remove the material precipitable by H141-31, and vice versa (Mierau and Cramer 1984).

There is only a single class I gene found in the D region of the b haplotype (Weiss et al. 1984). To further prove that the COB6-3 epitope was dependent on this single gene, cosmids containing the cloned $D^b$ gene (Weiss et al. 1984; Schmidt et al. 1985) were transfected into $Ltk^-$ cells, and a panel of transfectants was analyzed by cytotoxicity assays and by flow cytometry using the antibodies COB6-3 and H141-31. The two antigenic determinants were both present or absent together on any given cell clone (Table 1), and even cosegregated in one transfectant line that stopped expressing $D^b$ after prolonged cell culture. Thus, an explanation had to be found for the existence of two serologically different molecular entities encoded by a single H-2 class I gene.

**Table 1.** Reactivity pattern of monoclonal antibodies with L cells transfected with the cloned H-$2D^b$ gene

| | Cytolysis by | | |
|---|---|---|---|
| | H141-31[a] anti-$D^b$ | COB6-3 | H100-5/28 anti-$K^k$ |
| Transfectants: [b] | | | |
| Series 1: 4 clones | + [c] | + | + |
| Series 2: 2 clones | + | + | + |
| Series 3: 2 clones | + | + | + |
| Transfectants: | | | |
| Series 1: 5 clones | – | – | + |
| Series 2: 2 clones | – | – | + |
| Series 3: 5 clones | – | – | + |
| $Ltk^+$ (H-$2^k$) | – | – | + |
| EL4 (H-$2^b$) | + | + | – |

[a] Monoclonal antibodies H141-31 and H100-5/28 were described by Lemke et al. (1979)
[b] $Ltk^-$ cells were cotransfected by the calcium phosphate mediated DNA transfer technique (Wigler et al. 1979), with the herpes tk gene containing plasmid pOPF (Grosveld et al. 1982), together with cosmids containing the cloned H-$2D^b$ gene: In series 1 and 2, cosmids LS-Aw-1 (Weiss et al. 1984) and B1-28 (Schmidt et al. 1985), respectively, were used. In series 3, a 10-kb HindIII subfragment of cosmid LS-Aw-1, subcloned in pBR327, was transfected. The $Ltk^+$ line was constructed by transfection with pOPF alone. Transfectants were selected in HAT medium
[c] Results of microcytotoxicity assays (see legend to Fig. 2): specific lysis above 70% is scored as +, below 20% as –. Results were confirmed by flow cytometry with three of the $D^b$ positive and one of the $D^b$ negative transfectants. One of the positive cell clones from series 1 became negative for H141-31 and COB6-3 simultaneously after some weeks of continuous cell culture

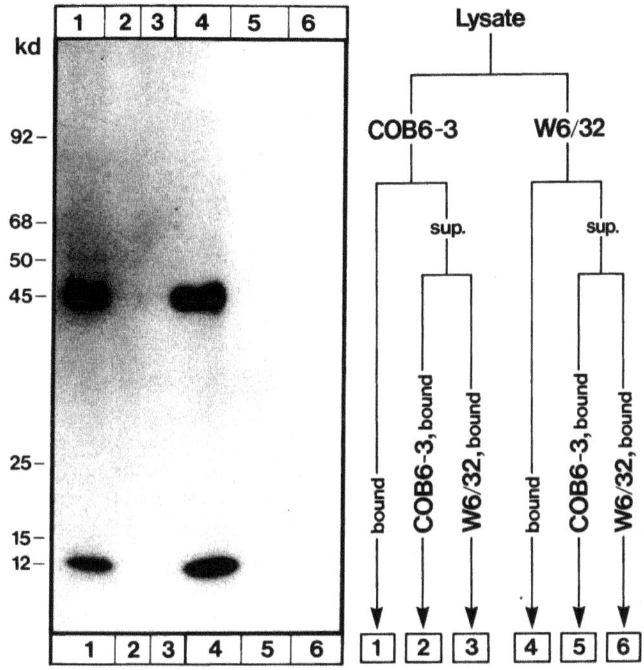


**Fig. 1.** Autoradiograph of a 7.5%–15% polyacrylamide gradient SDS-gel electrophoresis of immunoprecipitates (Mierau and Cramer 1984). Human lymphocytes were surface radio-iodinated, lysed with Nonidet-P 40 detergent, and sequentially precipitated with ascites fluid containing monoclonal antibodies (COB6-3 or W6/32) and formalin-fixed *Staphylococcus aureus,* according to the scheme depicted in the *right half* of the figure; *sup.,* supernatant of first immunoprecipitation. Molecular mass markers were phosphorylase (92 000), bovine serum albumin (68 000), immunoglobulin $\gamma$ and light chains (50 000 and 25 000, respectively), ovalbumin (45 000), ribonuclease (15 000), and cytochrome C (12 000). *kd,* kilodalton

## Monomorphic Reaction with Human MHC Class I Antigens

We next extended our analysis of the reactivity of COB6-3 to human cells. We were prompted to perform these experiments by reports on a cross-reaction between human MHC class I antigens and the H-2D$^b$ molecule as detected by the monoclonal antibody W6/32 (Ivanyi and van de Meugheuvel 1984; Maziarz et al. 1986). Indeed, we found that the reactivity pattern of COB6-3 looked very similar to, if not identical with, the reactivity of W6/32: both monoclonal antibodies are monomorphic anti-HLA class I antibodies. In our hands, peripheral blood lymphocytes from 141 individual human donors as well as the human cell line WI-L2, but not the MHC class I negative cell line DAUDI, were killed by COB6-3 plus complement (Mierau et al. 1987). Sequential immunoprecipitations showed that all molecules detected by COB6-3 are also detected by W6/32, and vice versa (Fig. 1). Complete reciprocal binding inhibition by the two antibodies on human cells was observed by flow cytometry (Mierau et al. 1987). Thus, the epitopes detected by COB6-3 and W6/32 on HLA molecules are situated close together and may be identical.

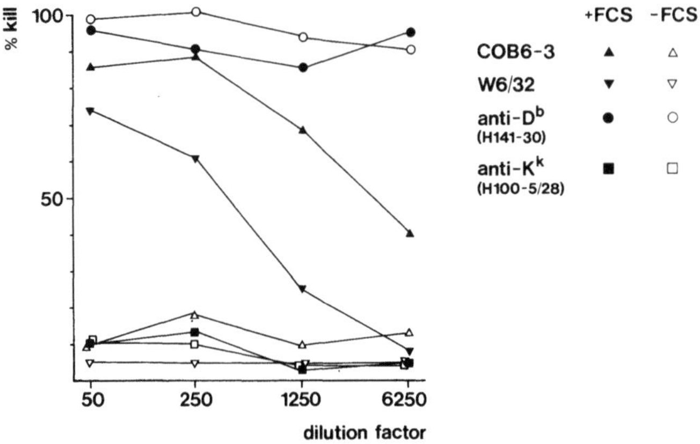


**Fig. 2.** Two-step complement-dependent microcytotoxicity assay. Freshly prepared murine C57BL/6 spleen cells (H-2D$^b$) were used as targets for monoclonal antibodies with or without preincubation of the cells with 10% fetal calf serum *(FCS)*. Ascites fluid or, in the case of W6/32, purified antibody, in different dilutions (2 μl) were incubated with the cells (3 μl) for 40 min at 37 °C, followed by incubation with 2 μl of selected rabbit complement. Percent kill was estimated by trypan blue exclusion. Results are given as means of duplicate determinations. The antibody H100-5/28 has been described by Lemke et al. (1979). It was used as a negative control

## Bovine $\beta_2$-Microglobulin is Part of the COB6-3 Epitope

The above-mentioned problem concerning the molecular nature of the COB6-3 epitope on mouse cells – a molecule differing from the authentic H-2D$^b$ molecule but being strictly dependent on the D$^b$ gene – was eventually solved when we modified our experimental protocol by performing cytotoxicity assays in the presence or absence of fetal calf serum (FCS). The result of a typical cytotoxicity titration is shown in Fig. 2. Whereas the anti-D$^b$ antibody 141-30 was cytolytic for C57BL/6 spleen cells regardless of whether FCS was present or absent, COB6-3 killed in the presence of FCS only. The same was true for the antibody W6/32. On human cells, on the other hand, the anti-class I activities of COB6-3 and W6/32 were not FCS dependent (data not shown).

We further analyzed the FCS dependence of COB6-3 reactivity on H-2D$^b$ positive cells by incubating C57BL/6 spleen cells with purified bovine or human $\beta_2$-microglobulin ($\beta_2$m), kindly supplied by Dr. A. R. Sanderson, Mill Hill. As shown in Fig. 3, in a cellular protein A binding assay, binding of COB6-3 to H-2D$^b$ cells was only detected after incubation with bovine $\beta_2$m. Binding of the classical anti-D$^b$ antibodies B22-249 (Fig. 3) and H141-31 (Mierau et al. 1987), on the other hand, was independent of previous $\beta_2$m incubation. It is concluded that the epitope defined by COB6-3 is formed by a murine class I antigen, D$^b$ in this case, in association with bovine $\beta_2$m which replaced its murine counterpart. Such an exchange mechanism is well known from previous reports (Schmidt et al. 1981; Bernabeu et al. 1984). The results of the cellular binding assays represented by Fig. 3 were corrobo-

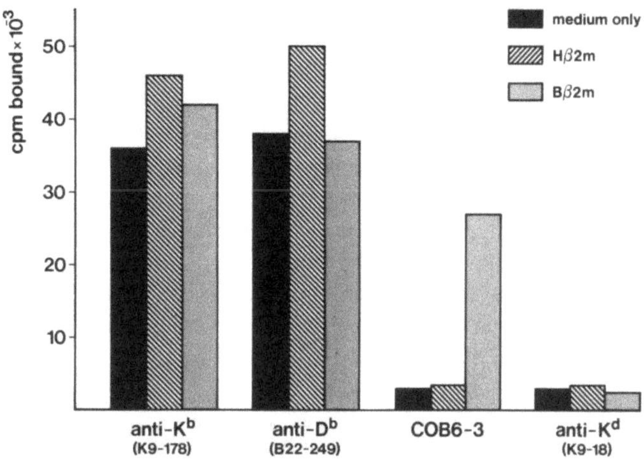


**Fig. 3.** Cellular protein A binding assay. Spleen cells of C57BL/6 mice (H-$2^b$) were incubated for 2 h (37 °C) with culture medium with or without purified human (*H*$\beta_2$m) or bovine (*B*$\beta_2$m) $\beta_2$-microglobulin (20 µg/ml), washed, and treated afterwards by two incubations (4 °C) with antibody-containing ascites dilutions and with $^{125}$I-labeled protein A, respectively. Cell-bound radioactivity was determined in a gamma counter. Monoclonal antibodies K9-178, B22-249, and K9-18 have been described previously (Hämmerling et al. 1982; Lemke et al. 1979; Arnold et al. 1985)

rated by polyacrylamide SDS-gel electrophoresis of immunoprecipitates (Mierau et al. 1987).

Using the antibody W6/32 instead of COB6-3 in the binding studies with $D^b$ antigens and heterologous $\beta_2$m, very similar results were obtained by others (Kahn-Perles et al. 1987) as well as by us (Mierau et al. 1987): H-$2D^b$ molecules were detected by W6/32 only after preincubation with bovine $\beta_2$m. One difference between the specifities of COB6-3 and W6/32 on mouse cells was detected, however: W6/32 but not COB6-3 was able to bind the $D^b$ positive cells modified by human instead of bovine $\beta_2$m (Mierau et al. 1987).

## Structural and Serological Homology of MHC Class I Antigens

In conclusion, COB6-3, originally thought to represent the monoclonal counterpart of the anti-class I alloantisera that can be produced after syngeneic immunizations (Ivanyi et al. 1980), instead defines a cross-reacting antigenic determinant shared by most or all HLA class I molecules and by only certain H-2 class antigens modified by bovine $\beta_2$m. Monoclonal antibodies behaving polymorphically in one species and monomorphically in another have occasionally been found previously (Parham et al. 1979; Smilek et al. 1980). The similarity of the reactivity patterns of COB6-3 and W6/32, despite their very different origins (Mierau and Cramer 1984; Barnstable et al. 1978), is particularly remarkable. The antigenic stimulus giving rise to COB6-3 most probably was created because the ConA blasts used for immunization of a

CAL.20 mouse carried H-2K$^d$ molecules which had been modified by bovine $\beta_2$m during the mitogen stimulation in FCS-containing culture medium.

The reaction pattern of COB6-3 can be viewed as a reflection of the remarkable structural homologies of MHC class I molecules of various species – mouse, cow, and man in this case. Such antibodies may in the future be useful to determine the biochemistry of the interacting surface areas of class I chains and $\beta_2$m. Furthermore, they may lead us to an understanding of the genetic mechanisms that create an enormous polymorphism in the MHC class I genes of every individual species, but at the same time guarantee enough structural interspecies homology to be easily detectable by, for example, antibodies like COB6-3. This homology in turn may reflect stringent structural requirements for the physiological effector functions of MHC class I molecules.

*Acknowledgments.* This work was supported by the Deutsche Forschungsgemeinschaft through SFB 74, SFB 217, and grant Ro 692.

## References

Arnold B, Horstmann U, Kuon W, Burgert HG, Hämmerling GJ, Kvist S (1985) Alloreactive cytolytic T-cell clones preferentially recognize conformational determinants on histocompatibility antigens: analysis with genetically engineered antigens. Proc Natl Acad Sci USA 82: 7030–7034

Barnstable CJ, Bodmer WF, Brown G, Galfre G, Milstein C, Williams AF, Ziegler A (1978) Production of monoclonal antibodies to group A erythrocytes, HLA and other human cell surface antigens. New tools for genetic analysis. Cell 14: 9–20

Bernabeu C, van de Rijn M, Lerch PG, Terhorst CP (1984) $\beta_2$-Microglobulin from serum associates with MHC class I antigens on the surface of cultured cells. Nature 308: 642–645

Grosveld FG, Lund T, Murray EJ, Mellor AI, Dahl HHM, Flavell RA (1982) The construction of cosmid libraries which can be used to transform eukaryotic cells. Nucleic Acids Res 10: 6715–6732

Hämmerling GJ, Rüsch E, Tada N, Kimura S, Hämmerling U (1982) Localization of allodeterminants on H-2K$^b$ antigens determined with monoclonal antibodies and H-2 mutant mice. Proc Natl Acad Sci USA 79: 4737–4741

Ivanyi D, van de Meugheuvel W (1984) A monomorphic HLA-specific monoclonal antibody, W6/32, reacts with the H-2D$^b$ molecule of normal mouse lymphocytes. Immunogenetics 20: 699–703

Ivanyi P, van Mourik P, Breuning M, Melief CJM (1980) Anti-H-2 antibodies induced by syngeneic immunization. Immunogenetics 10: 319–322

Kahn-Perles B, Boyer C, Arnold B, Sanderson AR, Ferrier P, Lemonnier FA (1987) Acquisition of HLA class I W6/32 defined antigenic determinant by heavy chains from different species following association with bovine $\beta_2$-microglobulin. J Immunol 138: 2190–2196

Lemke H, Hämmerling GJ, Hämmerling U (1979) Fine specificity analysis with monoclonal antibodies of antigens controlled by the major histocompatibility complex and by the Qa/TL region in mice. Immunol Rev 47: 175–206

Maziarz RT, Fraser J, Strominger JL, Burakoff SJ (1986) The human HLA-specific monoclonal antibody W6/32 recognizes a discontinuous epitope within the $\alpha$2 domain of murine H-2D$^b$. Immunogenetics 24: 206–208

Mierau R, Cramer M (1984) Serological and immunochemical analysis of H-2 class I molecules encoded by the D$^b$ region. Immunogenetics 20: 341–345

Mierau R, Robinson PJ, Sanderson AR, Genth E, Cramer M (1987) Antigenic determinants shared between HLA-A, -B, -C antigens and H-2 class I molecules modified by bovine beta-2 microglobulin. Immunogenetics 26: 351–355

Parham P, Sehgal PK, Brodsky FM (1979) Anti-HLA-A,B,C monoclonal antibodies with no alloantigenic specificity in humans define polymorphisms in other primate species. Nature 279: 639–641

Schmidt W, Festenstein H, Ward PJ, Sanderson AR (1981) Interspecies exchange of $\beta_2$-microglobulin and associated MHC and differentiation antigens. Immunogenetics 13: 483–491

Schmidt W, Fergula J, Festenstein H, Weiss E (1985) Identification of murine H-$2D^b$ histocompatibility antigens in cells transfected with cloned H-2 genes. Mol Immunol 22: 1151–1158

Smilek DE, Boyd HE, Wilson DB, Zmijewski CM, Fitch FW, McKearn TJ (1980) Monoclonal rat anti-major histocompatibility complex antibodies display specificity for rat, mouse, and human target cells. J Exp Med 151: 1139–1150

Weiss EH, Golden L, Fahrner K, Mellor AL, Devlin JJ, Bullman H, Tiddens H, Bud H, Flavell RA (1984) Organization and evolution of the class I gene family in the major histocompatibility complex of the C57BL/10 mouse. Nature 310: 650–655

Wigler M, Pellicer kA, Silverstein S, Axel R, Urlauss G, Chasin L (1979) DNA-mediated transfer of the adenine phosphoribosyltransferase. Proc Natl Acad Sci USA 76: 1373–1376

# A Problem Perplex, MHC + X

PETER PARHAM

Stanford University, Department of Cell Biology, Sherman Fairchild Building, Stanford, CA 94305, USA

There is a radio-sketch in which Sir Walter Raleigh, having voyaged to the New World, makes a telephone call back to a friend or sponsor. In it he enthuses about the custom of smoking tobacco, the pleasure it brings, and even suggests it might be a thing to import. Trying hard to convince, he never penetrates his friend's incredulity, formed by the belief that anyone who submits to putting rolled, dried leaves in their mouth and setting fire to them in the pursuit of happiness has clearly been in the colonies too long. In first encountering MHC restriction I responded in a similar vein. It was a Wednesday seminar at Harvard Medical School and the message was that you take cells, treat them with trinitrophenyl (TNP), and then immunize a mouse. Then, abracadabra, out come T cells with specificity for H-2 and TNP. Surely the vicissitudes of comprehending inbred strains of mice and MHC serology were bad enough without this molecular improbability! However, it did not go away, and in fact became more complex with the eventual formulation of MHC + X.

Now we understand the mystery of MHC + X. Three recent advances, fortuitously coming in close succession, have allowed a simple and coherent picture to emerge. First the demonstration that class I, as well as class II restricted T cells, recognize peptides (Townsend et al. 1986), derived from cellular processing of protein antigens. Second the discovery that class II molecules bind antigenic peptides in vitro with specificity that correlates with the patterns of MHC restriction (Babbitt et al. 1985). Third the solution of the HLA-A2 crystal structure and its revelation of a tailor-made binding site and an accommodating peptide (Bjorkman et al. 1987a). Thus, MHC molecules are peptide binding proteins, in which X is the bound peptide and the complex so formed then serves as the ligand for T cell receptors. Their binding specificity is broad and analogous to that of a protease. For example, trypsin will bind to many different proteins although it has a basic requirement for lysine or arginine residues. Unlike trypsin, an MHC molecule does not enzymatically modify and eject its captive. In contrast, the relationship is long lived and may in physiological terms be a lifelong commitment. A molecular lifetime that is.

When peptide is bound by MHC it forms an integrated structure in which side chains of the long $\alpha$-helices of the $\alpha_1$ and $\alpha_2$ domains, as well as those of the peptide, project from an extensive surface (Bjorkman et al. 1987b). If the interaction between this complex ligand and the T cell receptor is in any way similar to that between antibody and protein antigens then it is highly probable that the receptor will contact both MHC and the peptide. Residues showing high variability between

MHC alleles are found in positions lining the binding site, where they may interact with both peptide and the T cell receptor. Therein lies the molecular basis for the fundamental observation of MHC restriction, that polymorphic differences between MHC molecules alter T cell reactivity. This can be effected by substitutions that alter either binding of peptides and/or of T cell receptors.

Selection of the T cell repertoire and establishment of the pattern of MHC restriction takes place in the thymus. It is easy to envisage both positive and negative selection arising from interactions between T cell receptors and the class I and II MHC molecules expressed in the thymus. The product is a collection of receptors having affinity for MHC within a prescribed window. T cell clones with too high an affinity for self-MHC, and potentially autoreactive, are eliminated, and those with insufficient affinity to be functional are also deleted. An important question is whether and how X is involved. Available biochemical, structural, and immunological evidence is consistent with the notion that a bound peptide is as important for the structural and functional integrity of MHC molecules as the association of $\beta_2$-microglobulin is for class I (Lancet et al. 1979). The major site for peptide association is likely to be inside the cell, although in experimental situations the reaction can be driven at the cell surface (Townsend et al. 1986; Clayberger et al. 1987). Peptides derived from self-proteins and exogenous or foreign proteins compete for MHC molecules and the heterogeneous mixture of immature lymphocytes, epithelical cells, dendritic cells, and macrophages in the thymus will potentially present peptides derived from all the proteins made by those cells. The actual array of peptides presented will depend on many variables, including the abundance of the protein, the rate of degradation, the site of degradation, and the relative affinities of the peptides for the expressed MHC molecules. Selection of the T cell repertoire would in this manner be against MHC + X where X would be the products of thymic gene expression and would clearly include all of the major cellular components, housekeeping enzymes, the machinery of DNA replication and cell division, etcetera, etcetera. Tolerance of these components could thus be accomplished by clonal deletion.

An outstanding question concerns how tolerance is produced to tissue specific proteins such as thyroglobulin, which, despite its name, is not made in the thymus. One answer is to invoke the same mechanism as for thymic proteins and to argue that sufficient amounts of these proteins or their fragments are carried by blood to the thymus, traverse the blood/thymus barrier, and bind in sufficient quantities to the MHC molecules of thymic cells. A second is to postulate a period during development in which there is a mechanism that establishes tolerance to self-antigens presented by cells in the periphery, i. e., all tissues except the thymus are operative. This could be through clonal deletion or by active mechanisms that would phenotypically be suppressive. A third answer is that tolerance to these molecules is not produced by an active process. They may be nonimmunogenic for various reasons: a lack of immunogenic T cell epitopes, a resistance to degradation, or a predilection to be degraded through pathways where they do not encounter MHC molecules. Alternatively, they are expressed in cells that do not express class II molecules and in consequence never stimulate sufficient help to initiate a response. Autoimmunity is when all this goes wrong.

In speculating about cellular interactions one is constantly asking whether a cell responds or not, whether a target is killed or not. These are threshold phenomena in which a complete effect is observed once a critical strength of interaction between receptors and ligands is reached. In molecular terms the situation is extremely complicated. Separate sets of ligand receptor pairs contribute to the overall interaction: MHC+X and the T cell receptor, CD4 and class II or CD8 and class I, as well as non-specific adhesion molecules such as LFA-1. As the interactions are between membranes in which motion is restricted, multivalent interactions contributed by sets of identical ligand/receptor pairs will increase effective avidities. Clustering induced either by ligand receptor zippering or by the action of cytoplasmic elements will have similar effects. Within such a complex framework it is unclear how continuously varying one variable, the affinity of the T cell receptor for MHC plus peptide, will affect the threshold phenomenon that is being observed. For example, in the classic situation (Zinkernagel and Doherty 1974) a given clone of cytotoxic T cells (CTLs) kills cells having $MHC^A$+X but not $MHC^B$+X or $MHC^A$+Y. When numbers of T cells are titrated one might find a 10- to 20-fold difference in the ratio of effector to target cells that is required for a given level of lysis. This might reflect a twofold, 20-fold, or 1000-fold difference in the affinity of a single (CTL) receptor for a single complex of $MHC^A$+X compared to $MHC^B$+X or $MHC^A$+Y. Such experiments have yet to be done and await reliable methods for producing soluble T cell receptors. With a complex mechanism one can envisage situations in which small differences can be amplified or in which large differences are of little consequence.

Some idea of the potential complexity with multivalent attachment emerged from studies on the binding of monoclonal antibodies to cells (Mason and Williams 1980; Ways and Parham 1983). In these systems one is only dealing with monovalent or divalent binding, the hydrogen atom of cellular interactions. The reactions are commonly controlled by the dissociation rate of the antigen-antibody complex and this becomes manifest during the washing steps at the end of an assay that serve to remove unbound reagents. If a single antibody combining site has an affinity of $\geqslant 10^9 M^{-1}$ then dissociation is low and the extent of measured binding is not significantly increased by using bivalent IgG or $F(ab')_2$ over Fab. On the other hand, if a single antibody combining site has an affinity of $\sim 10^7 M^{-1}$ then dissociation will be rapid and measured binding for Fab will be undetectable. In this case the use of bivalent antibody increases the effective strength of binding by a factor $>100$ and the effective binding from nothing to saturation. A hybrid molecule was made from two such monoclonal antibodies that are dependent upon bivalent binding for stable attachment to cells (Parham 1985). One antibody was specific for HLA-A2 and the other for HLA-B7. As predicted, the hybrid showed considerably greater binding to cells that expressed both HLA-A2 and HLA-B7 than to cells expressing only one or the other. In a formal sense this hybrid antibody can be considered as being MHC restricted, in that it only shows strong binding to HLA-A2 in the context of HLA-B7 or vice versa, if you prefer. Although the binding to HLA-A2 or HLA-B7 is not easily detected in the cell binding assay, it being the choice of the investigator, it does not imply no interaction. Similar considerations will hold for T cell receptors. The observation that our hypothetical CTL does not kill $MHC^B$+X or $MHC^A$+Y does not mean there is no significant affinity between its receptor and these ligands.

These considerations will be of importance for authentic MHC restricted antibodies, in which a single combining site preferentially binds MHC + X, and the properties we expect of them. An ideal MHC restricted antibody should show a high affinity for peptide X bound to $MHC^A$ and a low affinity for complexes of other peptides with $MHC^A$ or for peptide X with $MHC^B$. From the structure of HLA-A2 we see that only a restricted part of the bound peptide is exposed and available for contacting either a T cell receptor or antibody combining site (Bjorkman et al. 1987 a). Thus we must expect that any antibody made against the complex will have extensive interactions with both MHC and the peptide and will inevitably show a significant, albeit lower, affinity for the separate components. Use of the criterion that an MHC restricted antibody shows no affinity for MHC or X separately would in all likelihood eliminate all antibodies, including those that show some degree of restriction. For example, let us consider a situation in which antibodies were initially selected for binding to cells with $MHC^A$ + viral peptide X and not to $MHC^A$ cells without viral peptide X. If, as a second screen, one then eliminated all antibodies binding to a column of peptide X, it is unlikely that any antibodies would survive the test as column binding will generally detect much lower affinities than cell binding assays.

Another potential problem in deciding when an MHC antibody is restricted is that one generally demands a greater degree of specificity from antibodies than T cells. A specificity ratio of 20 in a cellular assay is considered good, in a serological assay disappointing.

Devising strategies of immunization and screening for MHC restricted antibodies is difficult. All evidence suggests that the MHC molecules of a cell, even when infected with virus, present a large and heterogeneous collection of peptides. The abundance for any given complex is low and this is likely to decrease its immunogenicity. Problems of detectability in screening exist for the same reason. Expression of less than a few thousand copies per cell is unlikely to be detected in most commonly used assays. With the total expression of MHC molecules in the range of $10^4$–$10^6$ and the number of potential peptides being very large, it is going to be important to manipulate the system in order to increase the representation of a particular complex. This is going to be easier for class II than class I where demonstrable binding in vitro of pure peptide to a significant proportion of purified Ia and HLA-DR molecules has been possible (Babbitt et al. 1985; Bjorkman et al. 1987 b). Such in vitro formed complexes could then be used in immunization and in the screening of hybridoma supernatants. Other approaches are to grow cells in the presence of peptide or to transfect cells with constructs that overproduce a particular peptide.

Initial interest in MHC restricted antibodies was raised over questions of B and T cell repertoire (Wylie et al. 1982; Froscher and Klinman 1986). Now they are needed as critical reagents for examining the intracellular processes and pathways which bring MHC molecules and peptides together. Tracking the bulk movement of peptides or MHC molecules through the cell will not be sufficient to answer the critical and increasingly cell biological questions about MHC + X. It seems likely that it is a minority of MHC molecules binding a minority of peptides that give the complexes which are important for an immune response. One cannot rely on the pathways leading to these encounters being representative of the bulk movements, and probes specific for the complexes are required. MHC restricted antibodies are the logical choice.

## References

Babbitt BP, Allen PM, Matsueda G, Haber E, Unanue ER (1985) Binding of immunogenic peptides to Ia histocompatibility molecules. Nature 317: 359–361

Bjorkman PJ, Saper MA, Samraoui B, Bennett WS, Strominger JL, Wiley DC (1987a) Structure of the human class I histocompatibility antigen, HLA-A2. Nature 329: 506–512

Bjorkman PJ, Saper MA, Samraoui B, Bennett WS, Strominger JL, Wiley DC (1987b) The foreign antigen binding site and T cell recognition regions of class I histocompatibility antigens. Nature 329: 512–518

Clayberger C, Parham P, Rothbard J, Ludwig DS, Schoolnik GK, Krensky AM (1987) HLA-A2 peptides can regulate cytolysis by human allogeneic T lymphocytes. Nature 330: 763–765

Froscher BG, Klinman NR (1986) Immunization with SV40-transformed cells yields mainly MHC-restricted monoclonal antibodies. J Exp Med 164: 196–210

Lancet D, Parham P, Strominger JL (1979) Heavy chain of HLA-A and HLA-B antigens is conformationally labile: A possible role for $\beta_2$-microglobulin. Proc Natl Acad Sci USA 76: 3844–3848

Mason DW, Williams AF (1980) The kinetics of antibody binding to membrane antigens in solution and at the cell surface. Biochem J 187: 1–20

Parham P (1985) In vitro production of a hybrid monoclonal antibody that binds to cells that express both HLA-A2 and HLA-B7. Hum Immunol 12: 213–221

Townsend ARM, Rothbard J, Gotch FM, Bahadur G, Wraith D, McMichael AJ (1986) The epitopes of influenza nucleoprotein recognized by cytotoxic T lymphocytes can be defined with short synthetic peptides. Cell 44: 959–968

Ways JP, Parham P (1983) The binding of monoclonal antibodies to cell surface molecules: a quantitative analysis with IgG against two alloantigenic determinants of HLA-A2. Biochem J 216: 423–432

Wylie DE, Sherman LA, Klinman DR (1982) The participation of the major histocompatibility complex in antibody recognition of viral antigens expressed on infected cells. J Exp Med 155: 403–414

Zinkernagel RM, Doherty PC (1974) Restriction of in vivo T cell-mediated cytotoxicity in lymphocytic choriomeningitis within a syngeneic or semiallogeneic system. Nature 248: 701–702

# MHC-Restricted Antibodies: Facts and Interpretation

Pavol Ivanyi and Femia Kievits

Central Laboratory of the Netherlands Red Cross Blood Transfusion Service and Laboratory for Experimental and Clinical Immunology, University of Amsterdam, Amsterdam, The Netherlands

T cell recognition of foreign antigen is MHC restricted. This restriction holds for helper as well as cytolytic T cells. A major part of the problem of MHC restriction is connected with the structure and binding site specificity of the T cell receptor (TcR). The serendipity of discovery and changing views about TcR specificity have been reviewed by various authors. The most typical situation can still be characterized by the original description of the MHC restriction phenomenon by Zinkernagel and Doherty (1974, 1979). They found that antiviral cytotoxic T-lymphocytes (CTLs) recognize antigen only on target cells together with self-MHC molecules. The phenomenon has been amply confirmed, although various observations have extended the horizon: (a) viral-specific CTLs frequently recognize MHC alloantigens without virus, (b) determinants shared by various MHC molecules can restrict antigen recognition (cross-reactions among MHC molecules), (c) determinants shared by various foreign antigens can be recognized by the same CTL (cross-reactions among foreign antigens), (d) allo-(MHC)-restricted clones are induced when responder cells are stimulated with alloantigen and virus, and (e) exceptionally, autoreactive T cell clones have been described that have recognized not only allogeneic but also syngeneic cells without antigen.

To which category of the above-mentioned possibilities a CTL belongs depends on the reactivity pattern against a broad panel of target cells. Therefore, CTL reactivity must be "defined" for any discussion about whether MHC-restricted antibodies exist. In this discussion we shall consider as MHC restriction what was found as most typical for bulk CTL cultures derived from individuals sensitized in vivo and therefore probably appears most frequently: thus a typical antiviral MHC-restricted CTL is antigen specific and restricted exclusively by self-MHC molecules (self-MHC+X). Such an oversimplification is needed as a *starting* point to answer the question of whether MHC-restricted antibodies do indeed exist. In this way the very existence of MHC-restricted antibodies has also been discussed by various authors (Van Leeuwen et al. 1979; Wylie et al. 1982; Levy et al. 1984; Sherman et al. 1983; Froscher and Klinman 1986; Opolski et al. 1986; Schmidt 1987; Kievits et al. 1987; Tamminen et al. 1987; Pestalozzi et al. 1987).

The second part of the problem of MHC restriction is connected with the nature of MHC+X. What does the TcR actually see? Whereas this problem remains unsolved, present theory says that protein antigens are degraded into peptides by intracellular processing. These peptides are presented in the groove of the MHC molecules. However, peptide selection and presentation is dependent on the specificity of the MHC molecule (Allen et al. 1987) formed by the configuration of the groove that is created by two long $\alpha$-helices and some parts of the $\beta$-sheets (Bjorkman et al. 1987). A "motif" – not precisely understood – is the basis that allows certain peptides to bind into the respective grooves of the MHC molecule (Sette et al. 1987). Either the TcR recognizes the peptide alone or it recognizes both the peptide and the presenting MHC molecule as a complex (MHC + X). Because we do not yet know what is the exact nature of the MHC + X "neodeterminant," neither can we know exactly what to expect from an MHC-restricted antibody. However, we can examine whether antibodies – MHC-restricted antibodies – exist that exert a reactivity pattern similar to that of a typical MHC-restricted CTL.

Antibody recognizes antigen alone. This cannot be disputed. However, on cell membranes, antigens occur in the context of MHC molecules as peptides bound in the groove of MHC molecules. Structural data provide firm support for the hypothesis that histocompatibility antigens can bind peptides and that the complex of MHC + X is recognized by CTLs. The question can be asked: Do antibodies exist that recognize the antigen in a similar way as the TcR, i.e., in the context of (self)-MHC? If yes, then in analogy to the above-given reactivity pattern of most typical CTLs, the MHC-restricted antibodies should not react with the antigen only, but they should recognize the antigen in the context of self-MHC. Furthermore, they should not bind to autologous MHC without the antigen, thereby being antigen specific, and might recognize alloantigens without foreign antigen. However, the only really specific target is self-MHC + X, the structure that originally triggered the T cell proliferation.

In experiments, mice were injected with syngeneic virus infected cells (or other forms of MHC + X such as H-Y, tumor antigen, TNP-hapten, MuLV-transfected cells) and MHC-restricted antibodies were sought for the criteria delineated above.

Were monoclonal or polyclonal serum antibodies induced that follow the same reactivity pattern as CTLs induced by injection of syngeneic cells infected by the same virus and examined on the same panel of target cells? Asking for a parallel in the reactivity pattern of antibody and CTL is a condition sine qua non, otherwise it would not be clear what kind of antibody we are looking for.

A search for MHC-restricted antibodies might lead to one of three situations:

1. MHC-restricted antibodies are a regular and frequent component of the immune response to injections of syngeneic virus infected cells.
2. MHC-restricted antibodies are not regular or frequent but occur exceptionally.
3. MHC-restricted antibodies do not exist as there is no analogy in antigen recognition and/or binding between the T cell (TcR) and the B cell (Ig).

We believe that the answer to this important question is still not clear and we intend to discuss why it is not clear even though some relevant experiments have been performed.

We have performed various experiments (Kievits et al. 1987, 1988a, b, c, this volume; Opolski et al. 1986) all of which, with the models we have used, gave the answer that MHC-restricted antibodies do not exist at all. Recently, similar negative results have been published by others (Tamminen et al. 1987; Pestalozzi et al. 1987). This is at variance with the systemastic work of the group of Klinman (Wylie et al.

**Table 1.** Reactivity patterns of two spleen fragment supernatants isolated from one B6 spleen. Both antibodies are interpreted as anti-H-$2^{b,d}$ with cross-reactivity to H-$2^{s}$ and H-$2^{q}$. The antibodies were examined by the cytotoxic test on lymphocytes of eight H-2 congeneic strains, Sendai virus negative (SV−), and Sendai virus infected (SV+)

| Spleen fragment supernatant | H-2 haplotypes | | | | | | | | | | | | | | | |
|---|---|---|---|---|---|---|---|---|---|---|---|---|---|---|---|---|
| | b | | d | | f | | k | | p | | q | | r | | s | |
| no. | SV− | SV+ | SV− | SV+ | SV− | SV+ | SV− | SV+ | SV− | SV+ | SV− | SV+ | SV− | SV+ | SV− | SV+ |
| 1 | 14[a] | 14 | 11 | 15 | 0 | 0 | 0 | 0 | 0 | 0 | 2 | 5 | 0 | 0 | 3 | 8 |
| 2 | 16 | 18 | 12 | 14 | 0 | 0 | 0 | 0 | 0 | 0 | 2 | 4 | 0 | 0 | 4 | 10 |

[a] Lymphocytotoxicity scoring grade (CSG) obtained from titration of each antibody in geometrical dilutions against the respective target cells

1982; Froscher and Klinman 1986) for anti-influenza and anti-SV40 antiviral MHC (H-2) restricted antibodies as well as with the occasional observation of the MHC (HLA) restricted anti-H-Y antibody (Van Leeuwen et al. 1979; Goulmy et al., this volume).

We believe that the models we have explored, although not identical (qua foreign antigen used) are nevertheless analogous to those of others. Moreover, Sendai virus infected (SV+) and trinitrophenyl (TNP)-treated cells are classical models for the induction of MHC-restricted CTLs. In our experiments, the humoral response to injections of SV+ cells resulted in the production of two types of antibodies: (a) SV specific antibody, reactive with free virus, and (b) lymphocytotoxic antibodies specific for H-2 class I molecules, and therefore not SV specific. No MHC-restricted antibodies were found that conformed with the above-given definition (Kievits et al. 1987). The response to syngeneic TNP-treated cells resulted in the production of one type of antibody which was TNP specific (Kievits et al., this volume). In these experiments the secondary humoral immune response was examined after repeated injections with syngeneic SV+ cells. Therefore we recently explored the primary response in the spleen fragment cultures as described by Wylie and Klinman (1981). The reaction patterns of two spleen fragment supernatants are shown in Table 1 as selected illustrative examples. In short (Kievits et al., 1988c), the same type of anti-H-2 class I specific antibody was obtained as in the experiments with multiple injections in vivo. Therefore we had to conclude again that MHC-restricted antibodies do not exist.

However, there are certain aspects of our work where interpretation of the results plays a major role.

First, it is clear that injections of syngeneic cells (Ivanyi et al. 1980) with viral antigen (Kievits et al. 1987) can in some situations induce H-2 specific antibodies. This phenomenon, like the existence of "naturally" occurring MHC specific antibodies (Ivanyi et al. 1982; Cerny-Provaznik and Ivanyi, this volume), was not known before the early 1980s. How MHC+X induces allo- and autoreactive MHC specific antibodies is not clear. All in all, three possibilities compete for triggering of MHC specific antibody production: (a) polyclonal stimulation, (b) triggering by MHC+X neodeterminant, and (c) triggering by modified exposure of MHC molecules. We have discussed several times why polyclonal stimulation does not appear as a plausible explanation (Kievits et al. 1987; Černý-Provaznik and Ivanyi, this volume), and we have constantly argued for the third alternative, i.e., not MHC+X=ALLO but X→MHC=ALLO. In other terms, the foreign antigen modifies the exposure of MHC determinants so that they trigger the proliferation of alloreactive B cells.

Secondly, after injection of syngeneic SV+ cells, we regularly obtained sera from individual mice or prepared serum pools which contained H-2 specific cytotoxic antibodies binding more intensely to SV+ than to SV− cells. Binding to SV+ syngeneic cells often occurred (Kievits et al. 1987, 1988c) but was never exclusive. These findings might be easily interpreted as due to MHC-restricted anti-SV antibodies (see Table 2). However, enhanced binding to SV+ cells was not specific for SV and could also be seen after infection of cells with other viruses. Similarly, mitogen-stimulated lymphoblast cells, or INF-treated cells, bound the SV-induced antibodies more intensely thanSV+ cells. Evidently, MHC exposure on the cell membrane was a critical factor for the quantity of antibody bound. A more detailed analysis of the specificity of the immune response, characterization of the target structure for antibodies, absorption studies, and some other tests (Kievits et al. 1987) led us to conclude that the induced antibodies are not anti-SV (MHC restricted) but anti-MHC.

We consider a monoclonal antibody (McAb) isolated from a mouse with a typical serum reactivity pattern to be an example of antibodies induced by injection of syngeneic SV+ cells (Opolski et al. 1986). McAb-Q1 isolated from a B10.Q H-$2^q$ mouse after injections of syngeneic SV+ cells is a typical anti-H-2.25 (a short public K-end determinant shared by H-$2^k$ and H-$2^r$) alloreactive antibody with some weaker cross-reactions to other antigens; binding with relatively highest affinity to $K^k$ targets was not dependent on the presence of SV. However, some other H-2 molecules

**Table 2.** Schematic presentation of antiviral MHC-restricted CTLs and H-2 specific antibodies induced by injection of syngeneic virus infected cells

| | Syngeneic | | Syngeneic | | Allo-A | | Allo-B | | Allo-C | | Allo-D | | Interpretation |
|---|---|---|---|---|---|---|---|---|---|---|---|---|---|
| | V1− | V1+ | V2− | V2+ | V1− | V1+ | V1− | V1+ | V1− | V1+ | V1− | V1+ | |
| CTL | | | | | | | | | | | | | |
| 1 | 0 | 4 | 0 | 0 | 0 | 0 | 0 | 0 | 0 | 0 | 0 | 0 | Anti-V1 |
| 2 | 0 | 4 | 0 | 0 | 0 | 0 | 3 | 3 | 2 | 2 | 0 | 0 | Anti-V1 + Allo-B,C |
| McAb | | | | | | | | | | | | | |
| 1 | 0 | 0 | 0 | 0 | 0 | 0 | 4 | 4 | 0 | 0 | 0 | 0 | Anti-B |
| 2 | 0 | 0 | 0 | 0 | 0 | 0 | 4 | 4 | 1 | 1 | 0 | 0 | Anti-B |
| 3 | 0 | 1 | 0 | 1 | 0 | 0 | 4 | 4 | 0 | 0 | 4 | 4 | Anti-B + D |
| 4 | 0 | 1 | 0 | 1 | 4 | 4 | 0 | 0 | 1 | 3 | 0 | 0 | Anti-A |
| 5 | 0 | 2 | 0 | 2 | 1 | 2 | 4 | 4 | 0 | 0 | 0 | 0 | Anti-B |
| 6 | 1 | 3 | 1 | 3 | 0 | 0 | 0 | 0 | 0 | 0 | 3 | 3 | Anti-D |
| 7 | 2 | 4 | 2 | 4 | 2 | 3 | 1 | 2 | 2 | 3 | 0 | 0 | Anti-? |
| 8 | 4 | 4 | 4 | 4 | 3 | 4 | 0 | 0 | 0 | 0 | 3 | 3 | Anti-auto |
| 9 | 1 | 3 | ND | | 0 | 0 | 0 | 0 | 0 | 0 | ND | | Anti-?[a] |
| 10 | 1 | 2 | 1 | 2 | 4 | 4 | 2 | 3 | 1 | 3 | 1 | 3 | Anti--A[b] |
| Serum pool | | | | | | | | | | | | | |
| 1:10 | 1 | 3 | 1 | 3 | 1 | 2 | 2 | 3 | 0 | 1 | 1 | 3 | Polyvalent |
| 1:100 | 0 | 2 | ND | | ND | | ND | | 0 | 0 | ND | | Anti-? |

[a] Might be anti-D, see McAb, no. 6
[b] Broadly cross-reactive

V1 and V2 are two different viruses. V1− (V2−) and V1+ (V2+) are normal and virus-infected lymphocytes. Allo-A, B, C, D are H-2 congeneic mouse strains with four different H-2 haplotypes. A schematic cytotoxicity scoring grade (or titer) is shown as 0, 1, 2, 3, 4.

*Interpretations:*

CTL: Two anti-V1 MHC-restricted clones are illustrated; one without and the second with alloreactivity

McAb: Ten H-2 specific antibodies are illustrated, all induced by injections of syngeneic V1+ cells. The ten McAbs have various reactivity patterns on the panel of syngeneic and allogeneic target cells. The serum pool is shown (in two dilutions) as a mixture of all antibodies. None of the antibodies, or the serum pool, can be classified as viral specific (anti-V1) but MHC restricted. However, the reactivity pattern of the antibodies or serum can be easily misinterpreted if the panel of targets is not representative and/or antigen expression on various target cells is not similar or was not compared. It can be noticed that McAb no. 9 and the serum pool tested at dilution 1:100 might easily be interpreted as MHC-restricted anti-V1 because insufficient target cells were tested were recognized (with lower affinity) only after cell infection with SV (or influenza virus). Finally, autologous cells (H-$2^q$), without SV almost negative, became clearly positive only after SV or after influenza virus infection. Similar reactivity patterns, although without autoreactivity, were obtained with other McAbs isolated from B6 mice after injections of syngeneic SV+ cells (Kievits et al. 1988 a). Wide variation was observed in H-2 specificity among sera of individual mice and McAbs isolated from injected mice with SV+ cells.

We explain the reaction pattern of H-2 specific antibodies induced by SV+ cells schematically in Table 2. This interpretation implies that antibodies suspected to be antiviral MHC restricted should be examined for specificity on an extensive panel of target cells (a) with a comparable expression of MHC molecules, (b) comprising a number of H-2 haplotypes, and (c) expressing various foreign antigens. The schematic list of ten antibodies in Table 2 illustrates the complexity of the situation and the difficulties of interpretation. The scheme also shows why ordinary serum pools might be very difficult to interpret.

Another source of confusion in our search for MHC-restricted antibodies might be the reactivity pattern of an antibody against a non-MHC cell membrane antigen whose expression is somehow regulated by the MHC. Three examples of such a situation will be discussed here:

1. Antibodies to $\beta_2$-microglobulin ($\beta_2$m) might simulate MHC-restricted antibodies. Bovine $\beta_2$m specific antibodies were induced by injections of syngeneic cells on which bovine $\beta_2$m partially replaced endogenous (mouse) $\beta_2$m (Kievits et al. 1988 b). Such antibodies are cytotoxic to mouse cells dependent on the presence of bovine $\beta_2$m and certain H-2 haplotypes. The H-2 haplotype dependent variation in binding was surprising and indicated that antibody binding (to bovine $\beta_2$m) is restricted by the H-2 haplotype. However, the correct interpretation is that (for unknown reasons) bovine $\beta_2$m is bound by H-2 molecules of various haplotypes to a rather different extent (Kievits and Ivanyi 1987, 1988 b). Thus, B10.S (H-$2^s$) FCS coated cells bind much less anti-bovine $\beta_2$m antibodies than B10 (H-$2^b$) FCS coated cells simply because the latter exchange much more murine $\beta_2$m for bovine $\beta_2$m.
2. Antibodies to Thy.1 (theta) might simulate MHC-restricted antibodies in a similar way, but by another mechanism. On a series of H-2 congeneic strains (all of the same Thy.1 allotype), binding of anti-Thy.1 antibody was greatly dependent on the H-2 haplotype (Mickova and Ivanyi 1973). Thus, antibody binding to a non-MHC cell membrane antigen appeared as MHC restricted. However, Thy.1 expression was found to be hormone (testosterone) dependent (Mickova 1975) and there are some genes in the MHC region involved in hormone metabolism. So, although the H-2 restricted variations in binding of anti-Thy.1 antibodies are not fully explained, they are most probably due to hormone-regulated – and H-2 associated – differences in cell membrane expression of Th.1 antigen.
3. A somewhat similar explanation might hold for the references of Ohno and Stapleton (1981) to the reactivity pattern of an anti-H-Y serum of Scheid et al. (1972). This anti-H-Y serum recognized H-$2^b$ and H-$2^d$ but not H-$2^k$ male cells. Therefore Ohno and Stapleton interpreted the results as follows: "The fact is that neither T cell receptor nor humoral antibodies of the mammalian immune system recognize H-Y antigen per se. Instead they invariably recognize H-Y+ altered MHC

antigen complexes." To our knowledge, there are no data to support these statements. The rather limited data of Scheid et al. (1972) show that AKR H-$2^k$ cells bind the anti-H-Y antibody efficiently. However, binding to C3H H-$2^k$ cells is only slightly above the (female) background values. The original work does not refer to an H-2 influence (restriction), which was actually not tested, but to strain differences in susceptibility to complement-mediated lysis or strain differences in the amount of H-Y antigen present on the cell surface. Hence the data quoted by Ohno and Stapleton do not substantiate their statement that "it should be pointed out that T cell receptors as well as humoral antibodies of the mammalian immune system never recognize H-Y antigen itself but in association with ubiquitously expressed MHC antigens."

Many more data are needed for an understanding as to whether and why antibodies to non-MHC cell membrane antigens bind differently, dependent on the H-2 haplotype of the target cell.

All these data show that there is no possibility of drawing a conclusion on the occurrence of MHC-restricted antibodies to non-MHC antigens. Our data do not show their existence and they even point to various aspects of misinterpretation. This holds at least for the occurrence of MHC-restricted antibodies in analogy to the classical description of MHC-restricted CTLs, which are antigen specific, self-MHC restricted, and do not bind to autologous MHC without antigen.

In extension of our view that virus-specific self-MHC-restricted antibodies do not exist at all, the question emerges of how we interpret the observation that B cells (Ig) do not see a whole universe of MHC + X determinants. We consider two possibilities:

1. MHC + X is not seen by the B cell receptor (Ig) owing to the location and/or shape of the MHC groove with its peptide(s) to which antibody does not attain access.
2. MHC + X does not exist as a specific composite immunogen. The TcR recognizes only antigen (peptide); however, the exposure of the antigen epitope is modified by the MHC molecule. Alternatively, the TcR recognizes only MHC; however, the exposure of MHC epitope is modified by the antigen (peptide). The latter possibility also implies that the TcR and Ig do not recognize the same antigenic determinants.

*Acknowledgments.* This investigation was supported in part by the Foundation for Medical Research MEDIGON, which is subsidized by the Netherlands Organization for the Advancement of Pure Research (ZWO) (grant no. 509-087), and in part by the Deutsche Forschungsgemeinschaft (grant no. 5/1-2).

## References

Allen PM, Matsueda GR, Evans RJ, Dunbar JB, Marshall GR, Unanue ER (1987) Identification of the T cell and Ia contact residues of a T cell antigenic epitope. Nature 327: 713–715

Bjorkman PJ, Saper MA. Samraoui B, Bennett WS, Strominger JL, Wiley DC (1987) Structure of the human class I histocompatibility antigen, HLA-A2. Nature 329: 506–512

Froscher BG, Klinman NR (1986) Immunization with SV40-transformed cells yields mainly MHC-restricted monoclonal antibodies. J Exp Med 164: 196–210
Ivanyi P, van Mourik P, Breuning MH, Melief CJM (1980) Anti-H-2 antibodies induced by syngeneic immunization. Immunogenetics 10: 319–332
Ivanyi P, van Mourik P, Breuning MH, Kruisbeek AM, Kröse CJM (1982) Natural anti-H-2 antibodies in sera of aged mice. Immunogenetics 15: 95–102
Kievits F, Ivanyi P (1987) Monomorphic anti-HLA monoclonal antibody (W6/32) recognizes polymorphic H-2 heavy-chain determinants exposed by association with bovine or human but not murine beta-2-microglobulin. Hum Immunol 20: 115–126
Kievits F, Rocca A, Opolski A, Limpens J, Leupers T, Kloosterman T, Boerenkamp WP, Pla M, Ivanyi P (1987) Induction of H-2 specific antibodies by injections of syngeneic Sendai virus-coated cells. Eur J Immunol 17: 27–35
Kievits F, Opolski A, Boerenkamp WJ, Pla M, Ivanyi P (1988 a) Monoclonal H-2 class I specific antibodies isolated after immunization of C57B1/6 mice with syngeneic Sendai virus-coated cells. In: David C (ed) H-2 antigens: genes, molecules, function. Plenum, New York, pp 669–680
Kievits F, Boerenkamp WJ, Ivanyi P (1988 b) H-2 dependent binding of xenogeneic beta-2-microglobulin from culture media. J Immunol 140: 4253–4255
Kievits F, Boerenkamp W, Ivanyi P (1988 c) Immunization with syngeneic sendaivirus-infected cells induce no MHC-restricted antibodies but antibodies specific for H-2 class-I determinants. Immunogenetics, in press
Levy RG, Dower SK, Shearer GM, Segal DM (1984) Trinitrophenol modification of $H\text{-}2^k$ and $H\text{-}2^b$ spleen cells results in serological detection of $K^k$-like determinants. J Exp Med 159: 4164–1472
Mickova M (1975) The effect of orchiectomy on the manifestation of $Thy\text{-}1^b$ (theta-C3H) antigen in H-2 congeneic mouse strains. Folia Biol (Praha) 20: 343–345
Mickova M, Ivanyi P (1973) Genetic differences in the manifestation of theta-C3H-antigen. Association with the H-2 system. Transplant Proc 5: 1421–1425
Ohno S, Stapleton DW (1981) Associative recognition of testis-organizing H-Y antigen and immunological confusion. In: Jagielle G, Vogel A (eds) Bioregulators of reproduction. Academic, New York, pp 25–42
Opolski A, Kievits F, Ivanyi P (1986) Polymorphic and autoreactive H-2 specific monoclonal antibodies isolated after injections of syngeneic Sendai virus-coated lymphocytes. Immunogentics 24: 402–408
Pestalozzi B, Stitz L, Zinkernagel RM (1987) Monoclonal antibodies against viral determinants are not restricted to the K/D end of the MHC. J Exp Med 166: 295–299
Scheid M, Boyse EA, Carswell EA, Old LJ (1972) Serologically demonstrable alloantigens of mouse epidermal cells. J Exp Med 135: 938–955
Schmidt W (1987) H-2 specific antibodies induced by injection of syngeneic leukemia cells. Immunogenetics 25: 215–221
Sette A, Buus S, Colon S, Smith JA, Miles C, Grey HM (1987) Structural characteristics of an antigen required for its interaction with Ia and recognition by T cells. Nature 328: 395–399
Sherman LA, Vitiello A, Klinman NR (1983) T-cell and B-cell responses to viral antigens at the clonal level. Annu Rev Immunol 1: 63–86
Tamminen WL, Wraith D, Barber BH (1987) Searching for MHC restricted antiviral antibodies: antibodies recognizing the nucleoprotein of influenza virus dominate the serological response of C57B1/6 mice to syngeneic influenza-infected cells. Eur J Immunol 17: 999–1006
Van Leeuwen AE, Goulmy A, van Rood JJ (1979) Major histocompatibility complex restricted antibody reactivity mainly, but not exclusively, directed against cells from male donors. J Exp Med 150: 1075–1083

Wylie DE, Klinman NR (1981) The murine B cell repertoire responsive to an influenza infected syngeneic cell line. J Immunol 127: 194–198

Wylie DE, Sherman LA, Klinman NR (1982) Participation of the major histocompatibility complex in antibody recognition of viral antigens expressed on infected cells. J Exp Med 155: 403–414

Zinkernagel RM, Doherty PC (1974) Restriction of in vitro T cell mediated cytotoxicity in lymphocytic chorionmeningitis within a syngeneic or semi-allogeneic system. Nature 248: 701

Zinkernagel RM, Doherty PC (1979) MHC restricted cytotoxic T cells: studies on the biological role of polymorphic major transplantation antigens determining T cell restriction specificity. Adv Immunol 27: 51–177

# Biochemistry of HLA-B27 in Transfected Cells and Transgenic Mice

Erik J. Baas, Paul J. F. M. Derhaag, and Hidde L. Ploegh

Department of Cellular Biochemistry, The Netherlands Cancer Institute, Antoni van Leeuwenhoekhuis, Plesmanlaan 121, 1066 CX Amsterdam, The Netherlands

To study the functional properties of HLA molecules, genes encoding these proteins have been isolated and used to transfect mouse cells. We have isolated the genes encoding the HLA-B27k and -B27w proteins (Seemann et al. 1986). A number of diseases are associated with certain HLA alleles. Of these associations, the one linking the HLA-B27 alleles with ankylosing spondylitis is the most pronounced. Several models have been presented over the past few years to explain this fact. However, the lack of a proper model system to study this HLA disease association has restricted the experimental basis for these theories. Transgenic mice may provide us with a better defined model system to study the immunobiology of HLA-B27 and – together with biochemical analysis of the HLA-B27 gene product – may give us better insight into HLA disease association.

Transfection of mouse cells with the HLA-B27k or -B27w gene does not result in any surface expression of the HLA heavy chain as detected by FACS analysis using the monoclonal antibody W6/32, reactive with human heavy chain associated with the HLA light chain, $\beta_2$-microglobulin ($\beta_2$m) (Parham et al. 1979). This is in contrast to transfection experiments with the genes for HLA-B7 or -A2, where HLA-B7 or -A2 heavy chains are expressed on the cell surface of the transfected mouse cells associated with murine $\beta_2$m (m$\beta_2$m) as detected with W6/32 (Rein et al. 1987) (Fig. 1). The level of expression of HLA on these transfectants is comparable to the level of expression on human B lymphoblastoid cell lines. The HLA-B27 transfectants do express HLA heavy chain intracellularly, as indicated by immunoprecipitations of metabolically labeled proteins with monoclonal antibody HC-10, reactive with the human free heavy chain (Stam et al. 1986). This pool of free HLA-B27 heavy chain is not sialylated and is endo-H sensitive, which is indicative of accumulation of the protein in the pre-Golgi stage of biosynthesis. Transfected cells having received both the HLA-B27k and the HLA-A2 gene further confirmed this clear difference in biochemical behavior of the two proteins: HLA-A2 is expressed at the cell surface, while HLA-B27k is arrested intracellularly. HLA-B7 and -B27 are identical in amino acid sequence of their $\alpha 3$ domain down to their carboxy terminus. Differences in m$\beta_2$m binding capacity therefore must be contributed by the $\alpha 1$ or $\alpha 2$ domain. Exon shuffling experiments have confirmed this notion (Hochstenbach et al., personal communication), in agreement with the recently published three-dimensional structure of the HLA-A2 protein (Bjorkman et al. 1987).

To overcome the lack of surface expression of HLA-B27 in mouse transfected cell

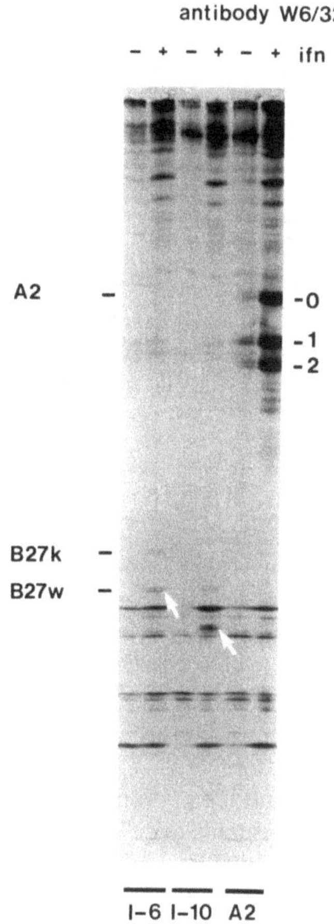


**Fig. 1.** IEF analysis of class I antigens precipitated with W6/32 from metabolically labeled transfected mouse cells indicates the virtual absence of HLA-B27k *(I-6)* and HLA-B27w *(I-10)* but the abundant presence of HLA-A2 *(A2)* as a complex with murine $\beta_2$m. Only minute quantities of HLA-B27 are visible in the absence or presence of interferon *(arrows)*. For HLA-A2 the different sialylated forms are indicated on the *right* (0, 1, and 2 sialic acids) (Rein et al. 1987)

lines, we have cloned the gene encoding the human $\beta_2$m (h$\beta_2$m) and used this gene in a second round of transfection on the HLA-B27 transfected mouse cells (Guessow et al. 1987). Normal surface expression of HLA-B27 is the result, with h$\beta_2$m as light chain. There is thus a striking resemblance with the situation in the human cell line DAUDI, where no heavy chain is detectable on the cell surface due to a mutation in the $\beta_2$m gene, resulting in a defective mRNA and the inability to synthesize h$\beta_2$m protein. Association of the heavy chain with a light chain (human or murine) thus appears necessary for cell surface expression. Inability to form the complex results in absence of surface expression of human class I antigens. The double transfected cells (having received both HLA-B27 and h$\beta_2$m) express a correctly assembled protein, having h$\beta_2$m as light chain as shown by immunoprecipitations with W6/32. The heavy chain is sialylated and endo-H resistant. FACS data fully substantiate the surface expression of HLA-B27 (Fig. 2). On the basis of the results with the HLA-B27 gene in transfection experiments we can make a clear prediction: in order to obtain expression of HLA-B27 in a transgenic mouse model, both human subunits will be required to produce authentic HLA-B27. Only by introducing both the HLA-B27

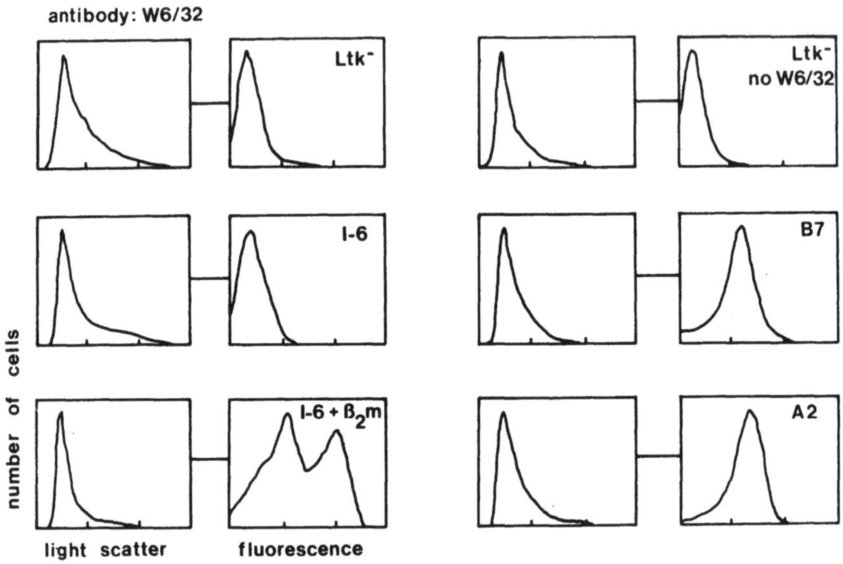


**Fig. 2.** FACS analysis of transfected mouse cells, using W6/32 as antibody. Indicated are non-transfected cells *(Ltk –)*, and cells transfected with HLA-B27k *(I-6)*, HLA-B7 *(B7)*, HLA-A2 *(A2)*, and HLA-B27k together with h$\beta_2$m. As control, fluorescence in the absence of W6/32 is shown (Rein et al. 1987)

gene and the h$\beta_2$m gene in the germline can we expect to obtain HLA-B27 expressing transgenic mice.

To study the in vivo properties of HLA-B27k, the pI6 plasmid containing the HLA-B27k gene was microinjected into fertilized eggs of (CBA × C57Bl/LiA) mice to generate transgenic mice (Krimpenfort et al. 1987). The offspring shown to have incorporated the gene, as indicated by Southern blot analysis of genomic mouse DNA, was analyzed for transcription of the gene and expression of the human heavy chain. Of 12 mice transgenic for HLA-B27k, all were shown to produce the correct mRNA and, intracellularly, the heavy chain protein.

In addition, a second set of transgenic mice was generated, with an identical murine background, using the h$\beta_2$m gene containing plasmid p$\beta_2$m-13. Southern blot analysis indicated the 12 mice to have incorporated the h$\beta_2$m gene in their genome. However, only two appeared to produce the appropriate mRNA and to express the h$\beta_2$m protein, probably due to the presence of bacterial vector sequences flanking the microinjected gene. These sequences are not present in the microinjected HLA-B27k gene.

On crossing an HLA-B27k positive mouse with a transgenic expressing h$\beta_2$m, offspring were produced of which some individuals were positive for both human genes (Fig. 3). Immunoprecipitations and FACS analysis show the obligatory requirement of h$\beta_2$m for HLA-B27k expression (data not shown), in concordance with the data of the transfection experiments. Mice transgenic for h$\beta_2$m *only* express the human light chain on splenocytes, associated with certain murine H-2 heavy chains. These H-2 class I heavy chains were not characterized further. On splenocytes of

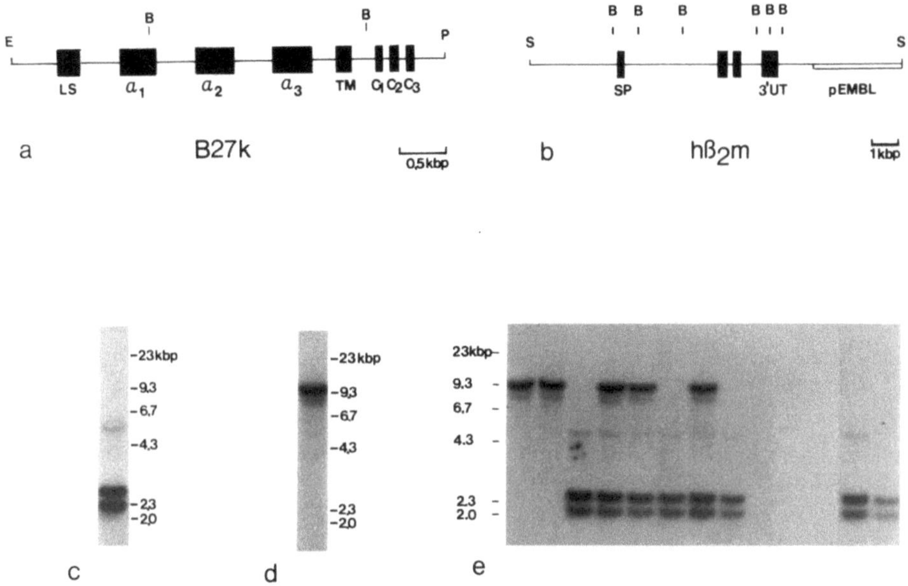


**Fig. 3. a** The HLA-B27k gene used to generate transgenic mice. *Solid bars* represent exon sequences. *E,* EcoRI; *B,* BamHI; *P,* PstI. **b** The $h\beta_2m$ gene containing construct used for microinjection. The *open bar* represents vector sequences. *S, SalI.* **c** Detection of HLA-B27k on Southern blot analysis in BglII digested genomic mouse DNA, using an HLA-B7 cDNA fragment as probe. This mouse was used in the cross shown in **e. d** Detection of $h\beta_2m$ on Southern blot analysis in BglII digested genomic mouse DNA, using pEMBL9 as probe. This mouse was used in the cross shown in **e. e** Southern blot analysis of BglII digested genomic DNA from the offspring of an HLA-B27k and an $h\beta_2m$ transgenic mouse. The upper hybridizing fragment indicates the presence of the HLA-B27k gene; the lower hybridizing fragments indicate the presence of the $h\beta_2m$ gene. Both transgenes are transmitted independently (Krimpenfort et al. 1987)

mice transgenic for both HLA-B27k and $h\beta_2m$ (double transgenic mice, dTGM) a level of HLA class I expression comparable with the level of expression found on human EBV transformed B cell lines was seen.

Human alloreactive cytotoxic T-lymphocytes (CTLs) raised against human HLA-B27k expressing cells are able to specifically lyse dTGM splenocytes, indicating the expressed human class I molecule indeed to be conformationally identical to the HLA-B27k molecule as expressed on human cells (Kievits et al. 1987). In order to analyze the in vivo function of the HLA-B27k/$h\beta_2m$ complex, the ability of the dTGM to raise HLA-B27k restricted CTLs was studied. After in vivo priming with either Sendai virus or influenza A/X79, the ability of dTGM splenocytes to lyse virus-coated (Sendai) or virus-infected (influenza) autologous cells or human HLA-B27k or -B27w positive cells was measured in a chromium release assay. HLA-B27 negative human lymphocytes served as controls (Fig. 4) (Kievits et al. 1987). The results indicate that the murine T cell repertoire is capable of specifying HLA-B27k restricted CTLs, and antigen can be presented in the context of HLA-B27k in these transgenic mice. This antigen restriction is HLA-B27k specific: HLA-B27w, another HLA-B27

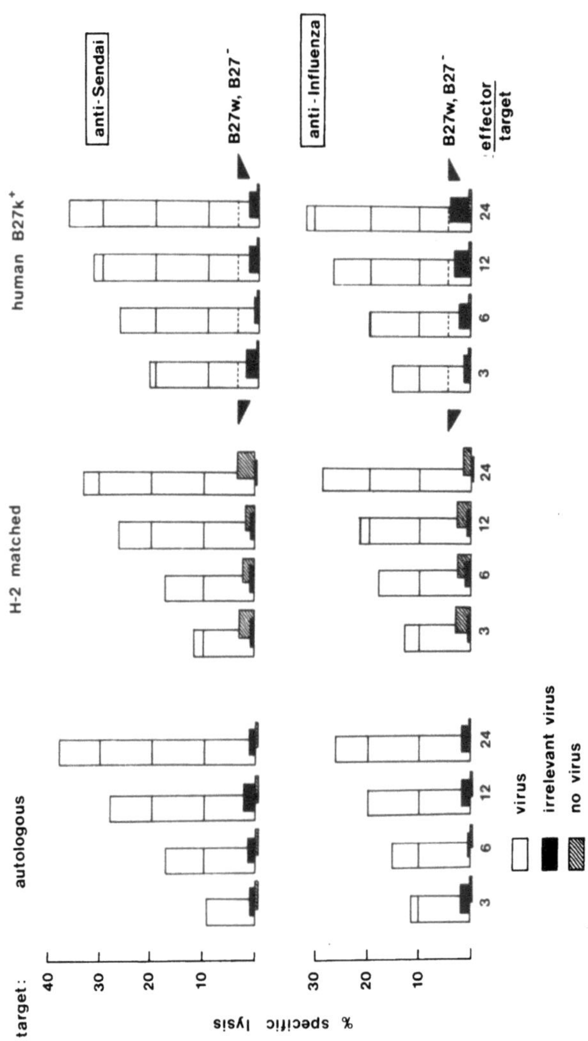


**Fig. 4.** Chromium release of target cells induced by CTLs specific for Sendai virus *(upper panel)* or influenza virus *(lower panel). Dashed line* indicates Cr release of human HLA-B27w or -B27 negative target cells, both virus infected and virus uninfected. *Open bars:* target cells infected with the virus against which the CTLs were raised. *Hatched bars:* uninfected target cells. *Solid bars:* irrelevant virus (Sendai in the case of influenza specific CTLs, influenza in the case of Sendai specific CTLs) (Kievits et al. 1987)

subtype, does not serve as a restriction element. This ability to discriminate between HLA-B27 subtypes differing in only three amino acids accurately reflects the situation observed in the human system.

In conclusion, the HLA-B27k transgenic mice offer a suitable model system for in vivo HLA studies. Association of the HLA-B27 heavy chain with murine $\beta_2$m appears insignificant, and in dTGM the presence of h$\beta_2$m prevents the formation of such hybrid molecules: none were detectable with the W6/32 antibody. The use of these transgenics in disease association studies now focuses on the immunological effects of expression of HLA-B27k: dTGM are tolerant for HLA-B27k and h$\beta_2$m, thereby possibly also creating other "holes" in the T cell repertoire; further, the mode of presentation of specific antigen by HLA-B27k molecules results in an HLA-B27k restricted murine T cell response. It remains to been seen whether a challenge with arthritogenic microorganisms will result in ankylosing spondylitis-like symptoms in HLA-B27 transgenic mice.

## References

Bjorkman PJ, Saper MA, Samraoui B, Bennet WS, Strominger JL, Wiley DC (1987) Structure of the human class I histocompatibility antigen, HLA-A2. Nature 329: 506–512

Guessow D, Rein R, Ginjaar I, Hochstenbach F, Seemann G, Ploegh HL (1987) The human $\beta_2$-microglobulin gene, primary structure and definition of the transcription unit. J Immunol 139: 3132–3138

Kievits F, Ivanyi P, Krimpenfort P, Berns A, Ploegh HL (1987) HLA restricted recognition of viral antigens in HLA transgenic mice. Nature 329: 447–449

Krimpenfort P, Rudenko G, Hochstenbach F, Guessow D, Berns A, Ploegh H (1987) Crosses of two independently derived transgenic mice demonstrate functional complementation of the genes encoding heavy (HLA-B27) and light ($\beta_2$-microglobulin) chains of HLA class I antigens. EMBO J 6: 1673–1676

Parham PR, Barnstable CJ, Bodmer WF (1979) Use of a monoclonal antibody (W6/32) in structural studies of HLA-A, B, C, antigens. J Immunol 123: 342–349

Rein RS, Seemann GHA, Neefjes JJ, Hochstenbach FMH, Stam NJ, Ploegh HL (1987) Association with $\beta_2$-microglobulin controls the expression of transfected human class I genes. J Immunol 138: 1178–1183

Seemann GHA, Rein RS, Brown CS, Ploegh HL (1986) Gene conversion-like mechanisms may generate polymorphism in human class I genes. EMBO J 5: 547–552

Stam NJ, Spits H, Ploegh HL (1986) Monoclonal antibodies raised against denatured HLA-B locus heavy chains permit biochemical characterization of certain HLA-C locus products. J Immunol 137: 2299–2306

# Interaction of the Qa Region with $\beta_2$-Microglobulin and with Non-Qa Genes in Determination of Qa Phenotype

MARGRIET OUDSHOORN-SNOEK[1], SURISH RAGHOEBIER[1], LORRAINE FLAHERTY[2], and PETER DEMANT[1]

[1] Department of Molecular Genetics, The Netherlands Cancer Institute (Antoni van Leeuwenhoekhuis), Plesmanlaan 121, 1066 CX Amsterdam, The Netherlands
[2] Wadworth Center for Laboratories and Research, New York State Department of Health, Albany, NY 12201, USA

The class I histocompatibility antigens are cell surface glycoproteins with a heavy chain of 40000–49000 and a light chain of 12000 daltons, representing the associated $\beta_2$-microglobulin molecule ($\beta_2$m). The heavy and light chains are encoded on different chromosomes.

In the mouse the class I antigens are encoded by the *H-2* complex on chromosome 17 and the $\beta_2$m gene is located on chromosome 2. The classical transplantation antigens encoded by the *K* and *D* regions are expressed on almost all somatic cells, while the class I products encoded by the *Qa-Tla* region show a tissue-specific expression. The *Qa-Tla* gene products appear to be under complex genetic control, as is for example shown by the quantitative trait of Qa-2: high and low levels of Qa-2 expression exist and this appears to be controlled by a gene in the *D-Tla* segment (Oudshoorn-Snoek et al. 1981; Rucker et al. 1983).

Another example of complex genetic regulation is the Qa-11 antigen. This determinant, detected by immunoprecipitation of radiolabeled cell membrane preparations, is controlled by at least two components: one that maps to the right of *Qa-2* on chromosome 17, and another that is identical or closely linked to the *b*-allele of the $\beta_2$m gene of chromosome 2 (Van de Meugheuvel et al. 1985). The gene controlling Qa-11 expression has been mapped to the *Tla* region on the basis of the B6.K2 strain (*Qa-2*$^a$-*Tla*$^b$). However, we suggested, because of the indistinguishable

**Table 1.** Mapping of Qa-11 using Qa-Tla recombinants

| Strain | $\beta_2$m | Genotype | | | | Phenotype | |
|---|---|---|---|---|---|---|---|
| | | K | D | Qa | Tla | Qa-2 | Qa-11[a] |
| C57BL | b | b | b | b | b | + | + |
| B6.K2 | b | b | b | b | k | + | − |
| B6.K1 | b | b | b | k | k | − | − |
| B6-Tla[a] | b | b | b | b | a | + | + |
| B6.K3 | b | k | k | b | a | + | + |
| B6.K4 | b | k | k | k | a | − | − |

[a] Qa-11 was detected by SDS-Page determination of immunoprecipitation of $^{125}$I-labeled lymph node cells (Van de Meugheuvel et al. 1985)

biochemical characteristics of Qa-11 with Qa-2 (apparent molecular weight, isoelectric point, Cleveland peptide maps), that the *Tla* region gene actually regulates the expression of a *Qa-2* region structural gene or modifies its product (Van de Meugheuvel et al. 1985). The same kind of regulation has been subsequently proposed for the expression of Qa-6 (Widacki et al. 1985). Now, using two new recombinant inbred strains, B6.K3 and B6.K4 (Flaherty et al. 1985), which differ in the *Qa-2* region, we can show that the *Qa-2* region is indispensable for the expression of Qa-11, suggesting that it may indeed carry the relevant structural gene (Table 1).

$F_1$ complementation tests (Table 2) showed that the gene, located in the *Tla* region and controlling the Qa-11 expression, is not acting in trans position, while hybrid complementation of $\beta_2$m *b*-allele and a suitable *Qa-Tla* haplotype leads to expression of Qa-11 in $F_1$ hybrids of two Qa-11-negative strains.

Tissue distribution studies showed that Qa-11 is expressed on T cells and not on B-lymphocytes (Table 3).

We did not demonstrate a 40000-dalton band on immunoprecipitations of lysates of concanavalin A (ConA) or lipopolysaccharide (LPS)-stimulated spleen cells. The negative result on T-lymphoblasts is most probably partly due to the low sensitivity of the immunoprecipitation assay, since serologically (cytotoxic assay) in B6.K3 mice we could demonstrate low but definite amounts of Qa-11 on the cell surface of activated T cells (Table 3). This was done with the Qa-11 determining antiserum S4(A.SW × C3H)$F_1$ anti-B10.A(4R) absorbed with AKR lymphocytes to remove the contaminating Ly-6.2 antibodies. Another reason for the low amount of Qa-11 on T-lymphoblasts may be substitution of murine b-type $\beta_2$m chain with the bovine $\beta_2$m, as it has been shown that $\beta_2$m from fetal calf serum associates with MHC class I antigens on the surface of cultured cells (Bernabeu et al. 1984). Since Qa-11 coprecipitates with Qa-2 (Van de Meugheuvel et al. 1985), it is also likely that Qa-11

**Table 2.** $F_1$ complementation for Tla and $\beta_2$m

| *a. No complementation with Tla* Strain | $\beta_2$m | K | D | Qa | Tla | Qa-2 | Qa-11 |
|---|---|---|---|---|---|---|---|
| B6-Tla[a] | b | b | b | b | a | + | + |
| B6.K2 | b | b | b | b | k | + | – |
| B6.K4 | b | k | k | k | a | – | – |
| $F_1$ hybrid B6.K2 × B6.K4 | b | bk | bk | bk | ka | + | – |
| *b. Complementation with $\beta_2$m* Strain | $\beta_2$m | K | D | Qa | Tla | Qa-2 | Qa-11 |
| C57BL/10 | b | b | b | b | b | + | + |
| BALB.B | a | b | b | b | b | + | – |
| B6.K2 | b | b | b | b | k | + | – |
| $F_1$ hybrid B6.K2 × BALB.B | ab | bb | bb | bb | bk | + | + |

**Table 3.** Tissue distribution of Qa-11

| Cell type | 40-kd Qa-11 molecule (C57BL/10) | Cytotoxic titer[a] with anti-Qa-11 (B6.K3) |
|---|---|---|
| Lymph node cells | Present | 64 |
| T-lymphocytes[b] | Present | 250 |
| B-lymphocytes | Absent | 0 |
| ConA lymphoblasts[c] | Doubtful | 10 |
| PHA lymphoblasts[c] | Not done | 25 |
| LPS lymphoblasts[c] | Absent | 0 |

[a] Complement-dependent cytotoxicity assay was performed as described previously (Snoek et al. 1979). Cytotoxic end titer is 75% of maximal lysis

[b] For immunoprecipitation studies enriched T cell populations were obtained by panning (Wysocki and Sato 1978). For cytotoxicity, T cells were purified using nylon wool filtration

[c] Mitogen stimulation was performed as follows: Splenic lymphocytes were cultured at a concentration of $2.5 \times 10^6$ cell/ml in $\alpha$-RPMI-1640 (Flow, Irvine, Scotland), supplemented with 10% fetal calf serum, $2 \times 10^{-5}$ M $\beta$-mercaptoethanol, and 2.5 µg/ml ConA (Sigma, St. Louis, Mo.) or 1 µg/ml PHA (Welcome, Beckenham, England) or 100 µg/ml LPS (Sigma, St. Louis, Mo.) for 72 h at 37 °C

is secreted by activated T cells, as was shown for Qa-2 by Soloski et al. (1986). This might explain the low level of Qa-11 on 3-day T-lymphoblasts. We observed that during mitogen activation of T-lymphocytes the level of expression of the membrane-bound form of Qa-2 gradually decreases with time (Oudshoorn-Snoek, unpublished data), which is in accordance with low levels of Qa-2 on day 3 of T cell stimulation (Rabinowitz et al. 1986).

The genetic determination of Qa phenotype is still largely unresolved. Although evidence is growing that the different Qa-2-related specificities are epitopes of one or more Qa-2-positive products, these antigenic specificities show more limited tissue distribution than the Qa-2 specificity itself. For example, only Qa-2 was detected on B cells of Qa-2 strains with high levels of Qa-2 expression (Rucker et al. 1983). Divergence in cellular expression was also found between Qa-2 and Qa-m208, a T cell marker defined by a $K^b$-Qa-Tla cross-reactive monoclonal antibody (Oudshoorn-Snoek et al. 1984). Studies with activated T cells showed that T-lymphoblasts of Qa-m208-positive B10.A mice do react with the anti-Qa-m208 monoclonal antibody, while Qa-m208-positive B10.A(2R) T cells fail to react after stimulation. Our preliminary data suggest that also in this case *Qa* and *Tla* region interaction is responsible for the expression of the Qa-m208 determinant (Oudshoorn-Snoek, in preparation).

In summary, in accordance with the previously proposed model (Van de Meugheuvel et al. 1985), we can now show that Qa-11 is a pure T cell marker that is controlled by several elements, i.e., a gene (most likely the structural gene) in the *Qa-2* region and a (regulatory) gene in the *Tla* region as well as the $\beta_2$m *b*-allele. Preliminary data suggest that the expression of Qa-m208 on activated T-lymphocytes is also controlled by genes in both *Qa-2* and *Tla* regions. The analysis of the mechanisms of action of genes regulating the expression of Qa antigens may provide information relevant also for the biology of all class I antigens.

*Acknowledgments.* We thank Mr. Max Zomerdijk for technical assistance and Mrs. Marlijn Sonne for preparation of the manuscript.

## References


Bernabeu C, van de Rijn M, Lerch PG, Terhorst CT (1984) $\beta_2$-Microglobulin from serum associates with MHC class I antigens on the surface of cultured cells. Nature 308: 642–645

Flaherty L, DiBiase K, Lynes A, Seidman JG, Weinberger O, Rinchik EM (1985) Characterization of a Q subregion gene in the murine major histocompatibility complex. Proc Natl Acad Sci USA 82: 2844–2848

Oudshoorn-Snoek M, Ivanyi D, Demant P (1981) Qualitative and quantitative aspects of anti-H-2L$^D$ sera. Transplantation 32: 128–138

Oudshoorn-Snoek M, Mellor AL, Flavell RA, Demant P (1984) A new determinant, Qa-m208, detected on T lymphocytes and transfected L cells by a K$^b$-specific monoclonal antibody. Immonogenetics 19: 461–474

Rabinowitz R, Sharrow SO, Chatterjee-Das S, Rogers MJ, Sachs DH (1986) Qa alloantigen expression on functional T lymphocytes from spleen and thymus. Immunogenetics 24: 391–401

Rucker J, Horowitz M, Lerner EA, Murphy DB (1983) Monoclonal antibody reveals *H-2*-linked quantitative and qualitative variation in the expression of a Qa-2 region determinant. Immunogenetics 17: 303–316

Snoek M, Ivanyi D, Nusse R, Demant P (1979) A new H-2.1-positive D region allele, D$^{dx}$, controlling two molecules, H-2D$^{dx}$ and H-2L$^{dx}$. Immunogenetics 8: 109–125

Soloski MJ, Vernachio J, Einhorn G, Lattimore A (1986) Qa gene expression, biosynthesis and secretion of Qa-2 molecules in activated T cells. Proc Natl Acad Sci USA 83: 2949–2953

Van de Meugheuvel W, van Seventer G, Demant P (1985) A new *Tla* region antigen Qa-11, similar to Qa-2 and associated with b-type $\beta_2$-microglobulin. J Immunol 134: 2507–2512

Widacki SM, Flaherty L, Cook RG (1985) Biochemical characterization of the molecules reactive with Qa-6 antiserum and the monoclonal antibody 20-8-4: Evidence for structural similarity with the Qa-2 molecule. J Immunol 135: 3333–3339

Wysocki LJ, Sato VL (1978) Panning for lymphocytes: A method for cell selection. Proc Natl Acad Sci USA 75: 2844–2848

# $\beta_2$-Microglobulin and $\beta$-Galactosidase: A Real and a Surrogate Minor Histocompatibility Antigen

HANS-GEORG RAMMENSEE[1] and ULRICH THEOPOLD[2]

[1] Basel Institute for Immunology, Grenzacherstrasse 487, 4005 Basel, Switzerland
[2] Max-Planck-Institut für Immunbiologie, Stübeweg 51, 7800 Freiburg, Federal Republic of Germany

## Introduction

Minor H antigens are products of many different loci scattered throughout the genome, detected by graft rejection in vivo or by T cell recognition in vitro (Graff 1978; Loveland and Simpson 1986). The products of most of these genes are unknown. Based on cytotoxic T-lymphocyte (CTL) reactivity patterns, it was suggested that the *H-3* region on chromosome 2 of the mouse consists of several minor *H* genes and that one of these genes is identical to $\beta_2$m (Michaelson 1983; Rammensee and Klein 1983; Kurtz et al. 1985; Click 1985), which has been mapped to the *H-3* region (Michaelson 1981; Robinson et al. 1981). We and others have shown that CTLs specific for an $H\text{-}3^a$ product as expressed in C57BL/10 and restricted to $K^b$ are actually specific for $\beta_2$M-B, thereby identifying $\beta_2$M ($\beta_2$-microglobulin) as a minor H antigen as far as CTL recognition is concerned (Rammensee et al. 1986; Kurtz et al. 1987). In this paper, we show by fluorescence analysis that $\beta_2$M-B can bind to B10.C-$H\text{-}3^c$ cells which are genetically $\beta_2m^a$ and that a CTL clone specific for $\beta_2$M-B does not cross-react to $\beta_2$M-W1, $\beta_2$M-W2, and $\beta_2$M-W4, which are derived from wild mice, confirming the allele specificity of $\beta_2$M-B specific CTLs.

The nature of other minor H antigens is not known. Since some virus-specific CTLs recognize peptides derived from intracellularly expressed viral proteins in the context of class I (Townsend et al. 1985, 1986), it was suggested that minor H antigens might also be fragments of intracellular proteins (Kourilsky and Claverie 1986; Germain 1986; Bevan 1987). This hypothesis requires that degradation products of such normal self-proteins would be available on the cell surface to be recognized by CTLs in the context of class I. The own immune system would not produce a CTL response against these fragments because of self-tolerance. Upon transfer of cells to a genetically disparate recipient some of the fragments, the polymorphic ones, would then be recognized by recipient CTLs as minor H antigens. Some of the internal, polymorphic antigens might be encoded for by endogenous viral sequences (Colombo et al. 1987; for review see Simpson 1987), but others might be housekeeping proteins of the cell, like enzymes. To learn something about antigen presentation of nonviral intracellular proteins, we tested whether CTLs could recognize a cell expressing foreign protein in the cytoplasm, but not on the cell surface. Such a foreign internal protein introduced into a cell by transfection of the gene could be regarded as a surrogate minor H antigen. We transfected the gene for the bacterial enzyme $\beta$-galactosidase ($\beta$-gal) into BALB/c tumor cells and raised CTLs from BALB/c mice against the transfectants.

## Materials and Methods

### *Animals*

B10.C-*H-3*$^c$ ($\beta_2m^a$), BALB.B ($\beta_2m^a$), and B10.BR ($\beta_2m^b$) mice were bred and maintained in the animal facility of the Basel Institute for Immunology. BALB/c and C57BL/6 ($\beta_2m^b$) mice were purchased from OLAC Ltd., Bicester, Oxon, G.B.

### *Cell Lines*

181B is a CTL line of (BALB.K × B10.C-*H-3*$^c$)$F_1$ origin stimulated with C57BL/10, as described elsewhere (Rammensee et al. 1986). The CTL clone 181B-3E1 was derived from this line by limiting dilution. It is maintained by weekly stimulation with irradiated C57BL/10 or C57BL/6 cells in I1-2→I4-2 containing medium. IgM 662 is a mutant line derived from the BALB/c B cell hybridoma Sp6 (Baumann et al. 1985). IgM 662 differs from Sp6 by a mutation in the $\mu$ gene. C5-C6, C3-Ce→C4, and $C_{DEL}$ are IgM 662 cells transfected with different constructs containing the *E. coli* $\beta$-gal gene and part of the Sp6 $\mu$ gene, inserted into a PAG 60 vector as depicted in Fig. 3. 662 PAG was derived by transfecting IgM 662 with the vector alone. Transfectants were maintained in culture medium containing 1.5 mg/ml G418. Modes of $\beta$-gal expression of the transfectants are shown in Table 1.

0805A is a CTL line derived from a BALB/c mouse immunized i.p. with $1 \times 10^7$ C3-C4 cells. It is maintained by weekly stimulation with irradiated BALB/c spleen cells and C3-C4 cells.

### *Exchange of $\beta_2m$*

Spleen cells were incubated for 2 days in medium supplemented with 5% mouse serum containing the desired allelic form of $\beta_2$M and 2.5 µg/ml concanavalin A.

**Table 1.** Expression of the fusion proteins[a]

| Transfected plasmid | Cell line | Glycosylation | Secretion | Cell surface |
|---|---|---|---|---|
| pC3-4 | C3-Ce→C4 | + | – | – |
| pC5-6 | C5-C6 | + | – | + |
| $pC_{del}$ | $C_{DEL}$ | + | + | – |

[a] Plasmids of Fig. 3 were transfected into the cell line IgM 662. Cell surface expression was assayed by cell radioimmunoassays, transport to the endoplasmic reticulum was assayed by inhibition of glycosylation by tunicamycin, and secretion was assayed by precipitation from radioactive labeled supernatants and radioimmunoassays

### *CTL Assays*

These have been done as described elsewhere (Rammensee et al. 1984). Briefly, $10^4$ $^{51}Cr$ labeled concanavalin A induced blast cells were incubated with titrated numbers of CTLs for 3.5–4 h in round bottom 96 well plates. Specific release was calculated as follows:

$$\text{specific release} = 100 \times \frac{E-S}{D-S}$$

where $E$ = cpm released by responders, $S$ = cpm released by medium alone, and $D$ = cpm released by detergent.

### *Monoclonal Antibodies and Sera*

$K^b$ specific antibodies 5F1.2 (Rüsch et al. 1983) were a gift from Linda Sherman; $\beta_2$M-B specific antibodies S19.8 (Tada et al. 1980) were a gift from U. Hämmerling. Sera from mice heterozygous for wild mice-derived $\beta_2$m alleles and $\beta_2m^a$ (w3/a and w4/a) and from mice homozygous for $\beta_2m^{w1}$ were supplied by K. Fischer-Lindahl.

### *Fluorescence Analysis*

Cells were incubated in 96 well plates for 45 min with $K^b$ or $\beta_2$M-B specific antibodies at 24 °C, washed, and incubated with FITC-labeled sheep anti-mouse Ig (Silenus) for another 45 min. Analysis was done on a FACS analyzer (Becton Dickinson).

## Results

### *Specificity of CTL Clone 181B-3E1*

The CTL line 181B is specific for $\beta_2$M-B and restricted to $K^b$, as described elsewhere (Rammensee et al. 1986). The clone 181B-3E1 was derived from this line by limiting dilution. 181B-3E1 kills C57BL/6 targets which express $K^b$ and $\beta_2$M-A (Fig. 1 a). However, BALB.B *($\beta_2m^a$)* targets preincubated in B10.BR serum, which contains $\beta_2$M-B, were killed as efficiently as C57BL/6 targets. This strong killing of the target with acquired $\beta_2$M-B is in contrast to the results obtained with the uncloned line 181B, where killing of the target with acquired $\beta_2$M-B prepared under similar conditions is weaker than killing of targets with intrinsic $\beta_2$M-B (Rammensee et al. 1986). A possible explanation is that 181B contains an additional specificity for another *H-3* region encoded H antigen (Kurtz et al. 1985).

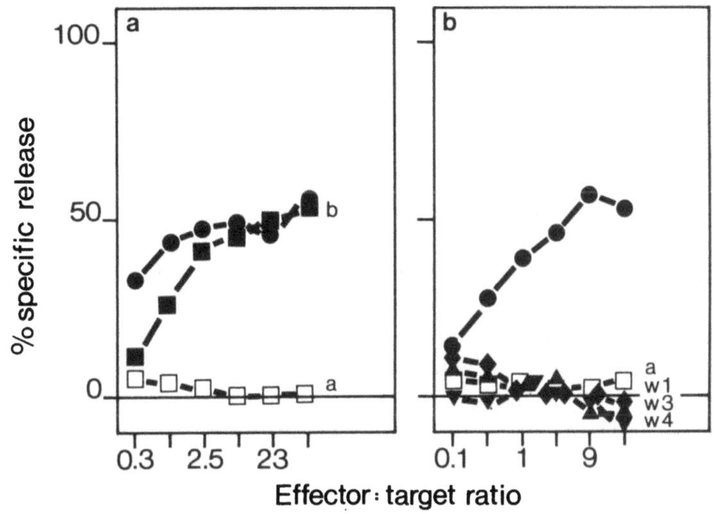


**Fig. 1a, b.** Cross-reactivity of 181-3E1 with the $\beta_2m$ allelic forms. Killing by 181-3B1 was tested in a $^{51}$Cr release assay on C57BL/6 targets (●) or on BALB.B targets *(all other symbols)* preincubated in 5% serum containing $\beta_2$M-B (■), $\beta_2$M-A (□), $\beta_2$M-W1 (◇), $\beta_2$M-W3 (▲), or $\beta_2$M-W4 (▼)

## *Demonstration of $\beta_2$M-B Acquired from Serum*

The binding of serum-derived $\beta_2$M-B to genetically $\beta_2m^a$ cells can be demonstrated readily with $\beta_2$M-B specific antibodies (Fig. 2). In this experiment, spleen cells of a $\beta_2m^a$, $K^b$ mouse (B10.C-*H-3*$^c$) were cultured with $\beta_2$M-B containing serum and concanavalin A under the same conditions as were used to produce targets for the CTL assay. These cells were positive for $\beta_2$M-B fluorescence analysis, whereas the same

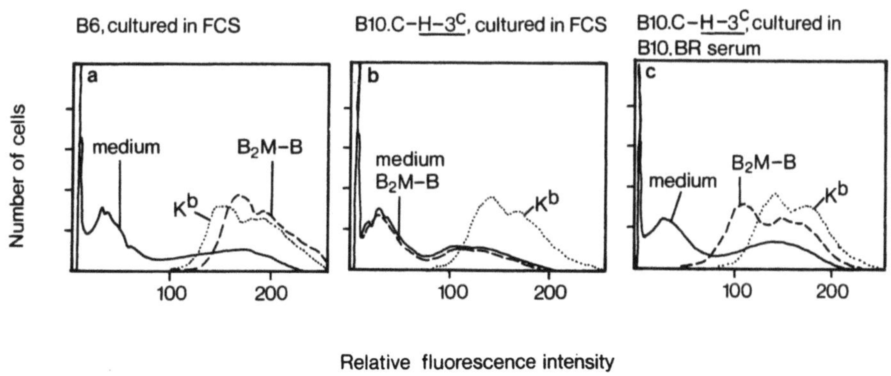


**Fig. 2.** Exchange of $\beta_2$M-B from serum to $\beta_2m^a$ cells. C57BL/6 (**a**) or B10.C-*H-3*$^c$ (**b, c**) spleen cells were cultured for 2 days with concanavalin A and either 10% fetal calf serum (**a, b**) or 5% serum of B10.BR containing $\beta_2$M-B specific antibodies and second step fluorescent antibody, and were analyzed by flow cytometry

kind of cells incubated with fetal calf serum instead of $\beta_2$M-B containing serum did not stain with $\beta_2$M-B specific antibody (Fig. 2 b, c). Both cell preparations stained well with $K^b$ specific antibodies used as positive control. C57BL/6 cells, producing intrinsic $\beta_2$M-B, stained with both antibodies.

### *Cross-reactivity with Other $\beta_2$m Allelic Forms*

The common laboratory mouse strains express either $\beta_2$M-A or $\beta_2$M-B (Michaelson 1983). Additional allelic forms have been described in wild mice (Robinson et al. 1984). To further analyze the specificity of 181B-3E1 CTLs, the lysis of BALB.B target cells preincubated in sera containing $\beta_2$M-W1, $\beta_2$M-W3, or $\beta_2$M-W4 was determined. As shown in Fig. 1 b, none of these targets were lysed, indicating that none of these additional allelic forms are recognized by 181B-3E1.

### *Rationale for the Production of CTLs Specific for Cells Transfected with $\beta$-Galactosidase*

The identification of $\beta_2$M as the molecule recognized by H-3-specific CTLs was possible by use of the reversible association of $\beta_2$M with class I heavy chains. Other minor H antigens might be degradation products from intracellular proteins, as was suggested by analogy with CTL recognition of fragments from viral proteins expressed intracellularly (Townsend et al. 1985; Kourilsky and Claverie 1986; Germain 1986; Bevan 1987). Thus, a method of identifying minor H antigens of this kind would be to transfect polymorphic genes coding for intracellular proteins into recipient cells and to raise CTLs against the transfected cells. To test the feasibility of this approach and to learn something about antigen presentation of intracellular proteins, we decided to transfect BALB/c tumor cells with a bacterial gene, *E. coli* $\beta$-galactosidase. This bacterial protein should be treated as foreign by a BALB/c mouse. Therefore, the occurrence of CTLs in BALB/c specific for $\beta$-galactosidase, or degradation products thereof, should be expected.

### *Design of Transfectants*

Since we wanted to study CTL recognition of protein expressed in different compartments of the cell, a versatile vector system was used, as depicted in Fig. 3. The pBR322 derived plasmid PAG60, which contains a G 418 resistance gene as a selectable marker, was used to insert the *$\beta$-gal* gene coupled to $\mu$ heavy chain gene segments at the sites indicated in Fig. 3. Deletion of different parts of the $\mu$ gene resulted in intracellular, membrane inserted, or secreted product in the transfectants, as summarized in Table 1. Note that all products expressed in the transfectants are expected to be fusion proteins between parts of the $\mu$ heavy chain and $\beta$-gal.

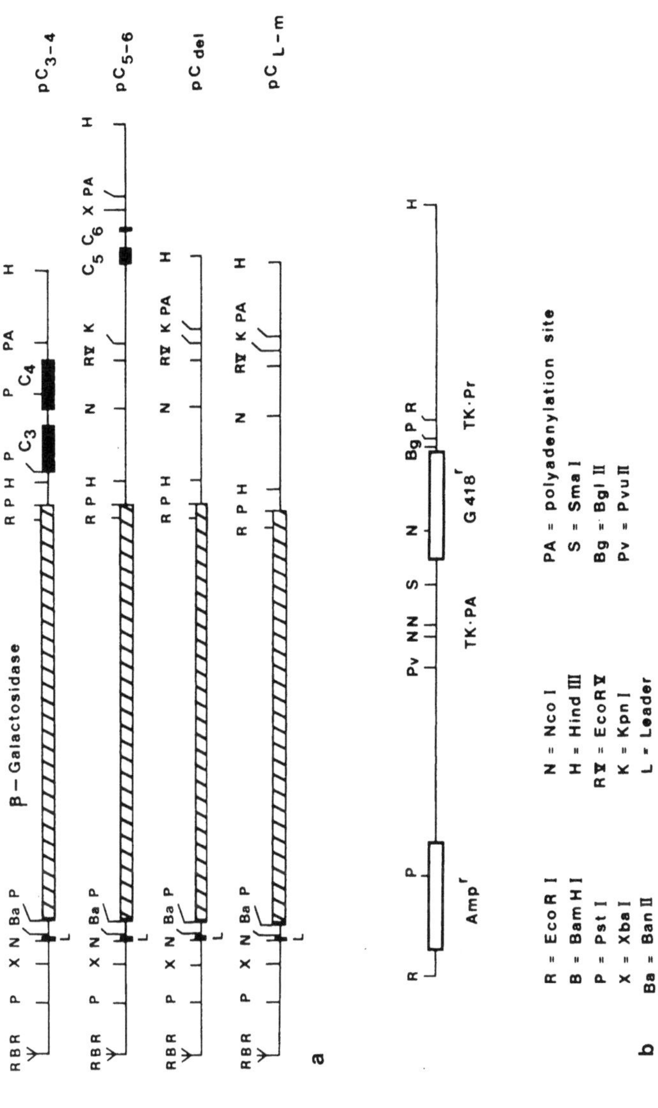


**Fig. 3.** Constructs used to transfect IgM 662 cells. All of the four constructs contain the regulatory sequences of the Ig heavy chain, including enhancer, octabox, and promoter (**a**). In the first three constructs, the leader of IgM is present as well as the first seven amino acids of V. For the last construct the leader was deleted in $pC_{del}$ between NcoI site spanning the initiation codon and the BAN II site and replaced by an oligonucleotide restoring the splice signals but containing only four amino acids in its coding region. $\beta$-Galactosidase was inserted into all four constructs with a splice donor site at its 3′ end. In $pC_{3-4}$, a Hind III fragment containing the 3rd and 4th exon of IgM was joined to the basic construct; in $pC_{5-6}$ a Hind III fragment containing the membrane spanning exons of IgM was joined. In $pC_{del}$ the membrane spanning exons were deleted between the Kpn I and Xba I site. **b** The part common to all of the four constructs, which is mainly the plasmid PAG 60 (Colbère-Garapin et al. 1981) containing G418 resistance as a selectable marker for eukaryotic cells

### *CTLs Specific for Cells Transfected with $\beta$-Galactosidase*

The CTL line 0805A was derived from a BALB/c mouse immunized with C3-C4 cells which express $\beta$-gal intracellularly. The specificity of this CTL line is shown in Fig. 4. IgM 662 and 662 PAG cells, both not producing $\beta$-gal, are not lysed, whereas C3-C4 cells are. C5-C6 cells expressing $\beta$-gal as membrane-inserted product, and $C_{DEL}$, secreting $\beta$-gal, are both lysed. These results indicate that CTLs are specific for the product common of all three constructs which is the V-$\beta$-gal fusion protein. Recognition by 0805A is class I restricted, since killing could be inhibited with a mixture of $K^d$, $D^d$, and $L^d$ specific antibodies (data not shown). Consistent with class I restriction of 0805A is that the line is $CD8^+$, $CD4^-$ and that the target cells are class II negative (data not shown).

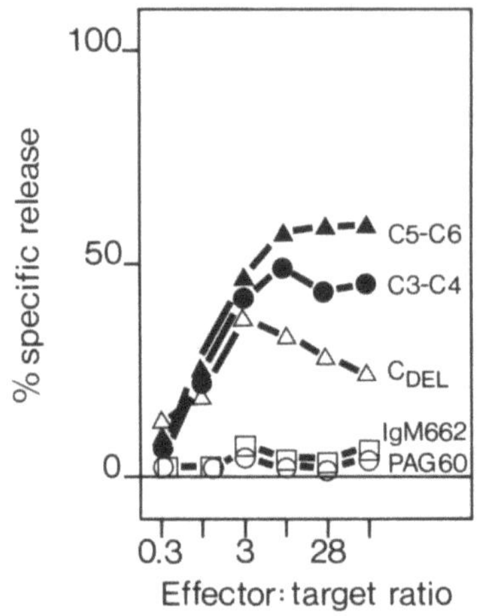


**Fig. 4.** Specificity of CTL line 0805A. 0805A CTL were tested in a $^{51}$Cr release assay on target cells C5-C6 (▲), C3-C4 (●), $C_{DEL}$ (△), IgM 662 (□), and PAG 60 (○)

## Discussion

The classical definition of minor H antigens is by graft rejection (Graff 1978). When mice identical at Mhc genes reject grafts of each other, they are said to differ at minor *H* genes. In most cases, the rejection of a minor *H* incompatible graft is accompanied by the activation of class I restricted, minor H specific CTLs. Therefore, a more accurate, however narrower, definition based on in vitro assays would be to call an intrinsic cellular antigen recognized by foreign class I restricted CTLs a minor H antigen. The in vitro assay can be more readily used to identify minor H molecules as compared to graft rejection in vivo. Once a molecule is identified as a CTL-defined minor H antigen, one can still test whether that also fulfils the criteria of the classical definition by graft rejection.

For these reasons we have used CTLs to address the question of which molecules might represent minor H antigens. In the first part of this paper CTLs are described which were raised across a pair of congeneic strains differing at *H-3* genes, to test the hypothesis that $\beta_2 m$ might be one of the *H* genes in the *H-3* region on chromosome 2 (Michaelson 1983; Rammensee and Klein 1983; Kurtz et al. 1985). The $\beta_2 m$ locus was mapped to that region and reactivity patterns of CTLs raised between several *H-3* congeneic strains provided evidence that (a) there are several *H* genes in the *H-3* region (Hauptfeld and Klein 1977; Roopenian and Click 1980; Click and Roopenian 1981) and (b) one of these *H* genes correlates with $\beta_2 m$ in the allelic distribution among *H-3* congeneic strains (Rammensee and Klein 1983; Kurtz et al. 1985). The H-3 specific, $K^b$-restricted CTL line 181B (Rammensee et al. 1986) and the clone 181B-3E1, which both distinguish between C57BL/10 ($H\text{-}3^a$, $\beta_2 m^b$) and B10.C-$H\text{-}3^c$, $\beta_2 m^a$), are specific for $\beta_2$M-B; other forms of $\beta_2$M, like $\beta_2$M-W1, $\beta_2$M-W3, $\beta_2$M-W4 (Robinson et al. 1984), $\beta_2$M-A, and bovine $\beta_2$M, are not recognized, as indicated by no lysis of BALB.B ($\beta_2 m^a$) targets incubated in fetal calf serum containing bovine $\beta_2$M (Fig. 1) which can associate with mouse class I heavy chains (Robinson, personal communication). These data indicate that the clone 181B-3E1 is specific for the single amino acid substitution of position 85, which distinguishes $\beta_2$M-A from $\beta_2$M-B

(Gates et al. 1981). Thus, $\beta_2$M is a minor H antigen, as far as defined by CTL recognition. Attempts to test whether recognition of $\beta_2$M is involved in skin graft rejection of B10.C-*H-3*$^c$ mice grafted with C57BL/10 skin have failed so far. We tried without success to induce nonresponsiveness or to immunize recipient mice by injections of $\beta_2$M.

The identification of $\beta_2$M as a CTL determinant was made possible by two or three fortuitous circumstances: (a) the correlation between H-3 specific CTL reactivity and allelic distribution of $\beta_2m$, (b) the relative ease with which $\beta_2$M could be purified (Graf et al. 1982), and (c) perhaps the reversible association of $\beta_2$M with class I heavy chains. The last point is expressed with a caveat, since it is not yet clear whether the CTLs recognize the complex of K$^b$ heavy chains and acquired $\beta_2$M-B, or rather a processed fragment of $\beta_2$M-B, presented by K$^b$. We are presently attempting to solve this issue.

Other minor H antigens cannot be expected to be identifiable by the same procedure. However, recent work on CTL recognition of viral antigens has, by analogy, brought about a new view as to what minor H antigens might be (Kourilsky and Claverie 1986; Germain 1986; Bevan 1987), which opens up new possibilities for their molecular identification. CTLs specific for influenza nucleoprotein have been described which recognize cells expressing nucleoprotein only intracellularly (Townsend et al. 1985, 1986). The CTLs recognize degradation products in the form of small peptides of nucleoprotein in the context of class I. By analogy, degraded products of any intracellular protein might be available on the cell surface for recognition by foreign CTLs in the context of class I.

Thus minor H antigens might be degradation products of normal intracellular proteins. To have a model for recognition of nonviral cellular proteins by CTLs, we raised CTLs from BALB/c mice against transfected BALB/c tumor cells expressing $\beta$-gal in their cytoplasm. These CTLs lyse the immunizing cells, C3-C4, but also cells expressing the transfected $\beta$-gal as membrane inserted product (C5-C6), and even cells which secrete $\beta$-gal ($C_{DEL}$). A summary of the characteristics of transfectants is provided in Table 1; the killer assay is shown in Fig. 4. The CTL recognition of cells expressing membrane inserted $\beta$-gal could be due to fragmentation of some $\beta$-gal molecules while still in the cytoplasm, as is probably the case for H-2-class I restricted CTL recognition of HLA products expressed in transfected mouse cells (Maryanski et al. 1986).

As is evident from the constructs shown in Fig. 3, all transfectants produce fusion proteins consisting of $\beta$-gal and parts of the $\mu$ heavy chain. Experiments are underway to define the peptide(s) recognized by 0805A CTL and to determine whether the CTLs are specific for $\beta$-gal or for a determinant on the fusion proteins. However, for the point to be made here, this issue is not relevant. We can conclude at this stage of the project that CTLs can recognize cells transfected with genes coding for cytoplasmatic, membrane inserted, or secreted proteins, suggesting that minor H antigens might also be derived from proteins expressed in these different compartments.

All the transfectants killed well by 0805A CTL have a leader sequence in front of the $\beta$-gal gene, resulting in protein expression and glycosylation in the endoplasmatic reticulum. We have also tested transfectants with deleted leader sequences (see Fig. 3), which should result in cytosolic protein expression. However, only weak killing was observed, allowing no conclusion to be drawn (data not shown). From the work on CTL recognition of viral products mentioned earlier, it is known that CTLs can recognize cells transfected with the nucleoprotein gene in the absence of a known leader sequence. If one also considers recent data showing that class I restricted CTLs can recognize fragments from a soluble protein taken up and processed by cells (Staerz et al. 1987), the following picture emerges. Class I restricted CTLs can recognize proteins, or fragments of proteins, (a) which are expressed in the cytosol or in the endoplasmatic reticulum, or are inserted into the cell membrane, (b) which are secreted, or (c) which are acquired from the extracellular compartment. Minor H antigens, therefore, might be expected in all these compartments.

## Summary

A special property of $\beta_2$-microglobulin ($\beta_2$M), namely its reversible association with MHC class I heavy chains, was used to identify $\beta_2$M as the molecule recognized by class I restricted cytotoxic T-lymphocytes (CTLs) raised across the incompatibility between the mouse strains C57BL/10 and B10.C-*H-3*$^c$/Sn. Both strains differ at a segment of chromosome 2, which includes *H-3* and $\beta_2m$ genes. A CTL clone, 181B-3E1, specific for $\beta_2$M-B, was derived and used to determine cross-reactivity with other $\beta_2$M-alleles obtained from wild mice. No cross-reaction with $\beta_2$M-W1, -W3, or -W4 was found. The data indicate that 181B-3E1 is specific for the one amino acid difference between $\beta_2$M-A and $\beta_2$M-B, and that $\beta_2m$ is one of the minor H genes in the *H-3* region on chromosome 2 of the mouse.

Based on work on CTL recognition of viral proteins it was suggested that other minor H antigens might be peptides derived from polymorphic intracellular proteins, presented to CTLs in the context of class I on the cell surface. To test the hypothesis that nonviral intracellular proteins, or fragments thereof, could serve as target structures for class I restricted CTLs, BALB/c mice were immunized with syngeneic tumor cells transfected with the *E. coli* $\beta$-galactosidase gene. CTLs specific for this surrogate minor H antigen were derived; they lyse cells transfected with different constructs engineered to result in intracellular, membrane inserted, or secreted product. The data suggest that real minor H antigens might also be derived from proteins expressed in these three ways.

*Acknowledgments.* We thank G. Köhler for providing the opportunity to produce the transfectants and for advice, D. Hügin for technical assistance, L. Sherman, U. Hämmerling, and K. F. Lindahl for providing reagents, C. Plattner and L. Yakes for secretarial help, and U. D. Staerz and B. Stockinger for reading the manuscript. The Basel Institute for Immunology was founded and is supported by F. Hoffmann-La Roche, Ltd. Co., Basel, Switzerland.

## References


Baumann B, Potash MJ, Köhler G (1985) Consequences of frameshift mutations at the immunoglobulin heavy chain locus of the mouse. EMBO J 4: 351–359

Bevan MJ (1987) Class discrimination in the world of immunology. Nature 325: 192–194

Click RE (1985) Complexity of minor histocompatibility loci. Hum Immunol 14: 220–223

Click RE, Roopenian DC (1981) A new minor histocompatibility locus linked to H-3. J Immunol 126: 2378–2381

Colbère-Garapin F, Horodniceau F, Kourilsky P, Garapin AC (1981) A new dominant hybrid selective marker for higher eukaryotic cells. J Mol Biol 150: 1–14

Colombo MM, Jaenisch R, Wettstein PJ (1987) Endogenous retroviruses lead to the expression of histocompatibility antigen detectable by skin graft rejection. Proc Natl Acad Sci USA 84: 189–193

Gates FT III, Coligan JE, Kindt TJ (1981) Complete amino acid sequence of murine $\beta_2$-microglobulin: structural evidence for strain related polymorphism. Proc Natl Acad Sci USA 78: 554–558

Germain R (1986) The ins and outs of antigen processing and presentation. Nature 322: 687–689

Graf L, Peterson PA, Sege K, Robinson PJ (1982) Isolation and characterization of mouse $\beta_2$-microglobulin allotypes. Mol Immunol 19: 877–884

Graff RJ (1978) Minor histocompatibility genes and their antigens. In: Morse HC III (ed) Origins of inbred mice. Academic, New York, pp 371–389

Hauptfeld V, Klein J (1977) A new histogenetic method for minor histocompatibility antigen typing. J Immunol 118: 423–426

Kourilsky P, Claverie JM (1986) The peptidic self model: A hypothesis on the molecular nature of immunological self. Ann Inst Pasteur Immunol 137D: 3–21

Kurtz ME, Graff RJ, Adelman A, Martin-Morgan D, Click RE (1985) CTL and serologically defined antigens of *$\beta_2 m$, H-3* region. J Immunol 135: 2847–2852

Kurtz ME, Martin-Morgan DM, Graff RJ (1987) Recognition of the $\beta$-2 microglobulin-B molecule by a CTL clone. J Immunol 138: 87–90

Loveland B, Simpson E (1986) The non-MHC transplantation antigens: neither weak nor minor. Immunol Today 7: 223–229

Maryanski JL, Pala P, Corradin G, Jordan BJ, Cerottini JC (1986) H-2 restricted cytolytic T cells specific for HLA can recognize a synthetic HLA peptide. Nature 324: 578–579

Michaelson J (1981) Genetic polymorphism of $\beta_2$-microglobulin *($\beta_2 m$)* maps to the *H-3* region of chromosome 2. Immunogenetics 13: 167–171

Michaelson J (1983) Genetics of $\beta_2$-microglobulin in the mouse. Immunogenetics 17: 219–259

Rammensee HG, Klein J (1983) Complexity of the histocompatibility-3 region in the mouse. J Immunol 130: 2926–2929

Rammensee HG, Fink PJ, Bevan MH (1984) Functional clonal deletion of class I-specific cytotoxic T lymphocytes by veto cells that express antigen. J Immunol 133: 2390–2396

Rammensee HG, Robinson PJ, Crisanti A, Bevan MJ (1986) Restricted recognition of $\beta_2$-microglobulin by cytotoxic T lymphocytes. Nature 319: 502–504

Robinson PJ, Lundin L, Sege K, Graf L, Wigzell H, Peterson PA (1981) Location of the mouse $\beta_2$-microglobulin *$\beta_2 m$* determined by linkage analysis. Immunogenetics 14: 449–452

Robinson PJ, Steinmetz M, Kazuo M, Lindahl KF (1984) Beta-2 microglobulin types in mice of wild origin. Immunogenetics 20: 655–665

Roopenian DC, Click RE (1980) A new cytotoxic lymphocyte-defined antigen coded by a gene closely linked to the H-3 locus. Immunogenetics 10: 333–341

Rüsch E, Kuon W, Hämmerling GJ (1983) Target sites for antibodies and cytotoxic T lymphocytes on H-2 molecules. Transplant Proc 15: 2093–2096

Simpson E (1987) Non-H-2 histocompatibility antigens: can they be retroviral products? Immunol Today 8: 176–178
Staerz UD, Karasuyama J, Garner AM (1987) CTL against a soluble protein. Nature 329: 449–451
Tada N, Kimura S, Hatzfeld A, Hämmerling U (1980) The *H-3* region of the mouse chromosome 2 controls a new surface alloantigen. Immunogenetics 11: 441–449
Townsend ARM, Gotch FM, Davey J (1985) Cytotoxic T cells recognize fragments of influenza nucleoprotein. Cell 42: 457–467
Townsend ARM, Rothbard J, Gotch FM, Bahadur G, Wraith D, McMichael AJ (1986) The epitopes of influenza nucleoprotein recognized by cytotoxic T lymphocytes can be defined with short synthetic peptides. Cell 44: 959–968

# Two Conformations of $L^d$ Distinguished by Their Abilities to Bind $\beta_2$-Microglobulin Are Synthesized, Processed, and Expressed on the Cell Surface

NANCY B. MYERS, DAVID R. LEE, and TED H. HANSEN

Department of Genetics, Washington University School of Medicine, St. Louis, MO 63110, USA

## Introduction

Many theories have emerged regarding the role of $\beta_2$-microglobulin ($\beta_2$m) in the synthesis and cell surface expression of class I molecules. Perhaps the most widespread theory is that $\beta_2$m is required for cell surface expression of class I molecules. It has been proposed that $\beta_2$m may act as a transport protein for class I molecules that facilitates their intracellular trafficking from the rough endoplasmic reticulum (ER) to the Golgi apparatus (Krangel et al. 1979). These conclusions are based primarily on studies using the human Burkitt's lymphoma Daudi. Daudi cells were found not to produce $\beta_2$m protein and not to express detectable class I molecules on the cell surface (Goodfellow et al. 1975). To establish the cause and effect of these observations, cytoplasmic class I proteins were detected in Daudi and surface expression of class I antigens could be rescued by fusion to $\beta_2m^+$ cell types (Fellous et al. 1977; Arce-Gomez et al. 1978).

In addition to its proposed role in transport and expression, $\beta_2$m is also thought to play a role in the conformation of class I molecules (Ploegh et al. 1979). This conclusion is supported by serological studies using the monoclonal antibody (McAb) W6/32 (Barnstable et al. 1978; Parham et al. 1979) as well as functional studies of anti-H-3 cytotoxic T cell responses (Rammensee et al. 1986). In both of these cases a conformational (or combinatorial) determinant is recognized that is dependent upon the interaction of class I with $\beta_2$m. In contrast to these studies that suggest the presence of $\beta_2$m is necessary for the conformation of class I, studies using murine systems suggest that this is *not* the case for the $L^d$ and $D^b$ molecules. The $L^d$ and $D^b$ molecules can be expressed on the cell surface as free heavy chains and appear serologically and functionally normal (Potter et al. 1984; Martinko et al. 1986). Finally, another line of investigation suggested that the role of $\beta_2$m is to stabilize the class I molecule, making it more resistant to proteolysis (Krangel et al. 1979; Ploegh et al. 1979).

A recent observation by Allen et al. (1986) raised several new questions regarding the role of $\beta_2$m in class I biosynthesis. They transfected the $D^b$ gene into the $\beta_2m^-$ murine cell line R1E and detected cell surface expression of $D^b$-like molecules. Therefore, this study appeared to negate the absolute requirement of $\beta_2$m for class I expression. However, the $D^b$ molecules expressed by R1E were not recognized by cytotoxic T-lymphocytes raised against native $D^b$, nor by most of the McAbs to $D^b$

antigens. The one exceptional antibody was 28-14-8, the only one of those antibodies tested that recognizes a determinant associated with the $\alpha 3$ domain. This study by Allen et al. (1986) did not determine whether $D^b$ molecules in a non-$\beta_2$m-associated conformation are normally expressed in $\beta_2m^+$ cells.

We demonstrate here that $L^d$ molecules like $D^b$ molecules can be detected in two different serological phenotypes that are also distinguished by their ability to associate with $\beta_2$m. Furthermore, the antigenic difference between these two types of $L^d$ appears to be solely attributable to the conformation of the heavy chain and not whether it is associated with $\beta_2$m. Surprisingly, the form of $L^d$ incapable of $\beta_2$m association was found to have oligosaccharide maturation as well as low levels of expression on the plasma membrane. Pulse-chase experiments indicated that these two forms of $L^d$ do not share a precursor–product relationship, but rather their distinguishing conformations are determined very early in their synthesis in the ER and remain fixed throughout maturation and expression. These findings are discussed as evidence that the primary role of $\beta_2$m is not as a transport molecule but may be to facilitate the folding of the $\alpha 1$ and $\alpha 2$ domains of the class I molecule during de novo synthesis.

## Materials and Methods

### *Cells*

Splenocytes were from BALB/c mice that were obtained from Dr. Donald Shreffler's animal facility (Washington University, St. Louis, MO). The $L^d$-gene transfectant T1.1.1 was obtained from Dr. David Margulies (NIH) and has been described previously (Evans et al. 1982 a).

### *Monoclonal Antibodies*

Murine McAbs used in this study have been previously described (Ozato et al. 1980; Evans et al. 1982 b). Antibody 30-5-7 was obtained from an immunization of dm2 mice with BALB/c cells and has been shown to recognize a determinant in the $\alpha 2$ domain of $L^d$ molecules. Antibody 28-14-8 was obtained from an immunization of C3H mice with C3H.SW (H-$2^b$) cells and has been shown to recognize a determinant in the $\alpha 3$ domain of several *D* region encoded molecules, including $L^d$ and $D^b$. Hybridomas producing both McAbs resulted from fusions with SP2/0 and both have the $\gamma$2a isotype. Ascites fluid containing high concentrations of McAb were used for immunoprecipitations.

### *Radiolabeling of Cells*

Pulse-chase studies were performed by preincubating splenocytes ($8 \times 10^7$ cells/ml) for 45 min in Earle's modified Eagle's medium (MEM) with 5% fetal calf serum (FCS; Hyclone, Logan, UT) but without methionine in a 37 °C $CO_2$ incubator. [$^{35}$S]Methionine (>1000 Ci/Mmol; Amersham, Arlington Heights, IL) was added to the cell culture for 15 min. The cells were then washed twice in complete Earle's MEM with 5% FCS before chasing for various lengths of time in the same medium. At each time point of chase, aliquots of cells were washed twice and the cell pellets were frozen at −70 °C before additional processing.

For continuously labeled cultures, splenocytes were resuspended in Earle's MEM (minus methionine) containing 5% FCS and [$^{35}$S]-methionine at the same cell and [$^{35}$S]methionine concentrations as above. The cells were then incubated at 37 °C in a humidified $CO_2$ incubator for 3–5 h. These cells were used for the endoglycosidase H (endo H) studies and as controls for the lactoperoxidase-$^{125}$I-labeled cells.

To examine cell surface expression of class I molecules, $2 \times 10^8$ splenocytes were surface radioiodinated by using Na $^{125}$I (5 mCi; Amersham), lactoperoxidase (Sigma Chemical Co., St. Louis, MO), *D*-glucose, and *D*-glucose oxidase (Sigma) for 15 min. The reaction was quenched with *L*-tyramine (6.95 µg/ml), and the cells were washed twice with phosphate-buffered saline containing 0.3 mg/ml ovalbumin (Sigma) and once with phosphate-buffered saline containing 60 ng/ml potassium iodide.

### *Isolation of Antigens*

Antigens were isolated based on a procedure from Cullen and Schwartz (1976). Cells were lysed in Tris-buffered saline, pH 7.0, containing 0.5% Nonidet P-40 (NP-40) and freshly added 0.2 mM (final) phenylmethylsulfonyl fluoride (PMSF; Sigma) for 30 min at 0 °C. Postnuclear supernatants were collected after centrifugation at $100\,000 \times g$ for 1 h. The preparations were then precleared with 1.5 ml 10% (v/v) IgGSorb (The Enzyme Center, Malden, MA) per $1 \times 10^8$ cell equivalents in the lysate. Before incubation of the precleared cell lysates with various McAbs, PMSF was added again to a final concentration of 0.2 mM to inhibit the proteolytic activity in the various ascites used as a source of McAbs. Aliquots of $5 \times 10^7$ cell equivalents were mixed with 100 µl of a McAb ascites and were allowed to incubate for 30 min at 0 °C, followed by incubation with 1 ml 10% IgGSorb for 30 min at 0 °C. The IgG-Sorb–McAb–antigen complexes were washed three times in cold Tris-buffered saline containing 0.25% NP-40 and 0.2 mM PMSF. The McAb–antigen complexes were eluted by incubation of the samples in 0.125 M Tris-HCl (ph 6.8), 12.5% glycerol, 2% sodium dodecyl sulfate (SDS), and 1% 2-mercaptoethanol [SDS–polyacrylamide gel electrophoresis (PAGE) sample buffer] at 90 °C for 2 min, followed by centrifugation to remove the IgGSorb.

### *Endo H Digestions*

Antigens to be digested with endo H were eluted from the IgGSorb by heating at 90 °C for 2 min with 50 mM sodium acetate (pH 5.5), 0.5% Triton X-100, 0.1% SDS, and 12.5 mM *β*-mercaptoethanol. The samples were divided and digested with 20 MU/ml endo H (ICN-Biochemicals, Naperville, IL) or were mock digested for 20 h at 37 °C. Incubations were terminated by the addition of one volume of SDS-PAGE sample buffer, followed by heating at 90 °C for 2 min.

*SDS-PAGE*

In all cases, 10%–15% linear polyacrylamide gradient slab gels (BioRad, Richmond, CA) were run by using the original Laemmli buffer system (Laemmli 1970). Prestained m.w. standards (BRL, Bethesda, MD) were included on every gel. Before drying down, the gels were treated with Enhance (NEN-Dupont, Boston, MA) as directed for fluorographic analysis. Tracings of fluorographs were performed on a Quick Scan Jr. densitometer (Helena Laboratories, Beaumont, TX).

## Results

*Detection of Two Serologically Different Forms of $L^d$ with Disparate $\beta_2$m Associations*

The molecular specificity of two different McAbs reactive with $L^d$ molecules was compared by immunoprecipitations that were resolved in 10%–15% linear polyacrylamide gradient slab gels. The McAbs used were 30-5-7, which is specific for a determinant in the $\alpha 2$ domain, and antibody 28-14-8, which is specific for a determinant in the $\alpha 3$ domain (Evans et al. 1982b). BALB/c spleen cells were metabolically labeled with $^{35}$S-met for 4 h and the glycoprotein pool was purified by lentil lectin affinity chromatography. Test aliquots of this BALB/c preparation were precipitated with either McAb 30-5-7 (Fig. 1, lane 1) or 28-14-8 (Fig. 1, lane 2). As expected, both McAbs precipitated a heavy chain of approximately 45000 kd associated with a $\beta_2$m light chain. It should be noted that the $L^d$ molecule as previously reported has a relatively low affinity for $\beta_2$m (Beck et al. 1986), and thus the $\beta_2$m-specific bands in these precipitates are weak. Interestingly, the 30-5-7 precipitates routinely contained more $\beta_2$m than the 28-14-8 precipitates. The reason for this became apparent later. To further compare the reactivities of antibodies 30-5-7 and 28-14-8, sequential precipitation analyses were performed. A pool of BALB/c antigen was sequentially treated with: (a) 30-5-7 (lane 3), (b) 30-5-7 (lane 4), and (c) 28-14-8 (lane 5). After complete clearance with 30-5-7, significant residual 28-14-8 reactive molecules were detected. When the reciprocal sequential immunoprecipitation experiment was performed (lanes 7–9), the antibody 28-14-8 was found to remove all residual 30-5-7 reactive molecules. Therefore, these sequential immunoprecipitation experiments defined two antigenically distinct molecules, one 30-5-7$^+$, 28-14-8$^+$ and the other 30-5-7$^-$, 28-14-8$^+$. When these two antigenic forms were assessed for their $\beta_2$m-association, a striking disparity was observed. As shown in lane 3, significant amounts of $\beta_2$m are associated with 30-5-7$^+$ molecules, whereas *no* detectable $\beta_2$m was observed with the 30-5-7$^-$ molecules (lane 5).

To determine whether both of the antigenic forms are products of the $L^d$ gene, sequential precipitation analysis was also performed on metabolically labeled antigen from the $L^d$-gene transfectant T1.1.1. Again, two antigenic forms were defined, one 30-5-7$^+$, 28-14-8$^+$ and the other 30-5-7$^-$, 28-14-8$^+$. This result demonstrated that both of these antigens are products of the $L^d$ gene. Based on densitometric comparisons of multiple independent experiments, the ratio of the 30-5-7$^+$ to

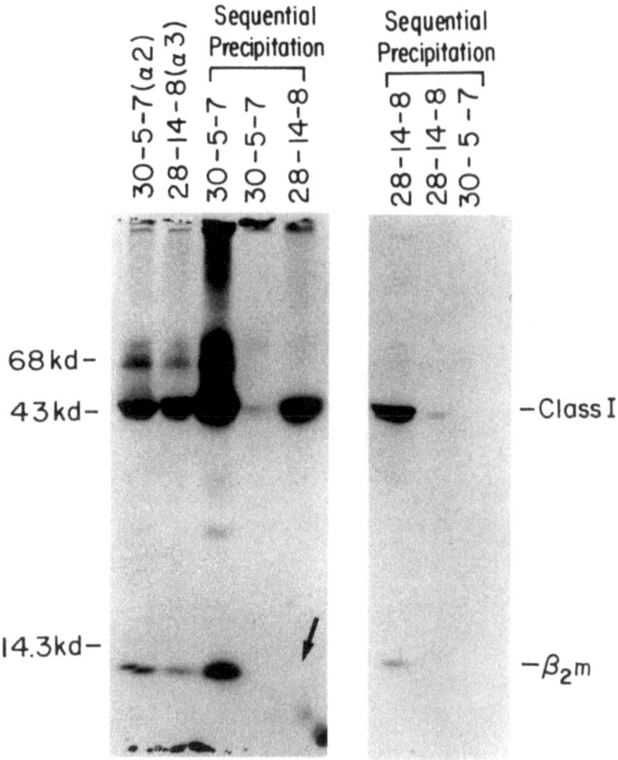


**Fig. 1.** Sequential precipitations of metabolically labeled BALB/c antigen with McAbs 30-5-7 and 28-14-8. Test precipitates with antibodies 30-5-7 and 28-14-8 are shown in the two left-hand lanes respectively. From left to right, the next three lanes contain precipitates from antigen treated sequentially with 30-5-7, 30-5-7, and 28-14-8. The last three lanes contain precipitates from antigen treated sequentially with 28-14-8, 28-14-8, and 30-5-7. These results define two antigenic forms of $L^d$, one 30-5-$7^+$, 28-14-$8^+$ and the other 30-5-$7^-$, 28-14-$8^+$. This latter form of $L^d$ has no detectable $\beta_2$m associated with it, as indicated by the *arrow*

30-5-$7^-$ $L^d$ molecules in the biosynthetically labeled preparation from spleen was approximately 1:0.3–0.4, whereas from T1.1.1 it was approximately 1:3–10. Furthermore, there appeared to be appreciably less $\beta_2$m associated with the $L^d$ molecules precipitated from T1.1.1 compared to BALB/c antigen (data not shown). This observation probably indicates that $L^d$ molecules are overproduced by the transfectant and thus exceed the available supply of $\beta_2$m.

### *Oligosaccharide Processing and Cell Surface Expression of $L^d$ Molecules in a Non-$\beta_2$m-Associated Conformation*

The oligosaccharide processing of the two antigenic forms of $L^d$ were compared. As commonly occurs with glycoproteins, newly synthesized class I molecules on the ER have N-linked high mannose side chains. After transport to the Golgi apparatus,

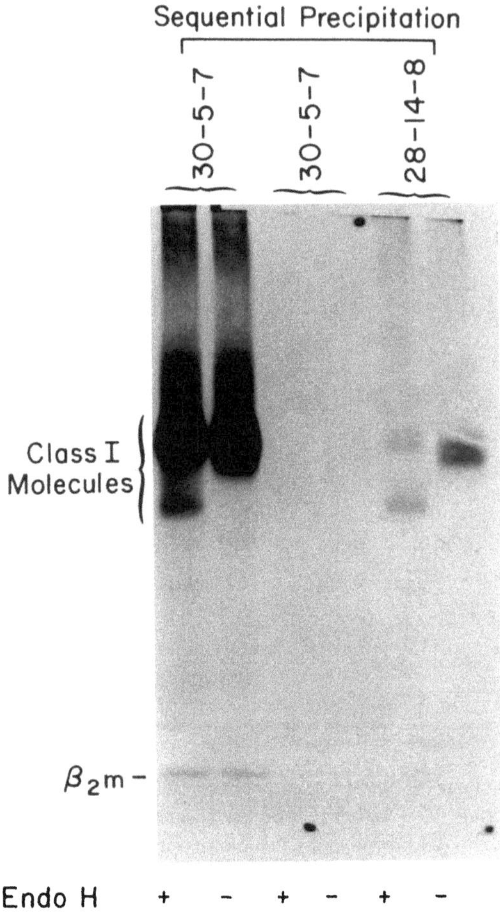


**Fig. 2.** Endo H treatment of the two antigenic forms of $L^d$ isolated from BALB/c antigen metabolically labeled for 4 h. The lanes contain sequential precipitates using 30-5-7 *(lanes 1, 2)*, 30-5-7 *(lanes 3, 4)*, and 28-14-8 *(lanes 5, 6)*. The odd numbered lanes contained aliquots that were treated with endo H, whereas the even numbered lanes contained aliquots that underwent mock digestions as indicated along the *bottom of the figure*

these high mannose residues are converted to complex-type oligosaccharides (Krangel et al. 1979). These two forms of glycosylation of the class I molecule can readily be distinguished using endo H. Endo H cleaves only the high mannose side chains, resulting in a difference that can be observed on SDS-PAGE. Therefore, resistance to endo H can be used as a marker to assess the intracellular transport of class I molecules from the ER to the Golgi apparatus. Since $\beta_2$m has been implicated in the intracellular transport of class I molecules, it was of interest to compare the oligosaccharide processing of the two antigenic forms of $L^d$ distinguished by their reactivity with McAb 30-5-7.

Biosynthetically labeled BALB/c antigen was precipitated sequentially with 30-5-7 (Fig. 2, panels 1, 2), 30-5-7 (panels 3, 4), and 28-14-8 (panels 3, 4). Half of each of

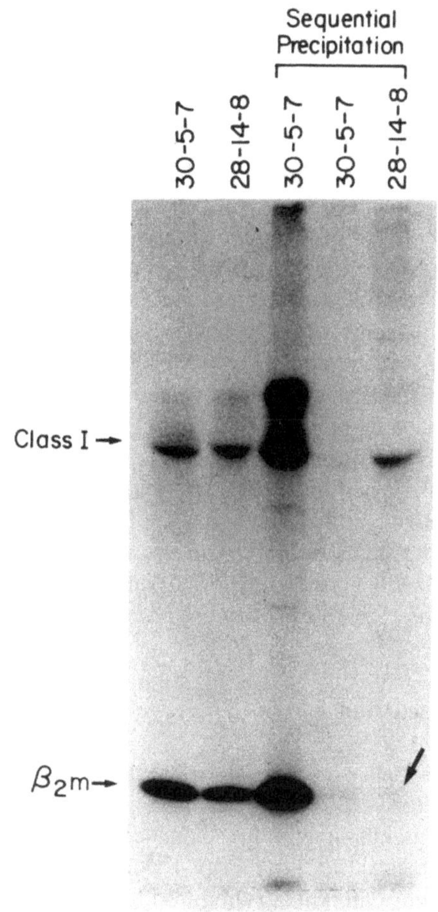


**Fig. 3.** Sequential precipitation of $^{125}$I surface labeled $L^d$ from T1.1.1 cells. The first two lanes contained test precipitates using 30-5-7 and 28-14-8 respectively, whereas the next three lanes contained sequential precipitates of 30-5-7, 30-5-7, and 28-14-8. These results indicate that both 30-5-7$^+$ and 30-5-7$^-$ $L^d$ molecules make it to the cell surface and their ratio is approximately 10:1 as measured by densitometry. As indicated by the *arrow,* the 30-5-7$^-$ form of $L^d$ has no detectable $\beta_2$m associated with it

these precipitates was treated with endo H (lanes 1, 3, 5) whereas the other half underwent a mock digestion (lanes 2, 4, 6). When either 28-14-8 or 30-5-7 was initially used for precipitation, most class I molecules were found to be endo H resistant. This finding indicated that during the culture period most of the $L^d$ molecules were transported from the ER to the Golgi apparatus in BALB/c spleen cells. When the non-$\beta_2$m-associated, 30-5-7$^-$, 28-14-8$^+$ $L^d$ molecules were digested with endo H, about half or more were found to be endo H sensitive. This observation suggested that 30-5-7$^-$ $L^d$ molecules were being transported to the Golgi apparatus, however less efficiently then the 30-5-7$^+$ $L^d$ molecules. This observation also indicated that at least some of the $L^d$ molecules incapable of $\beta_2$m association were processed and transported like conventional class I molecules. Thus these findings raised the question of whether 30-5-7$^-$ $L^d$ molecules are expressed on the cell surface. To address this question T1.1.1 cells were surface labeled with $^{125}$I using lactoperoxidase. This antigen preparation was precipitated with 30-5-7 (Fig. 3, lane 1), 28-14-8 (lane 2), or sequentially with 30-5-7 (lane 3), 30-5-7 (lane 4), and 28-14-8 (lane 5). As indicated in the first two lanes of Fig. 3, both McAbs 30-5-7 and 28-14-8 precipitated $\beta_2$m-associated forms of $L^d$. It should be pointed out that these T1.1.1 cells were grown in FCS prior to labeling, and under such conditions bovine $\beta_2$m is known to rapidly replace the mouse $\beta_2$m (Kahn-Perles et al. 1987; Kievits and Ivanyi 1987). This exchange probably reflects the curious phenomenon that murine class I molecules appear to have a higher affinity for bovine $\beta_2$m than endogenous mouse $\beta_2$m (Kievits and Ivanyi 1987). When sequential precipitation analysis was performed on the surface-labeled preparation, approximately 18% of the $L^d$ molecules were found to be 30-5-7$^-$, 28-14-8$^+$. Furthermore, whereas the 30-5-7$^+$ $L^d$ molecules showed significant $\beta_2$m association, the 30-5-7$^-$ $L^d$ molecules had no detectable $\beta_2$m. When the same experiment was performed on $^{125}$I-labeled BALB/c spleen cell antigen, approximately 10% of the total $L^d$ molecules were 30-5-7$^-$, 28-14-8$^+$, and non-$\beta_2$m associated (data not shown). Consistent with previous reports (Martinko et al. 1986; Beck et al. 1986; Coligan et al. 1980), most $L^d$ molecules on the cell surface, even those 30-5-7$^+$, were non-$\beta_2$m associated.

## Discussion

The data presented here provide a genetic and chemical explanation for the molecules we previously defined as $R^d$ (Hansen et al. 1981). As originally detected by sequential immunoprecipitation, $R^d$ was 30-5-7$^-$, 28-14-8$^+$ and was thought to be genetically distinct from $L^d$ or $D^d$. Subsequent studies reported that $R^d$ showed minor differences compared to $L^d$ by two-dimensional gel electrophoresis (Koch et al. 1983; Nichols et al. 1983), and by tryptic peptide mapping $R^d$ and $L^d$ were found to be indistinguishable (Lillehoj et al. 1985). Here we demonstrate that $R^d$ is a product of the $L^d$ gene and that it can be distinguished from $L^d$ by a total lack of $\beta_2$m association. We thus recommend dropping the designation $R^d$ and will refer to this molecule as 30-5-7$^-$ $L^d$ or non-$\beta_2$m-conformed $L^d$. The 30-5-7$^-$ forms of $L^d$ appear to be analogous to the serologically aberrant $D^b$ molecules expressed by transfection of the $D^b$ gene into the $\beta_2$m$^-$ cell line R1E (Allen et al. 1986). Both non-$\beta_2$m-associated forms of $L^d$ and $D^b$ were detected by antibody 28-14-8, which recognizes a shared $\alpha 3$ determinant. In contrast, antibodies to $\alpha 1$ and/or $\alpha 2$ determinants fail to react with either of these non-$\beta_2$m-associated forms. The $L^d$ and $D^b$ genes and their products have been found to be highly homologous by several criteria. Therefore, it will be of considerable interest to determine whether other class I molecules besides $L^d$ and $D^b$ have non-$\beta_2$m-associated forms that can be detected in the cytoplasm or expressed on the cell surface.

Taken together, studies of $D^b$ and $L^d$ permit new hypotheses to be made regarding the role of $\beta_2$m for class I synthesis and expression. Findings with $D^b$ and $L^d$ suggest that $\beta_2$m induces a conformational change early in the synthesis of class I molecules. Furthermore, the $\beta_2$m-induced conformation appears to affect only the $\alpha 1$

and $\alpha 2$ domains and not the $\alpha 3$ domain. It is intriguing to speculate that $\beta_2$m may act during de novo synthesis to properly fold the outer domains. This would be a vitally important function, since it is now clear that the outer domains have an intricate folding pattern that creates a pocket for binding immunogenic peptide antigens (Bjorkman et al. 1987a, b). If $\beta_2$m is responsible for the correct conformation, then one might predict that these non-$\beta_2$m-associated forms of $D^b$ and $L^d$ should not bind peptides and thus not function in T cell activation. Consistent with this hypothesis, $D^b$ molecules expressed in R1E were not recognized by alloreactive or virus-specific cytotoxic T-lymphocytes generated to $D^b$ antigens produced in $\beta_2m^+$ cells. The studies of $L^d$ and $D^b$ also suggest that once $\beta_2$m has imparted conformation, it is free to dissociate or be exchanged. This is particularly true with $L^d$, since most of these molecules even in the $\beta_2$m-associated conformation (i.e. 30-5-$7^+$) are detected as free heavy chains. Similarly, $L^d$ and $D^b$ molecules have been detected on the cell surface as free heavy chains (Potter et al. 1984; Martinko et al. 1986). Yet, when grown in FCS, $\beta_2$m binding to surface 30-5-$7^+$ $L^d$ molecules is readily detectable (Fig. 3). Interestingly, none of these dissociation or exchange events have been shown to affect the recognition of $L^d$ or $D^b$ by alloantibodies or MHC-specific T cells (Potter et al. 1984; Martinko et al. 1986; unpublished observation). Therefore, the continuous presence of $\beta_2$m is not required for conformation of class I molecules, or at least $L^d$ and $D^b$. Alternatively, $\beta_2$m appears to impart a conformation during de novo synthesis that is then fixed. Our pulse-chase experiments with T1.1.1 support this notion. Throughout a 0–24 h chase period, the ratio of 30-5-$7^+$/30-5-$7^-$ $L^d$ molecules in the whole cell lysate remained constant in metabolically labeled antigen preparations (in preparation). In a study of the assembly and maturation of human class I molecules, Krangel et al. (1979) also reported that $\beta_2$m induces a conformational change in the first 10–15 min after synthesis. However, in their system the non-$\beta_2$m-associated and the $\beta_2$m-associated conformations of class I shared a product–precursor relationship (Krangel et al. 1979). Furthermore, in their system they were unable to detect non-$\beta_2$m-associated forms on the cell surface. Some of the differences between our results and theirs likely stem from the reagents used. Krangel et al. (1979) used the mouse McAb W6/32 to detect class I $\beta_2$m complex. This reagent reacts with a determinant on all HLA-A, -B, and -C heavy chains only when they are complexed with $\beta_2$m. To detect non-$\beta_2$m forms a rabbit serum to guanidine-denatured, purified HLA-B7 was used (Krangel et al. 1979). By contrast, our study uses McAbs that detect only the product of one locus, *H-2L^d^*, and that further detect epitopes which have been assigned to specific domains.

If a class I molecule is not $\beta_2$m conformed during de novo synthesis, then subsequent association with $\beta_2$m does not occur. This conclusion is supported by the observation that 30-5-$7^-$ $L^d$ molecules lack $\beta_2$m association, not only when detected biosynthetically, but also after they have been transported to the cell surface and exposed to bovine $\beta_2$m. Consistent with this conclusion, Ploegh et al. (1979) separately produced human class I and $\beta_2$m molecules in a cell-free translation system. When these two preparations were then mixed together, association did not occur. Therefore, class I molecules synthesized in vitro without exposure to $\beta_2$m subsequently do not associate with $\beta_2$m. Therefore, this $\beta_2$m-induced conformation on class I molecules appears to be a fixed and irreversible event.

During the course of the investigations, we envisaged three potential caveats to the model proposed above. They are listed below along with evidence that makes each of them seem very unlikely to us. First, the results could be explained if 30-5-7 recognizes a determinant associated with the complex of $L^d$ and $\beta_2$m, whereas 28-14-8 is capable of recognizing free heavy chain. This is unlikely since 30-5-7 precipitates contain very little $\beta_2$m based on molar ratios especially from $L^d$ transfectants (Beck et al. 1986; unpublished observation), and surface $L^d$ heavy chain free of detectable $\beta_2$m can be recognized by 30-5-7 (Martinko et al. 1986). Also, by analogy to the results with $D^b$, all of the McAbs to $D^b$ that were found to be negative on $D^b$-R1E would have to recognize the complex of $D^b$ and $\beta_2$m. This would also seem very unlikely. Second, the results could be explained by McAb 28-14-8, and not 30-5-7, causing the dissociation of $\beta_2$m from $L^d$. This possibility was ruled out because precipitates using 28-14-8 prior to clearance by 30-5-7 contained significant $\beta_2$m (Figs. 1, 3). Third, the results could be explained by a structural difference resulting from alternative RNA splicing. One could hypothesize that there are two different splice sites that result in alternative forms of $D^b$ and $L^d$ which can be distinguished serologically and by their disparate $\beta_2$m associations. This possibility also seems quite remote, since in the case of our studies the 30-5-7$^+$ and 30-5-7$^-$ $L^d$ molecules were found to be peptide map indistinguishable (Lillehoj et al. 1985). Furthermore, in the case of the $D^b$-R1E transfectant (Allen et al. 1986) one would have to hypothesize that $\beta_2$m determines the RNA splice site. Thus, we feel that each of the alternative explanations is unlikely and that the data best fit the hypothesis that $\beta_2$m determines conformation during de novo synthesis in the ER.

Our data extend the aforementioned study by Allen et al. (1986) of the $D^b$-R1E transfectant in some important ways. They demonstrated that $D^b$ molecules can be expressed in a $\beta_2$m$^-$ cell line. Our data demonstrate that in a normal $\beta_2$m$^+$ cell, a fraction of $L^d$ molecules have a conformation that lacks the ability to associate with $\beta_2$m. Yet surprisingly, the non-$\beta_2$m-conformed (30-5-7$^-$) $L^d$ molecules showed oligosaccharide processing indicating transport from the ER to the Golgi apparatus. Furthermore, these non-$\beta_2$m-conformed $L^d$ molecules are expressed on the cell surface. However, their transport and surface expression do appear less efficient than in the case of their $\beta_2$m-conformed (30-5-7$^+$) counterparts. Based on the above evidence, we would suggest that it is the $\beta_2$m-induced conformation and not the actual physical association with $\beta_2$m that accounts for their preferential transport and expression. Since the ratio of these two forms of $L^d$ remains constant in the total cell lysate, in spite of the fact that the 30-5-7$^+$ form is preferentially transported and expressed, our findings suggest that these two forms have distinct intracellular trafficking. In any case, these data demonstrate that $\beta_2$m is not an absolute requirement for class I expression even in $\beta_2$m$^+$ cells. Furthermore, these results indicate that at least for the $L^d$ and $D^b$ molecules, $\beta_2$m determines the correct folding of $\alpha$1 and $\alpha$2 domains. A mechanism by which $\beta_2$m could facilitate the folding of class I molecules is suggested by the apparent structural homologies between class I and II molecules. The class I $\alpha$1 and $\alpha$2 domains appear homologous to the class II $\alpha$1 and $\beta$1 domains, whereas the class I $\alpha$3 and $\beta_2$m domains appear homologous to the class II $\alpha$2 and $\beta$2 domains (cf. Bjorkman et al. 1987a). Given this premise, then $\beta_2$m could be envisaged as providing a support structure on which to properly fold the class I $\alpha$1 and $\alpha$2 domains to form their functional binding site.

## Summary

Sequential precipitation analysis of BALB/c antigen revealed two distinct antigenic forms of the $L^d$ molecule distinguished by their reactivity with McAb 30-5-7. A similar analysis of antigen from the $L^d$-transfectant T1.1.1 confirmed that both forms of $L^d$ are products of the $L^d$ gene. The 30-5-7$^+$ form of $L^d$ was found to be capable of association with $\beta_2$m but could also exist as a free heavy chain, whereas the 30-5-7$^-$ form of $L^d$ was incapable of $\beta_2$m association. Unexpectedly, this latter form of $L^d$ showed oligosaccharide maturation as well as cell surface expression, although less than the 30-5-7$^+$ form of $L^d$. Pulse-chase experiments demonstrated that these two forms of $L^d$ do not share a precursor–product relationship, but rather their distinguishing structures are fixed during de novo synthesis in the endoplasmic reticulum and remain constant throughout maturation and expression. Thus, $\beta_2$m association is not an absolute requirement for intracellular transport and expression on the plasma membrane even in $\beta_2$m-positive cell types. Furthermore, in the context of other recent studies of $L^d$ and $D^b$, our results suggest that $\beta_2$m plays a key role in folding the outer domains of class I molecules during de novo synthesis. It is speculated that $\beta_2$m may provide a support structure analogous to a class II second domain, on which the class I binding site can be properly formed.

*Acknowledgments.* We thank Mrs. Karen Perks for excellent secretarial assistance, Dr. David Margulies (NIH) for the $L^d$ transfectant T1.1.1, and Drs. Janet Connolly, Wen-Rong Lie, Ronald Rubocki, and Susan Cullen for their critical review.

This work was supported by grants AI 19687 and AI 19993 from the National Institutes of Health.

## References

Allen H, Fraser J, Flyer D, Calvin S, Flavell R (1986) $\beta_2$-microglobulin is not required for cell surface expression of the murine class I histocompatibility antigen H-2D$^b$ or of a truncated H-2D$^b$. Proc Natl Acad Sci USA 83: 7447

Arce-Gomez B, Jones EA, Barnstable CJ, Solomon E, Bodmer WF (1978) The genetic control of HLA-A and B antigens in somatic cell hybrids: requirement for $\beta_2$-microglobulin. Tissue Antigens 11: 96

Barnstable CJ, Bodmer WF, Brown G, Galfre G, Milstein C, Williams AF, Ziegler A (1978) Production of monoclonal antibodies to group A erythrocytes, HLA and other human cell surface antigens – new tools for genetic analysis. Cell 14: 9

Beck JC, Hansen TH, Cullen SE, Lee DR (1986) Slower processing weaker $\beta_2$-m association and lower surface expression of H-2L$^d$ are influenced by its amino terminus. J Immunol 137: 916

Bjorkman PJ, Saper MA, Samraoui B, Bennett WS, Strominger JL, Wiley DC (1987 a) Structure of the human class I histocompatibility antigen, HLA-A2. Nature 329: 506

Bjorkman PJ, Saper MA, Samraouri B, Bennett WS, Strominger JL, Wiley DC (1987 b) The foreign antigen binding site and T cell recognition regions of class I histocompatibility antigens. Nature 329: 512

Coligan JE, Kindt TJ, Nairn R, Nathernson SG, Sachs DH, Hansen TH (1980) Primary structural studies of an H-2L molecule confirm that it is a unique gene product with homology to H-2K and H-2D antigens. Proc Natl Acad Sci USA 77: 1134

Cullen SE, Schwartz BD (1976) An improved method for isolation of H-2 and Ia alloantigens with immunoprecipitation induced by protein A-bearing staphylococci. J Immunol 117: 136

Evans GA, Margulies DH, Ozato K, Camerini-Otero RD, Seidman JG (1982a) Structure and expression of a mouse major histocompatibility antigen gene H-2L$^d$. Proc Natl Acad Sci USA 79: 1994

Evans GA, Margulies DH, Shykind B, Seidman JG, Ozato K (1982b) Exon shuffing, mapping polymorphic determinants on hybrid mouse transplantation antigens. Nature 300: 755

Fellous M, Kamoun M, Niels J, Dausset J, Clements G, Zeuthen J, Klein G (1977) Induction of HLA expression in Daudi cells after cell fusion. Immunogenetics 5: 423

Goodfellow PN, Jones EA, van Heyningen V, Solomon E, Bobrow M, Miggiano V, Bodmer WF (1975) The $\beta_2$ microglobulin gene is on chromosome 15 and not in the HL-A region. Nature 254: 267

Hansen TH, Ozato K, Melino MR, Coligan JE, Kindt TJ, Jandiski JJ, Sachs DH (1981) Immunochemical evidence in two haplotypes for at least three *D* region-encoded molecules, D, L, and R. J Immunol 126: 1713

Kahn-Perles B, Boyer C, Arnold B, Sanderson AR, Ferrier P, Lemonnier FA (1987) Acquisition of HLA class I W6/32-defined antigenic determinant by heavy chains from different species following association with $\beta_2$-microglobulin. J Immunol 138: 2190

Kievits F, Ivanyi P (1987) Monomorphic anti-HLA monoclonal antibody (W6/32) recognizes polymorphic H-2 heavy chain determinants exposed by association with bovine or human but nor murine $\beta_2$-microglobulin. Hum Immunol 10: 115–126

Koch S, Robinson PJ, Koch N, Hammerling GJ (1983) Separation of H-2D$^d$, H-2L$^d$ and H-2R$^d$ by two-dimensional gel electrophoresis. Immunogenetics 17: 215

Krangel MS, Orr HT, Strominger JL (1979) Assembly and maturation of HLA-A and HLA-B antigens in vivo. Cell 18: 979

Laemmli UK (1970) Cleavage of structural proteins during the assembly of the head of bacteriophage T4. Nature 227: 680

Lillehoj EP, Myers NB, Lee DR, Hansen TH, Coligan JE (1985) Structural definition of a family of L$^d$-like molecules distributed among four of seven haplotypes compared. J Immunol 135: 1271

Martinko JM, Anderson SJ, Potter TA (1986) A murine cell line defective in expression of several class I molecules. Immunogenetics 23: 64

Nichols EA, Krakauer T, Hansen TH (1983) Two dimensional gel comparisons of murine *H-2D* region-associated antigens of different H-2 haplotypes. J Immunol 131: 2440

Ozato K, Hansen TH, Sachs DH (1980) Monoclonal antibodies to mouse MHC antigens. II. Antibodies to the H-2L$^d$ antigen, the product of a third polymorphic locus in the mouse MHC. J Immunol 125: 2473

Parham P, Barnstable CJ, Bodmer WF (1979) Use of monoclonal antibody (W6/32) in structural studies of HLA-A, B, C antigens. J Immunol 123: 342

Ploegh HL, Cannon LE, Strominger JL (1979) Cell-free translation of the mRNAs for the heavy and light chains of HLA-A and HLA-B antigens. Proc Natl Acad Sci USA 76: 2273

Potter TA, Boyer C, Schmitt-Verhulst A-M, Golstein P, Rajan TV (1984) Expression of H-2D$^b$ on the cell surface in the absence of detectable $\beta_2$-microglobulin. J Exp Med 160: 317

Rammensee HG, Robinson PJ, Crisanti A, Bevan MJ (1986) Restricted recognition of $\beta_2$m by cytotoxic T lymphocytes. Nature 319: 502

# Selective Increase of the Expression of HLA Class I Molecules by HLA-A3, HLA-B7, and HLA-Cw3 Transfected P815 Mastocytoma Cells, with Subsequent Transfection by the Human $\beta_2$-Microglobulin Gene

B. PERARNAU, A. GILLET, M. BARAD, R. HAKEM, and F. A. LEMONNIER

Centre d'Immunologie INSERM-CNRS de Marseille-Luminy, Case 906, 13288 Marseille Cedex 9, France

## Introduction

Expression of HLA class I heavy chains relies upon their association with $\beta_2$-microglobulin. Transfection experiments (Barbosa et al. 1982; Lemonnier et al. 1983) and murine-human cell fusions (Fellous et al. 1977; Arce-Gomez et al. 1978) have indicated that, in most cases, class I molecules can be formed by association of heavy and light chains originating from different species. However, such heterologous molecules have an altered serological structure (Ferrier et al. 1985; Kahn-Perles et al. 1987) and, in the case of HLA-B27, expression is markedly reduced (Rein et al. 1987; Krimpenfort et al. 1987). The importance of the association with human $\beta_2$-microglobulin for cell surface expression of other allelic forms of human HLA class I heavy chain was investigated. P815 (H-$2^d$) mastocytoma cells were first transfected with either HLA-A3, -B7, or -Cw3 class I genes. A single transfected clone was isolated by limiting dilution in each case and was further transfected with the human $\beta_2$-microglobulin gene which we isolated from a human genomic library. Both parental single and daughter surtransfected cells were studied to compare their expression of HLA class I molecules.

## Serological Analysis of Transfected Cells

Cells were first serologically analyzed in an indirect immunofluorescence assay with anti-HLA class I monoclonal antibodies selected because their reactivity is independent of the human or murine origin of the light chain associated to the heavy one.

The results illustrated in Fig. 1 indicate that the presence of human $\beta_2$-microglobulin results in a 6- (HLA-Cw3), 11- (HLA-A3), or 40-fold (HLA-B7) increase in HLA class I molecules at the surface of P815 cells. In contrast, only a 1.1- to 2.9-fold increase was observed for endogenous H-$2^d$ molecules. These values were consistent whichever clone of surtransfected cells was studied. Therefore, biochemical analyses were undertaken to define which step of the biosynthesis of the HLA class I heavy chains was altered in the absence of human $\beta_2$-microglobulin.

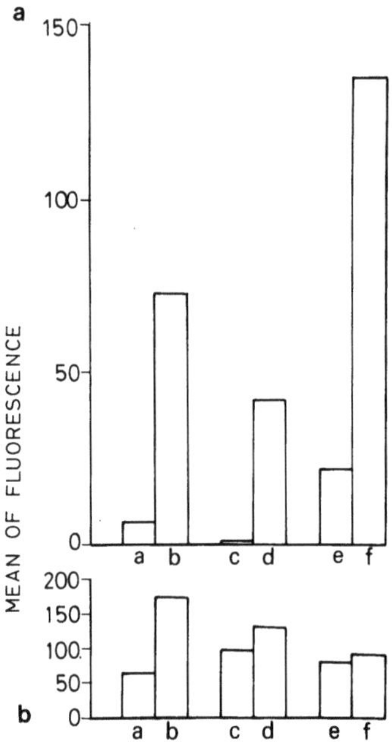


**Fig. 1 a, b.** Expression of HLA class I molecules by parental (HLA) and daughter (HLA x human $\beta_2$-microglobulin) transfected cells. HLA-A3 *(a)*, -B7 *(c)*, -Cw3 *(e)*, P815 parental single transfected cells, and the corresponding HLA x human $\beta_2$-microglobulin P815 surtransfected cells *(b, d, f)* were incubated first with saturating concentrations of B9.1.2.1 (HLA-A3) or B1.23.2 (HLA-B7; HLA-Cw3) anti-HLA class I monoclonal antibodies **(a)** or 20-8-4 S anti-2K$^d$ monoclonal antibodies **(b).** Following washings, cells were further incubated with fluoresceinated sheep anti-mouse immunoglobulin antibodies, washed, and analyzed with a fluoresceinated activating cell sorter. The height of the column is proportional to the mean of fluorescence, which was calculated by integration of the fluorescence curve

## Biochemical Analysis of Transfected Cells

RNA analysis indicated that HLA-A3, -B7, and -Cw3 parental and corresponding daughter human $\beta_2$-microglobulin surtransfected cells had the same amount of HLA class I heavy chain transcripts (data not shown). Immunoprecipitation studies of the various transfected cells were performed on $^{35}$S-methionine labeled cells using a rabbit anti-HLA class I heavy chain antiserum (which recognizes both isolated and $\beta_2$-microglobulin associated heavy chains). These studies are illustrated in Fig. 2.

In spite of some cross-reactivities with murine class I heavy chains and proximity of the actin band, the HLA-A3, -B7, and -Cw3 heavy chains could easily be identified. No noticeable quantitative differences were ever seen comparing the bands corresponding to the heavy chains from parental or surtransfected daughter cells. Furthermore, and of special interest, the human and murine $\beta_2$-microglobulins exhibited differential electrophoretic mobilities, the human light chain migrating slightly further than the murine one. Thus, by analyzing these immunoprecipitations, it became obvious that human heavy chains associate preferentially with human light chains since no detectable amount of murine $\beta_2$-microglobulin was coprecipitated by rabbit anti-HLA antibodies in the case of HLA-A3 x human $\beta_2$-microglobulin P815 double-transfected cells. Similar results were obtained with HLA-Cw3 and HLA-B7 transfectants (Fig. 2). In the case of P815 HLA-B7 single-transfected cells, under our experimental conditions no detectable amount of murine light chain could be coprecipitated with HLA-B7 heavy chains, suggesting an especially weak association. Since the increase in cell surface expression following surtransfection with the human $\beta_2$-microglobulin gene was particularly dramatic in the case of HLA-B7, these HLA-B7 transfectants were selected for further biochemical studies per-

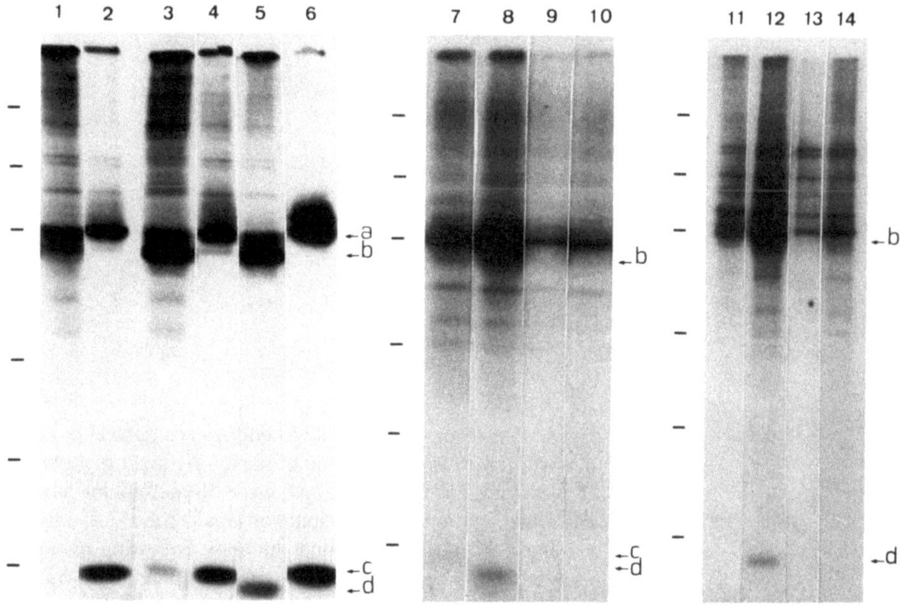


**Fig. 2.** Preferential pairing of human heavy and light chains in HLA x human $\beta_2$-microglobulin P815 double-transfected cells. Parental HLA-A3 *(lanes 3, 4)*, HLA-B7 *(lanes 11, 13)*, and HLA-Cw3 *(lanes 7, 9)* transfected cells, daughter HLA-A3 x human $\beta_2$-microglobulin *(lanes 5, 6)*, HLA-B7 x human $\beta_2$-microglobulin *(lanes 12, 14)*, and HLA-Cw3 x human $\beta_2$-microglobulin *(lanes 8, 10)* transfected cells, and untransfected P815 cells *(lanes 1, 2)* were $^{35}$S-methionine labeled. A rabbit antiserum which reacts with both isolated and $\beta_2$-microglobulin associated HLA class I heavy chains *(lanes 1, 3, 5, 7, 8, 11, 12)*, 20-8-4S, an anti-K$^d$ monoclonal antibody *(lanes 2, 4, 6)*, and 10-3.6, an anti-I-A$^k$ monoclonal antibody *(lanes 9, 10, 13, 14)* were used for immunoprecipitation studies. Murine H-2 (a), human HLA (b) heavy chains, and murine (c) and human (d) $\beta_2$-microglobulins were individualized following sodium dodecylsulfate-polyacrylamide gel electrophoresis using 94-, 66-, 45-, 29-, 24-, and 12-kd markers

formed with a monoclonal antibody (HC 10) which only reacts with isolated HLA class I heavy chains (Stam et al. 1986). Pulse-chase experiments first indicated that HC 10 reactive HLA-B7 molecules had a much longer life span in the absence of human $\beta_2$-microglobulin than in its presence (HC 10 reactive molecules could still be detected 2 h 30 min after the case in the absence of human $\beta_2$-microglobulin, whereas they had totally disappeared after 30 min in its presence). Thus, isolated heavy chains apparently accumulate intracytoplasmically in P815 HLA-B7 parental cells. The susceptibility of these molecules to endoglycosidase H was therefore evaluated.

As illustrated in Fig. 3, the totality of these heavy chains remained susceptible to this glycosidase, indicating that, in the absence of human $\beta_2$-microglobulin, HLA-B7 heavy chains have a reduced ability to translocate to the Golgi apparatus.

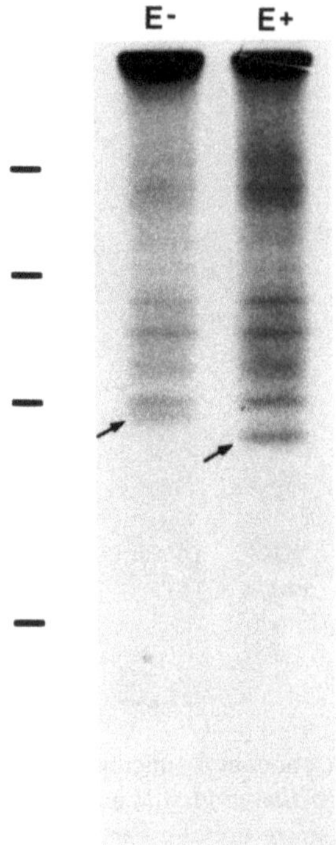


**Fig. 3.** Persistent susceptibility to endoglycosidase H of HLA-B7 synthesized in the absence of human $\beta_2$-microglobulin. HLA-B7 parental transfected cells were $^{35}$S-methionine labeled and lysed, and immunoprecipitation was carried out in the presence of HC 10 monoclonal antibody. Following elution from Sepharose beads, half of the immunoprecipitate was treated with endoglycosidase H *(E+)*, according to Sege et al. (1981), and electrophoresis was performed as indicated in the legend to Fig. 2. *Arrows* correspond to HC 10 reactive HLA-B7 heavy chains before and after endoglycosidase H treatment

## Conclusions

Pairing of HLA-B27 (Rein et al. 1987; Kimpenfort et al. 1987), -B7, -A3, and -Cw3 heavy with light chains exhibits some degree of species specificity, implying complementation between amino acid residues of both chains which might correspond to the species specific residues that have been identified by comparison of the amino acid sequences of the first, second, and third domains of H-2 and HLA class I heavy chains (Klein and Figueroa 1986). However, in spite of the fact that they share the amino acid residues associated to the third domain, the magnitude of increase of expression varies largely from one allelic product to another. This indicates that interactions which stabilize the HLA class I molecules develop between the two external domains of the heavy chains and the $\beta_2$-microglobulin.

*Acknowledgments.* This work was supported by grants from Institut National de la Santé et de la Recherche Médicale, Centre National de la Recherche Scientifique, and Association pour la Recherche sur le Cancer.

## References


Arce-Gomez B, Jones EA, Barnstable CJ, Salomon E, Bodmer WF (1978) The genetic control of HLA-A and B antigens in somatic cell hybrids: requirement for $\beta_2$ microglobulin. Tissue Antigens 11: 96

Barbosa JA, Kamarck ME, Biro PA, Weismann SM, Ruddle FH (1982) Identification of human genomic clones coding the major histocompatibility antigen HLA-A2 and HLA-B7 by DNA-mediated gene transfer. Proc Natl Acad Sci USA 79: 6327

Fellous M, Kaman M, Wiels J, Dausset J, Clements G, Zeuthen J, Klein G (1977) Induction of HLA expression in Daudi cells after cell fusion. Immunogenetics 5: 423

Ferrier P, Layet C, Caillol DH, Jordan BR, Lemonnier FA (1985) The association between murine $\beta_2$-microglobulin and HLA class I heavy chains results in serologically detectable conformational changes of both chains. J Immunol 135: 1281

Kahn-Perles B, Boyer C, Arnold B, Sanderson AR, Ferrier P, Lemonnier FA (1987) Acquisition of HLA class I W6/32 defined antigenic determinant by heavy chains from different species following associated with bovine $\beta_2$-microglobulin. J Immunol 138: 2190–2196

Klein J, Figueroa F (1986) The evolution of class I MHC genes. Immunol Today 7: 41–44

Krimpenfort P, Rudenko G, Hochstenbach F, Guesson D, Berns A, Ploegh H (1987) Crosses of two independently derived transgenic mice demonstrate functional complementation of the genes encoding heavy (HLA-B27) and light ($\beta_2$-microglobulin) chains of HLA class I antigens. EMBO J 6: 1673

Lemonnier FA, Dubreuil PA, Layet C, Malissen M, Bourrel D, Mercier P, Jakobsen BK et al. (1983) Transformation of LMTK$^-$ cells with purified HLA class I genes. II. Serological characterization of HLA-A3 and -Cw3 molecules. Immunogenetics 18: 65

Rein RS, Seemann GHA, Neffjes JJ, Hochstenbach FMH, Stam NJ, Pleogh HL (1987) Association with $\beta_2$-microglobulin controls the expression of transfected human class I genes. J Immunol 138: 1178

Sege K, Rask L, Peterson PE (1981) Role of $\beta_2$-microglobulin in the intracellular processing of HLA antigens. Biochemistry 20: 4523

Stam NJ, Spits H, Ploegh HL (1986) Monoclonal antibodies raised against denatured HLA-B locus heavy chains permit biochemical characterization of certain HLA-C locus products. J Immunol 137: 2299

# Exposure of the W6/32-Defined Determinant on Mouse Cells Is Dependent on the Interaction of Certain H-2 Heavy Chains with Human or Bovine $\beta_2$-Microglobulin

Femia Kievits, Walter J. Boerenkamp, and Pavol Ivanyi

Central Laboratory of the Netherlands Red Cross Blood Transfusion Service and Laboratory for Experimental and Clinical Immunology, University of Amsterdam, Amsterdam, The Netherlands

The HLA class I specific monoclonal antibody (McAb) W6/32 recognizes an antigenic determinant shared in humans by all products of the HLA-A, -B, and -C loci (Barnstable et al. 1978). The W6/32 determinant involves only amino acids of the HLA-A, B, C chain and is a product of their three-dimensional configuration stabilized by the association with $\beta_2$-microgobulin ($\beta_2$m) (Parham et al. 1979). This determinant has further been detected on the cells of some other primates and on bovine cells but not on cells from a number of other species, including mice (Brodsky et al. 1979, 1981; Brodsky and Parham 1982). It has, however, been reported that the W6/32 determinant is exposed on mouse cells expressing the H-2$D^b$ allele (Ivanyi and van der Meugheuvel 1984; Maziarz et al. 1986). Here we summarize the findings which show that the appearance of the W6/32 determinant on mouse cells results from the association of certain D or K gene products with either bovine or human $\beta_2$m.

Mouse spleen cells were incubated for 1 h at 37 °C in culture medium containing 10% fetal calf serum (FCS) or normal mouse serum (NMS). These cells were examined in a complement-dependent assay for lysis by W6/32. None of the target cells tested (including $D^b$ cells) were lysed by W6/32 after preincubation with NMS. However, when the cytotoxicity test was performed on cells incubated with FCS, the reactivity pattern of W6/32 was highly polymorphic. Cells expressing the H-2$D^b$, $K^f$, $D^s$, $K^v$, $K^d$, H-$2^p$, and H-$2^r$ alleles were lysed by W6/32 while the W6/32 determinant was absent from cells expressing the H-$2^q$ and H-$2^k$ alleles. Cells of the B10.D2/n and BALB/c (H-$2^d$) strains were only lysed after stimulation with concanavalin A (ConA). These results show that lysis of certain mouse cells by W6/32 is dependent on the H-2 haplotype and on the presence of FCS. The lysis was dose dependent on FCS and was diminished by short incubation periods or by incubation at low temperatures (Table 1). Maximal lysis was attained for H-$2^b$ cells after 40 min and for H-$2^f$ cells after 60 min of incubation at 37 °C in the presence of 10% FCS. The incubation temperature has to reach 37 °C to ensure maximal lysis of H-$2^b$ and H-$2^f$ cells. Table 1 also shows that H-$2^b$ cells, in contrast to H-$2^f$ cells, were already able to bind W6/32 after preincubation in a very low concentration FCS (0.5%). H-$2^k$ (B10.BR) cells were not lysed even after optimal incubation procedures in FCS. This FCS-dependent activity of W6/32 on mouse cells has not been described for the class I activity of W6/32 on human cells.

To examine whether the exposure of the W6/32 determinant on mouse cells was

**Table 1.** Lymphocytotoxic titers[a] of W6/32 for three cell types after monitoring the incubation time, the concentration of FCS, and the incubation temperature

| Parameters[c] | H-2 haplotypes[b] | | |
|---|---|---|---|
| | b | f | k |
| Incubation time (min) | | | |
| 0 | 0 | 0 | 0 |
| 5 | 320 | 0 | 0 |
| 10 | 640 | 0 | 0 |
| 20 | 1280 | 80 | 0 |
| 40 | 2560 | 320 | 0 |
| 60 | 2560 | 640 | 0 |
| 90 | 2560 | 640 | 0 |
| Concentration FCS (%) | | | |
| 0 | 0 | 0 | 0 |
| 0.5 | 320 | 0 | 0 |
| 2 | 640 | 80 | 0 |
| 5 | 1280 | 320 | 0 |
| 10 | 2560 | 640 | 0 |
| 20 | 2560 | 640 | 0 |
| Incubation temperature (°C) | | | |
| 0 | 0 | 0 | 0 |
| 4 | 0 | 0 | 0 |
| 20 | 40 | 0 | 0 |
| 28 | 320 | 20 | 0 |
| 37 | 2560 | 640 | 0 |

[a] Reciprocal titer shows the highest McAb dilution with positive (80%) cell lysis

[b] W6/32 was tested on spleen cells from H-2 congeneic mouse strains B10 (H-$2^b$), B10.M (H-$2^f$), and B10.BR (H-$2^k$)

[c] The standard incubation parameters were 10% FCS for 60 min at 37 °C

specific for FCS, spleen cells were preincubated (1 h, 37 °C) in culture medium containing 10% of human, rabbit, sheep, goat, or cow serum. We observed that murine class I molecules incubated in human sera (NHS) or in sera from bovine sources (cow, calf) do express the W6/32 determinant, whereas no expression of the W6/32 determinant was found after preincubation with rabbit, sheep, goat, or mouse serum. To further analyze the dependence on NHS for the reactivity of W6/32 on a panel of mouse cells, spleen cells were incubated (1 h, 37 °C) with various concentrations of purified human $\beta_2$m. Figure 1 shows that the incubation of H-$2^b$ and H-$2^r$ but not H-$2^k$ cells with human $\beta_2$m resulted in the exposure of determinants recognized by W6/32. It is concluded that the W6/32 defined determinant is formed by the association of certain mouse class I heavy chains with human or bovine $\beta_2$m. Exchange of endogenous $\beta_2$m from culture medium and the serological consequences are well known from previous reports (Ferrier et al. 1985; Jefferies and

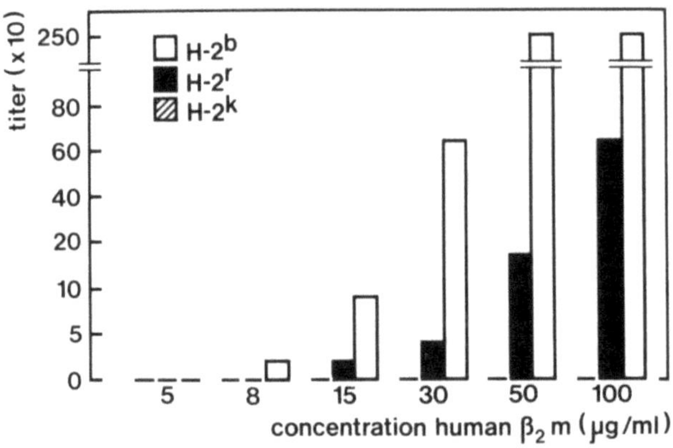


**Fig. 1.** Lymphocytotoxic titers of McAb W6/32 on B10 (H-$2^b$), B10.RIII (H-$2^r$), and B10.BR (H-$2^k$) spleen cells after 1 h of preincubation at 37 °C with various concentrations of purified human $\beta_2$m. Titers express the highest McAb dilution with positive (80%) cell lysis

MacPherson 1987; Mierau et al. 1987; Kievits and Ivanyi 1987). However, the question arose of whether differences in association of H-2 heavy chains with human or bovine $\beta_2$m are the reason for the polymorphic reactivity pattern of W6/32. Therefore, quantitative measurements were performed for binding of human $\beta_2$m to spleen cells of seven H-2 congeneic strains. Spleen cells were incubated in 10% NHS (1 h, 37 °C) and after washing, the binding of $\beta_2$m to the cells was estimated by indirect fluorescence using rabbit anti-human $\beta_2$m antiserum. Figure 2 shows that binding of human $\beta_2$m to mouse cells, and thus the replacement of endogenous $\beta_2$m, is dependent on the H-2 haplotype. Spleen cells from the B10.M (H-$2^f$) strain bound twice as much human $\beta_2$m as cells from the B10.BR strain. However, W6/32

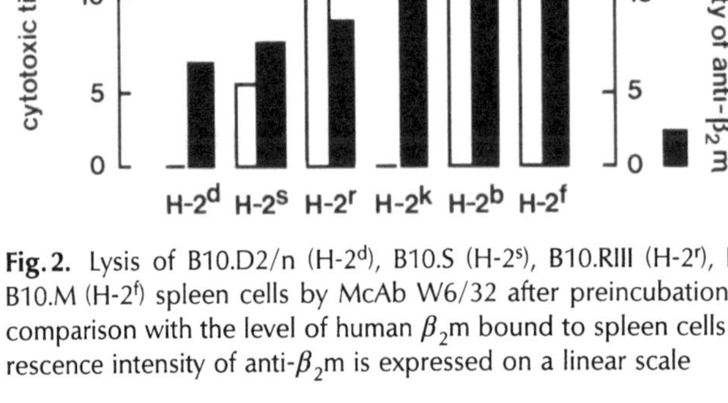


**Fig. 2.** Lysis of B10.D2/n (H-$2^d$), B10.S (H-$2^s$), B10.RIII (H-$2^r$), B10.BR (H-$2^k$), B10(H-$2^b$), and B10.M (H-$2^f$) spleen cells by McAb W6/32 after preincubation (1 h at 37 °C) in 10% NHS in comparison with the level of human $\beta_2$m bound to spleen cells of the same haplotypes. Fluorescence intensity of anti-$\beta_2$m is expressed on a linear scale

negative cells (H-$2^k$) bound more $\beta_2$m, as did the W6/32 positive cells H-$2^s$ and H-$2^r$. Therefore, W6/32 negativity is not due to the inability of H-$2^k$ cells to bind human $\beta_2$m but to the absence of the W6/32 determinant on H-$2^k$ heavy chains.

In conclusion, McAb W6/32 cross-reacts with H-2 antigens to a broader extent than was originally described (Ivanyi and van der Meugheuvel 1984). However, binding of W6/32 to mouse cells is completely dependent on preincubation of the target cells in sera from bovine or human sources. These findings imply a special quality shared by human and bovine $\beta_2$m that is not present on mouse, rabbit, goat, and sheep $\beta_2$m. However, similarity in amino acid sequences of human and bovine $\beta_2$m did not explain our findings (Kievits and Ivanyi 1987). Binding of human $\beta_2$m to mouse spleen cells was largely dependent on the H-2 haplotypes, as already reported for the exchange of murine $\beta_2$m (Sege et al. 1979). The reason for and the consequences of variations in the exchange of $\beta_2$m are unclear. The variations might have implications as they appear to be related to the H-2 haplotype differences in binding of viral antigens (Grundy et al. 1987). No simple correlation was found between the amount of human $\beta_2$m bound to mouse cells and the cross-reactivity of W6/32. Only certain H-2 heavy chains will expose the W6/32 determinant after complexing with human or bovine $\beta_2$m.

*Acknowledgments.* This investigation was supported in part by the Foundation for Medical Research, MEDIGON, which is subsidized by the Netherlands Organization for the Advancement of Pure Research (ZWO) (grant no. 509-087), and in part by the Deutsche Forschungsgemeinschaft (grant no. 5/1–2).

## References

Barnstable CJ, Bodmer WF, Brown G, Galfre G, Milstein C, Williams AF, Ziegler A (1978) Production of monoclonal antibodies to group A erythrocytes, HLA and other human cell surface antigens – new tools for genetic analysis. Cell 14: 9–20

Brodsky FM, Parham P (1982) Evolution of HLA antigenic determinants: species cross-reactions of monoclonal antibodies. Immunogenetics 15: 151–166

Brodsky FM, Bodmer WF, Parham P (1979) Characterization of monoclonal anti-beta2-microglobulin antibody and its use in the genetic and biochemical analysis of major histocompatibility antigens. Eur J Immunol 9: 536–545

Brodsky FM, Stone WH, Parham P (1981) Of cows and men: a comparative study of histocompatibility antigens. Hum Immunol 3: 143–152

Ferrier P, Layer C, Caillol DH, Jordan BR, Lemmonier FA (1985) The association between murine beta2-microglobulin and HLA class-I heavy chains results in serologically detectable conformational changes of both chains. J Immunol 135: 1281–1287

Grundy JE, McKeating JA, Ward PJ, Sanderson AR, Griffiths PD (1987) Beta2-microglobulin enhances the infectivity of cytomegalovirus and when bound to the virus enables class–I HLA molecules to be used as a virus receptor. J gen Virol 68: 793–803

Ivanyi P, van der Meugheuvel W (1984) A monomorphic HLA-specific monoclonal antibody, W6/32, reacts with H-$2D^b$ molecules of normal mouse lymphocytes. Immunogenetics 20: 699–703

Jefferies WA, MacPherson GG (1987) Expression of the W6/32 HLA epitope by cells of rat, mouse, human and other species: critical dependence on the interaction of specific MHC heavy chains with human or bovine beta2-microglobulin. Eur J Immunol 17: 1257–1263

Kahn-Perles B, Boyer C, Arnold B, Sanderson AR, Ferrier P, Lemmonier FA (1985) Acquisition of HLA class I W6/32-defined antigenic determinant by heavy chains from different species following association with bovine beta2-microglobulin. J Immunol 138: 1281–1287

Kievits F, Ivanyi P (1987) Monomorphic anti-HLA monoclonal antibody (W6/32) recognizes polymorphic H-2 heavy-chain determinants exposed by association with bovine or human but not murine beta2-microglobulin. Human Immunol 20: 115–126

Maziarz RT, Fraser J, Strominger JL, Burakoff SJ (1986) The human HLA-specific monoclonal antibody W6/32 recognizes a discontinuous epitope within the alpha2 domain of murine H-2$D^b$. Immunogenetics 24: 206–208

Mierau R, Robinson PJ, Sanderson AR, Genth E, Cramer M (1987) Antigenic determinants shared between HLA-A, -B, -C antigens and H-2 class I molecules modified by bovine beta2-microglobulin. Immunogenetics 26: 351–355

Parham P, Barnstable CJ, Bodmer WF (1979) Properties of an anti-HLA, B, C, antibody. Use of a monoclonal antibody (W6/32) in structural studies of HLA-A, B, C antigens. J Immunol 123: 342–349

Sege K, Östberg L, Peterson PA (1979) Demonstration of a murine cell surface component with affinity for exogenous beta2-microglobulin. Eur J Immunol 9: 964–971

# The Role of $\beta_2$-Microglobulin and Class I HLA Heavy Chain in Cytomegalovirus Infection

JANE E. GRUNDY, JANE A. MCKEATING, and PAUL D. GRIFFITHS

Department of Virology. The Royal Free Hospital School of Medicine, London NW 3 2QG, England

## Introduction

Human cytomegalovirus (CMV) is a member of the herpes virus group and causes widespread infection in man. In individuals with normal immunity, infection with the virus is usually asymptomatic although occasionally a mononucleosis-like illness can occur (Weller 1971). However, when the host's immunity is compromised, CMV becomes an important cause of morbidity and mortality. Thus immunosuppressed individuals, such as recipients of bone marrow or renal transplants (Ho 1977), patients infected with the human immunodeficiency virus (Sonnabend et al. 1983), and the immunologically immature fetus (Stagno et al. 1983) are all at risk from symptomatic CMV infection.

In addition to primary CMV infection, immunocompromised individuals are at risk from reactivation of preexisting infection (Ho 1977). The host immune response fails to eradicate the virus after primary infection and CMV, a double-stranded DNA virus, can then integrate into the host cell genome, establishing a latent infection which can potentially be reactivated. In addition, immunocompromised hosts can also be reinfected with new strains of CMV (Grundy et al. 1987 d), despite having previously been infected with the virus. Thus CMV can be reactivated or can cause reinfection despite the presence of antibodies to the virus. Furthermore, the fetus can be infected by transplacental passage of the virus from the mother, even though passively acquired maternal antibodies can be found in the fetal circulation (Stagno et al. 1977, 1980). Thus antibodies are apparently ineffective in defence against CMV infection, although it is possible that they could ameliorate the severity of infection.

## Binding of $\beta_2$m to CMV

Our recent studies may provide an explanation for the failure of antibody to be effective against CMV. We have found that CMV purified directly from the body fluids of patients is "coated" with the host protein $\beta_2$-microglobulin ($\beta_2$m) (McKeating et al. 1987 a). $\beta_2$m was shown to be bound to the viral envelope and the virus was bound specifically to an affinity column comprising a monoclonal antibody specific for $\beta_2$m linked to Sepharose (McKeating et al. 1987a), indicating that $\beta_2$m was exposed on the outside of the viral particle. Virus harvested from body fluids such as urine could not be neutralized by human hyperimmune globulin, human immune sera, or murine monoclonal antibodies which had good neutralizing activity against CMV grown in cell culture (McKeating et al. 1987a). Furthermore, urinary CMV could not be "captured" onto a solid phase by virus specific murine monoclonal antibodies which could bind cell culture grown virus (McKeating et al. 1986; Grundy et al. 1987b); however, it could be captured by monoclonal antibodies specific for $\beta_2$m (Grundy et al. 1987b). Thus we concluded that the binding of $\beta_2$m to CMV masked the viral antigenic sites recognized by murine monoclonal antibodies and, more importantly, the antigenic sites necessary for virus neutralization recognized by man's humoral immune response.

In vitro studies using the laboratory strain of CMV, strain AD169, showed that CMV bound $\beta_2$m from cell culture fluids after the virus was released from cells (Grundy et al. 1987a). Thus $\beta_2$m was not acquired as an integral part of the viral envelope as the virus budded through host cell membranes containing class I HLA antigens. Since most if not all body fluids contain free $\beta_2$m, we believe that in vivo CMV will be present in these fluids as $\beta_2$m-coated particles. As CMV is transmitted between individuals by direct contact with infected body fluids such as saliva, breast milk, and semen, we proposed that CMV has evolved this strategy of coating itself in a host protein as a means of evading neutralization and facilitating transmission between individuals (McKeating et al. 1987a).

## Role of $\beta_2$m in Infectivity of CMV

The above hypothesis supposed that the $\beta_2$m-coated urinary CMV particles are infectious, which is indeed the case in cell culture. Thus the question arose that if the glycoproteins on the surface of the viral envelope are masked by the binding of host $\beta_2$m, what does CMV use as a receptor to bind to host cells and initiate infection (Fig. 1)? We proposed that CMV used the bound $\beta_2$m to bind to host cells. If this was true, two predictions could be made: (1) $\beta_2$m should be required by CMV for infectivity and (2) CMV and $\beta_2$m should compete for the same binding sites on host cells. Our subsequent experiments verified both of these predictions. $\beta_2$m enhanced the infectivity of CMV (Grundy et al. 1987c), and $\beta_2$m and CMV competed for the same binding sites on fibroblasts and lymphoblastoid cells (Grundy et al. 1987c).

## Class I HLA as a Receptor for CMV

$\beta_2$-microglobulin is the light chain of the class I HLA dimer (Cresswell et al. 1974) and the main binding site known for $\beta_2$m on the cell surface is the class I HLA heavy chain (Hyafil and Strominger 1979), suggesting that CMV used class I molecules as a cell receptor. Studies of the binding of CMV to cells expressing and lacking expres-

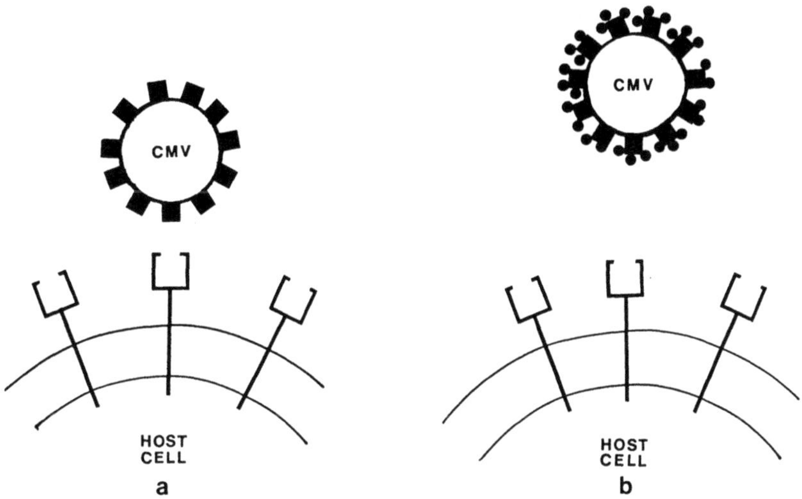


**Fig. 1 a, b.** Schematic representation of binding of CMV to host cells. In **a** CMV uses its surface glycoproteins (■) to bind to receptors on the host cell surface. In **b** these surface viral glycoproteins are masked by the binding of host $\beta_2$m (●), and binding to cellular receptors would be prevented. The distribution of host $\beta_2$m molecules on the viral surface is a schematic representation only; the exact number and distribution is presently unknown

sion of class I molecules showed that CMV could indeed use class I as a receptor (Grundy et al. 1987c), although it could not be shown in these studies that this was the only receptor. However, since $\beta_2$m substantially increased infectivity of the virus, we concluded that class I molecules were a major receptor for CMV.

## Characterization of the $\beta_2$m Binding Proteins of CMV

The proteins on the viral envelope responsible for the binding of $\beta_2$m were studied. Urinary CMV had two protein subunits of $M_r$ 55000 and 35000, which were found on the viral envelope as a disulfide-linked complex of approximate $M_r$ 200000 in a nondisulfide linkage with $\beta_2$m (McKeating et al. 1987b). The larger subunit may have a slightly higher $M_r$ (65000) for CMV strain AD169 from cell culture (Grundy et al. 1987a).

## Analysis of the Binding of $\beta_2$m to CMV

The binding of $\beta_2$m to CMV was compared with that to the class I HLA heavy chain and the results are summarized in Table 1. The monoclonal antibody BBM.1, which is specific for arginine at position 45 of $\beta_2$m (Parham et al. 1983), recognizes free $\beta_2$m, HLA-bound $\beta_2$m (Brodsky et al. 1979), and CMV-bound $\beta_2$m (McKeating et al.

**Table 1.** Reactivity of monoclonal antibodies specific for $\beta_2$m with free, HLA-bound, or CMV-bound $\beta_2$m

| Antibody | Reactivity with: | | |
|---|---|---|---|
| | Free $\beta_2$m | HLA/$\beta_2$m | CMV/$\beta_2$m |
| BBM.1 | + | + | + |
| C21 | + | − | + |
| C23 | + | + | + |
| B4 | + | + | + |

1987 a), indicating that this residue, which is on an external loop of the protein (Parham et al. 1983), is exposed in all of these situations. In contrast the antibody C21, which is specific for a determinant on free $\beta_2$m that is concealed when $\beta_2$m is bound to the class I heavy chain (Liabeuf et al. 1981), recognizes CMV-bound $\beta_2$m (Grundy et al. 1987 b). Thus the binding site on $\beta_2$m for CMV and class I heavy chain appear to be different.

## Model for the Binding of CMV to HLA

Based on the findings above we have proposed that CMV binds to class I HLA molecules on the cell surface by displacing $\beta_2$m from the class I dimer and binding to the heavy chain via the viral-bound $\beta_2$m (Grundy et al. 1987 c). This model is shown schematically in Fig. 2 a and is consistent with fact that CMV and the class I heavy chain bind $\beta_2$m at different sites, allowing $\beta_2$m to act as a bridge between the cell and the virus. Since CMV binds $\beta_2$m it is theoretically possible that "uncoated" CMV could bind directly to the class I dimer as illustrated in Fig. 2 b. However as all body fluids contain free $\beta_2$m it is unlikely that under physiological conditions CMV exists without surface-bound $\beta_2$m. Furthermore, the finding that the addition of $\beta_2$m enhances infectivity of the virus (Grundy et al. 1987 c) suggests that if such direct binding does occur it is not as favorable as the exchange mechanism proposed in Fig. 2 a:

Exchange between free $\beta_2$m and that on the cell surface is well documented (Hyafil and Strominger 1979: Kefford et al. 1984; Bernabeu et al. 1984), and since our studies have shown that CMV has a higher affinity of binding to cells than free $\beta_2$m (Grundy et al. 1987 c), we believe that displacement of $\beta_2$m from the class I dimer by CMV would be favored. This would be expected given the large number of molecules of $\beta_2$m likely to be bound to CMV in body fluids which would result in multipoint binding. Thus we propose that entry of CMV would probably be by receptor-mediated endocytosis.

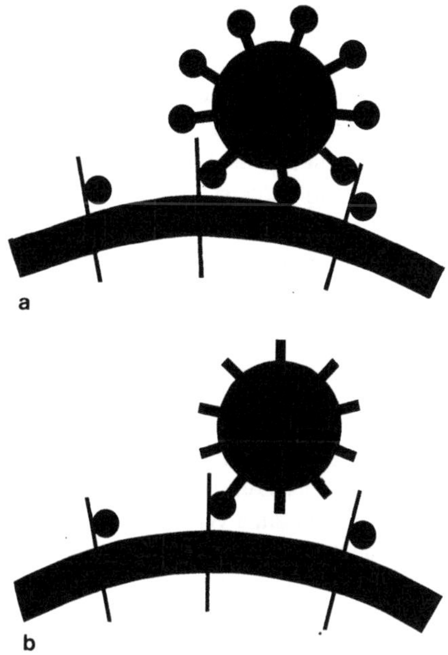


**Fig. 2a, b.** Models for the binding of CMV to class I HLA molecules on the cell surface. In **a** CMV has host $\beta_2$m (●) already bound to the viral envelope and uses this as a bridge to bind to the class I heavy chain, displacing the $\beta_2$m from the class I dimer on the cell surface. In **b**, CMV is not "coated" with host $\beta_2$m but binds directly to the $\beta_2$m present on the cell surface in the class I dimer

## Possible Role of HLA Haplotype in CMV Infection

The possibility exists that different class I HLA haplotypes might have different affinities of binding of free or CMV-bound $\beta_2$m which may affect the susceptibility of an individual to infection with the virus. It is of interest in this regard that in the murine model of CMV infection, the resistance of mice to lethal infection with murine CMV maps to the class I regions of the H-2 complex (Chalmer et al. 1977; Chalmer 1980; Grundy et al. 1981). Furthermore, in recombinant mice on the B10 genetic background possession of $D^b$ was associated with resistance (Grundy et al. 1981), and this class I molecule has been reported to have a reduced binding of $\beta_2$m (Potter et al. 1984, 1985), suggesting that the basis of the H-2 control of resistance is the affinity of binding of $\beta_2$m to the H-2 heavy chain. Differences in the strength of binding of $\beta_2$m to mouse splenocytes of different H-2 haplotypes have been reported (Sege et al. 1979). We are now testing this hypothesis and investigating the possibility that a similar mechanism operates in man.

## Conclusion

Our studies so far have led us to the conclusion that in vivo CMV binds $\beta_2$m from body fluids, that such binding masks the viral antigens preventing its neutralization by antibodies present in those fluids, and that CMV then uses its surface-bound $\beta_2$m to bind to class I HLA molecules on the host cell surface. We believe that the use of class I HLA as a receptor for CMV may indicate that an individual's HLA haplotype may affect his or her susceptibility to infection with the virus.

## References


Bernabeu C, van de Rijn M, Lerch PG, Terhorst CP (1984) $\beta_2$m from serum associates with class I antigens on the surface of cultured cells. Nature 308: 642–645

Brodsky FM, Bodmer WF, Parham P (1979) Characterization of monoclonal anti-$\beta_2$-microglobulin antibody and its use in the genetic and biochemical analysis of major histocompatibility antigens. Eur J Immunol 9: 536–545

Chalmer JE (1980) Genetic resistance to murine cytomegalovirus. In: Skamene E (ed) Genetic control of natural resistance to infection and malignancy. Academic, New York, pp 283–290

Chalmer JE, Mackenzie JM, Stanley NF (1977) Resistance to murine cytomegalovirus linked to the major histocompatibility complex of the mouse. J Gen Virol 37: 107–114

Cresswell P, Springer T, Strominger JL, Turner MJ, Grey HM, Kubo RT (1974) Immunological identity of the small subunit of HLA-A antigens as $\beta_2$m and its turnover on the cell membrane. Proc Natl Acad Sci USA 71: 2123–2127

Grundy JE, Mackenzie JS, Stanley NF (1981) Influence of H-2 and non H-2 genes on resistance to murine cytomegalovirus infection. Infect Immun 32: 277–286

Grundy JE, McKeating JA, Griffiths PD (1987 a) Cytomegalovirus strain AD169 binds $\beta_2$-microglobulin in vitro after release from cells. J Gen Virol 68: 777–784

Grundy JE, McKeating JA, Sanderson AR, Griffiths PD (1987 b) Cytomegalovirus and $\beta_2$-microglobulin ($\beta_2$m) in urine specimens: Reciprocal interference in their detection is responsible for artifactually high levels of urinary $\beta_2$m in infected transplant recipients. Transplantation 45: 1075–1079

Grundy JE, McKeating JM, Ward PJ, Sanderson AR, Griffiths PD (1987 c) $\beta_2$-microglobulin enhances the infectivity of cytomegalovirus and when bound to the virus enables class I HLA antigens to be used as a virus receptor. J Gen Virol 68: 793–803

Grundy JE, Super M, Lui S, Sweny S, Griffiths PD (1987 d) The source of cytomegalovirus infection in seropositive renal allograft recipients is frequently the donor kidney. Transplant Proc 12: 2129–2130

Ho (1977) Virus infection after transplantation in man. Arch Virol 55: 1–24

Hyafil F, Strominger JL (1979) Dissociation and exchange of the $\beta_2$-microglobulin subunit of HLA-A and HLA-B antigens. Proc Natl Acad Sci USA 76: 5834–5838

Kefford RF, Calabi F, Fearnley IM, Burrone OR, Milstein C (1984) Serum $\beta_2$-microglobulin binds to a T-cell differentiation antigen and increases its expression. Nature 308: 641–645

Liabeuf A, le Borgne de Kaquel C, Kourilsky FM, Malissen B, Manuel Y, Sanderson AR (1981) An antigenic determinant of human $\beta_2$-microglobulin masked by the association with HLA heavy chains at the cell surface: Analysis using monoclonal antibodies. J Immunol 127: 1542–1548

McKeating JA, Grundy JE, Varghese Z, Griffiths PD (1986) The detection of cytomegalovirus by ELISA in urine samples is inhibited by $\beta_2$-microglobulin. J Med Virol 18: 341–348

McKeating JA, Griffiths PD, Grundy JE (1987 a) Cytomegalovirus in urine specimens has host $\beta_2$-microglobulin bound to the viral envelope: A mechanism of evading the host immune response? J Gen Virol 68: 785–792
McKeating JA, Griffiths PD, Grundy JE (1987 b) Characterisation of the $\beta_2$-microglobulin binding proteins of cytomegalovirus (Abstr). 13th Herpesvirus Workshop, Philadelphia
Parham P, Androlewicz MJ, Holmes NJ, Rothenberg BE (1983) Arginine 45 is a major part of the antigenic determinant of human $\beta_2$-microglobulin recognized by mouse monoclonal antibody BBM.1. J Biological Chem 258: 6179–6186
Potter TA, Boyer C, Schmidt-Verhulst AM, Goldstein P, Rajan TV (1984) Expression of H-$2D^b$ on the cell surface in the detectable $\beta_2$-microglobulin. J Exp Med 160: 317–322
Potter TA, Zeff RA, Schmidt-Verhulst AM, Rajan TV (1985) Molecular analysis of an EL4 cell line that expresses H-$2D^b$ but not H-$2K^b$ or $\beta_2$-microglobulin. Proc Natl Acad Sci USA 82: 2950–2954
Sege K, Ostberg L, Peterson PA (1979) Demonstration of a murine cell surface component with affinity for exogenous $\beta_2$-microglobulin. Eur J Immunol 9: 964–971
Sonnabend J, Witkin SS, Purtilo DT (1983) Acquired immunodeficiency syndrome, opportunistic infections, and malignancies in male homosexuals. JAMA 249: 2370–2374
Stagno S, Reynolds DW, Huang E-S, Thames SD, Smith RJ, Alford CA (1977) Congenital Cytomegalovirus Infection: Occurrence in an Immune Population. N Engl J Med 296: 1254–1258
Stagno S, Reynolds DW, Pass RF, Alford CA (1980) Breast Milk and the Risk of Cytomegalovirus infection. N Engl J Med 302: 1073–1076
Stagno S, Pass RF, Dworfsky ME, Alford CA (1983) Congenital and perinatal infection cytomegalovirus. Seminars in Perinatology 7: 31–42
Weller TH (1971) The cytomegaloviruses: ubiquitous agents with protean clinical manifestations. N Engl J Med 285: 203–214, 267–274

# Structural Requirements of an Ovalbumin-Derived Immunogenic Peptide for T Cell Activation and Interaction with $IA^d$

ALESSANDRO SETTE, SØREN BUUS, and HOWARD M. GREY

Division of Basic Immunology, National Jewish Center for Immunology and Respiratory Medicine, 1400 Jackson Street, Denver, CO 80206, USA

## Introduction

As a general rule, T helper cells only recognize protein antigens that have undergone some process of physical alteration, such as fragmentation or denaturation of the polypeptide chain (Grey and Chesnut 1985) within an antigen presenting cell. Considerable evidence has accumulated to suggest that such a "processed" antigen is recognized by the T cell receptor in association with major histocompatibility complex (MHC) molecules (Buus et al. 1987 a; Watts and McConnell 1987; Allen et al. 1987). In the past 2 years we (Buus et al. 1986, 1987 a) and others (Babbitt et al. 1985) have been able to show that immunogenic peptides bind to Ia molecules, and that such binding correlates with known genetic restrictions of immune responses. Moreover, data obtained with several unrelated immunogenic peptides suggested that each Ia molecule possesses a single peptide binding site (Buus et al. 1987 a; Guillet et al. 1987). The chicken ovalbumin (Ova) peptide 323–339 has previously been shown to be immunodominant in H-$2^d$ mice and to bind specifically to its restriction element, $IA^d$. Using the $IA^d$ restricted presentation of Ova 323–339 as a model system and employing a large series of Ova 323–339 analogs, we here analyze the structural requirements for peptide-$IA^d$ binding and for T cell activation.

## Materials and Methods

### *Peptide Synthesis*

Peptides were synthesized with combined manual and automatic synthesis (Applied Biosystems 430A peptide synthesizer) and simplified by the use of the technique described by Houghton (1985). After the synthesis was completed, the peptide was cleaved from the resin, and the protecting groups were removed by treatment with hydrogen fluoride. The peptide was washed, extracted, lyophilized, desalted, and finally purified by reversed-phase high-performance liquid chromatography (HPLC). The purity of the peptides was substantiated by amino acid sequence and composition analysis. They were routinely >95% pure after HPLC. $^{125}I$ labeling of tyrosinated peptides was done by the chloramine T method (Greenwood et al. 1963).

### *Cells*

The B cell lymphoma A20-1.11 (A20) was used as a source of $IA^d$. A20 cells were maintained in vitro by culture in RPMI 1640 medium supplemented with 2 mM L-glutamine, 50 μM 2-mercaptoethanol, and 10% heat-inactivated fetal calf serum. Large quantities of cells were grown in spinner cultures (37 °C, 10% $CO_2$).

A20 cells were lysed at a concentration of $10^8$ cells/ml in phosphate buffered saline (PBS) containing 1% NP 40, 5 mM sodium orthovandate, and 25 mM iodoacetamide. The lysates were cleared of nuclei and debris by centrifugation at 10000 × *g* for 20 min.

### *Affinity Purification of Ia Molecules*

$IA^d$ molecules were purified as previously described (Buus et al. 1987a) by affinity chromatography using the monoclonal antibody MKD6 ($IA^d$ specific) coupled to Sepharose 4B beads.

### *The Binding of Radiolabeled Peptides to* $IA^d$

Purified $IA^d$ molecules (10–40 μM) were incubated with 0.2–0.6 μM $^{125}I$ radiolabeled Ova 323–339 for 48 h in the presence of a protease inhibitors cocktails [0.5% NP 40, 1 mM phenylmethylsulfonyl fluoride, 0.26 mg/ml 1,10-phenanthroline (50 μg/ml), pepstatin A, EDTA 3 mg/ml, and 0.1% $NaN_3$]. The Ia-peptide complexes were separated from free peptide by gel filtration on a Sephadex G50 (Pharmacia) column. The columns were eluted in 0.5 NP 40/PBS/0.1% $NaN_3$ and collected in 1-ml fractions, and each fraction was counted for 1 min. The fraction $\alpha$ of peptide bound to Ia relative to the total amount of peptide offered was calculated as the ratio of peptide in the void volume to the total peptide recovered. The binding capacity of the various peptide analogs of Ova 323–339 was evaluated by measuring the concentration of cold peptide necessary to inhibit by 50% the binding of the $^{125}I$-Ova 323–339 to the $IA^d$ molecules.

### *Structural Similarity Analysis*

To compare unrelated protein sequences and search for structural similarity we adapted a method described by Grantham (1974) and Padlan (1977). All the possible alignments between two sequences are generated by sliding one sequence along the other. For each particular alignment an average structural dissimilarity (ASD) value was calculated as follows: A dissimilarity value for each pair of corresponding amino acids was obtained from a table that takes into account differences in atomic composition, polarity, and molecular volume. The dissimilarity values are then normalized for mean dissimilarities of random substitutions and their average (ASD) is then calculated. The particular alignment yielding the lowest ASD value was selected as the best-fit alignment of the two sequences.

*T Cell Activation Assay*

T cell hybridomas 3DO-54.8 and 8DO-51.15, both specific for Ova 323–339, were used at a concentration of $2 \times 10^5$ cells/well together with various antigen concentrations, ranging from 0.1 to 100 μg/ml. $10^5$ A20 cells/well served as antigen presenting cells. After 24 h the IL-2 concentration in each culture supernatant was determined in the IL-2 dependent HT2 cell assay, and the amount of each peptide relative to the amount of the unsubstituted Ova 323–339 needed to induce the production of 80 units/ml of IL-2 was measured.

## Results

*N- and C-terminal Truncated Analogs of Ova 323–339 Define a Core Region Critical for IA$^d$ Binding*

To define the IA$^d$-interacting region within the peptide Ova 323–339 we synthesized a series of N- and C-terminal truncated analogs and tested these analogs for binding to purified IA$^d$ molecules (Table 1). Deletions from the C-terminal end of the molecule were without noticeable effect until the seventh ($E_{333}$) residue was removed, leading to a nearly three fold reduction in the IA$^d$ binding. Further removal of $A_{332}$ resulted in a complete loss of the IA$^d$-interacting capacity. From the N-terminal end of the molecule, the first two residues could be removed without noticeable effect, and removal of the next two residues ($Q_{325}$ and $A_{326}$) had a marginal but significant effect. A marked (20-fold) decrease was associated with the

**Table 1.** Truncations of the Ova peptide 323–339 and their binding to IA$^d$

| Ova peptide | Sequence | | | | | | | | | | | | | | | | | Relative IA$^d$ binding capacity[a] |
|---|---|---|---|---|---|---|---|---|---|---|---|---|---|---|---|---|---|---|
| 323–339 | I | S | Q | A | V | H | A | A | H | A | E | I | N | E | A | G | R | 1.0 |
| 323–336 | – | – | – | – | – | – | – | – | – | – | – | – | – | E | | | | 1.1 |
| 323–335 | – | – | – | – | – | – | – | – | – | – | – | – | N | | | | | 1.0 |
| 323–333 | – | – | – | – | – | – | – | – | – | – | E | | | | | | | 0.6 |
| 323–332 | – | – | – | – | – | – | – | – | – | A | | | | | | | | 0.2 |
| 323–331 | – | – | – | – | – | – | – | – | H | | | | | | | | | <0.005 |
| 325–339 | | | Q | – | – | – | – | – | – | – | – | – | – | – | – | – | – | 1.1 |
| 326–339 | | | | A | – | – | – | – | – | – | – | – | – | – | – | – | – | 0.5 |
| 327–339 | | | | | V | – | – | – | – | – | – | – | – | – | – | – | – | 0.3 |
| 328–339 | | | | | | H | – | – | – | – | – | – | – | – | – | – | – | 0.014 |
| 329–339 | | | | | | | A | – | – | – | – | – | – | – | – | – | – | <0.005 |

[a] The IA$^d$ interacting capacity of each peptide is expressed as the amount of unlabeled Ova 323–339 divided by the amount of truncated peptide needed to inhibit the binding of $^{125}$I radiolabeled Ova 323–339 to IA$^d$ by 50%. The data represent the average of two to four experiments

removal of the next residue ($V_{327}$), and a complete loss of binding activity was associated with removal of $H_{328}$. Thus, the critical residues for interaction with $IA^d$ molecules were contained within the sequence VHAAHA (Ova 327–332).

### *Effect of Single Amino Acid Substitutions on the $IA^d$ Interacting Capacity of Ova 323–336*

To further characterize the relative contribution to $IA^d$ binding of individual residues within the Ova peptide, we synthesized a series of 55 single substitution analogs of the peptide 323–336, which is highly stimulatory for certain T cells and binds equally well as Ova 323–339 to $IA^d$. For each of the 11 positions 325–335 we synthesized five analogs, each carrying a single amino acid substitution (two conservative, one semiconservative, and two nonconservative). When these analogs were tested for binding to $IA^d$, it was found that most of the substitutions introduced had little or no effect on Ia binding. Only 9 out of 55 (16%) had a significant (more than threefold reduction, $P<0.01$) effect, and only seven (13%) led to a more than fivefold reduction in $IA^d$ binding capacity. The most dramatic effects were seen at residues $V_{327}$ and $A_{333}$ (two and three, respectively, of five substitutions led to a greater than tenfold reduction in binding). These results are in agreement with those obtained with the truncated analogs, and point at $V_{327}$ and $A_{332}$ as being the most critical residues. A significant but less dramatic effect was detected at $H_{328}$ and $E_{333}$. Although we have interpreted the decrease in $IA^d$ binding capacity at these positions as caused by alterations or deletions of contact residues, we cannot exclude that the decreased binding could be secondary to a change in peptide conformation caused by the substitutions [e.g., a significant tenfold reduction in binding activity was apparent when $A_{330}$ was replaced by a G residue; however, since glycine is known to have a strong influence on peptide conformation (Chou and Fasman 1978), and this is the only substitution of those tested at this position showing

**Table 2.** The effect of single amino acid substitutions in Ova 325–335 on $IA^d$ binding and T cell activation

| Residue | $IA^d$ binding | T cell recognition | |
|---|---|---|---|
| | | 3DO-54.8 | 8DO-51.15 |
| $Q_{325}$ | – | – | ++ |
| $A_{326}$ | – | – | + |
| $V_{327}$ | ++ | – | – |
| $H_{328}$ | + | ++ | + |
| $A_{329}$ | – | ++ | ++ |
| $A_{330}$ | ? | + | ++ |
| $H_{331}$ | – | ++ | ++ |
| $A_{332}$ | ++ | – | + |
| $E_{333}$ | + | ++ | – |
| $I_{334}$ | – | + | – |
| $N_{335}$ | – | – | – |

a significant effect on $IA^d$ binding, we propose that this effect may be indirect and mediated by a secondary effect on the peptide structure]. A summary of the relative involvement of each residue is presented in Table 2. The complete data on the effect of each substitution on $IA^d$ binding and T cell activation have been presented elsewhere (Shimonkevitz et al. 1984; Sette et al. 1987).

### *Effect of Single Amino Acid Substitutions on the T Cell Activation Capacity of Ova 323–336*

The single substituted analogs of Ova 323–336 were then tested for their capacity to stimulate two Ova 323–339 specific T cell hybridomas, 3DO-54.8 and 8DO-51.15. The data summarized in Table 2 demonstrate that T cell activation is far more sensitive to single amino acid substitutions than $IA^d$ binding. Sixty-two percent of the substitutions led to a more than tenfold decrease in T cell activating capacity and 24% resulted in a complete loss of any stimulatory activity.

When the effects of the five single amino acid substitutions made at each position were analyzed in detail, different patterns emerged for the two T cell hybrids. The activation of 3DO-54.8 cells was affected only marginally by substitutions at residues $Q_{325}$, $A_{326}$, and $N_{335}$, while it was severely affected by any substitution made at residues $H_{328}$, $A_{329}$, $H_{331}$, and $E_{333}$. Somewhat intermediate effects were apparent at positions $A_{330}$ and $I_{334}$, where only some, usually conservative, substitutions were permissible. Finally, the effects of substitutions at residues $V_{327}$ and $A_{332}$ parallelled the effects observed on $IA^d$ binding.

A similar analysis of the activation requirements of 8DO-51.15 indicated that residues $E_{333}$, $I_{334}$, and $N_{335}$ were only marginally involved in recognition by this T cell hybrid, while $Q_{325}$, $A_{329}$, $A_{330}$, and $H_{331}$ were identified as crucial residues, in the sense that none of the five substitutions tested at each of these positions were found to be permissible. Substitutions at residues $A_{326}$, $H_{328}$, and $A_{332}$ were tolerated in some, but not other, instances.

When the effects on T cell activation of the 55 single amino acid substitutions tested were compared with the effects on $IA^d$ binding, a complex pattern became apparent. The effects on T cell activation of substitutions at position $V_{327}$ were readily accounted for by the decrease in binding capacity detected in the $IA^d$ binding assay. At position $A_{332}$, T cell recognition was impaired by the same substitutions that affected $IA^d$ binding. However, one substitution (A→V) which had bo effect on either $IA^d$ binding or 3DO-54.8 activation was found to affect recognition by 8DO-51.15. The only substitution affecting Ia binding at position $H_{328}$ was E→H; three of the four remaining substitutions affected the stimulatory capacity for 8DO-51.15, and all four affected 3DO-54.8. Likewise, the only substitution showing a significant effect on $IA^d$ binding at the residue $E_{333}$ was E→I, while five out of five substitutions affected 3DO-54.8 recognition. Thus, the data suggest that three out of the four residues involved in $IA^d$ binding were also involved in T cell receptor recognition. Furthermore, activation of either 3DO-54.8 or 8DO-51.15 T cells was also affected by substitutions at six out of the seven amino acid positions tested that apparently were not involved in $IA^d$ binding. For each of the residues examined (with the exception of $A_{329}$ and $H_{331}$) at least one of the substitutions that greatly affected

(>1000-fold) the activation of one of the T cell hybrids had either no effect or a much less drastic effects (10- to 100-fold) on the stimulation of the other hybrid, indicating that for these substitutions the reduced T cell stimulatory capacity was not caused by gross alterations of peptide conformation.

### *Structural Similarity of Good IA$^d$ Binders Derived from Unrelated Protein Antigens*

Work from our (Buus et al. 1987a) and other (Guillet et al. 1987) laboratories has suggested that each Ia molecule may possess a single peptide-binding site. A logical prediction based on this finding is that IA$^d$ binders derived from unrelated protein antigens contain some common structural characteristic dictating their interacting capacity. Using the method described by Pablan (1977) to evaluate structural similarities, the sequences of the good IA$^d$ binders identified so far were examined for structural similarities with the core region Ova 327–332.

The best-fit alignments between the core region Ova 327–332 and the other good IA$^d$ binders are shown in Table 3. A common structural motif seems to emerge from the present set of alignments. Position 1 appears to be the most preserved, being occupied by either V or I residues in all the six peptides. The next position is occupied by either a threonine or a basic residue, followed by two hydrophobic residues. Position 5 is less preserved, but tends to be occupied by polar or charged amino acids. Finally, position 6 is occupied either by A or by S or by G.

To provide some experimental support for this analysis, we analyzed two series of truncated peptides derived from influenza hemagglutinin (Ha) and sperm whale myoglobin (Myo). The results obtained indicate that the core regions Myo 102–117 and Ha 135–140 predicted on the basis of the best-fit alignments are indeed correct (Sette et al., in preparation). Overall, the homologies and similarities identified by this kind of analysis of unrelated good IA$^d$ binders support the observation made with the single substituted Ova 323–336 analogs, namely that peptide–Ia interaction is very permissive and that many variations on a common structural motif are compatible with good binding. This is exactly the type of specificity that would have been predicted on the basis of the determinant selection hypothesis (Buus et al. 1987b; Rosenthal 1978).

**Table 3.** Best-fit alignments of unrelated peptides with good I-A$^d$ binding capacity[a]

| | Residue number | | | | | |
|---|---|---|---|---|---|---|
| | 1 | 2 | 3 | 4 | 5 | 6 |
| Ova 327–332 | V | H | A | A | H | A |
| Ha 135–141 | V | T | A | A | C | S |
| Myo 66–72 | V | T | V | L | T | A |
| Myo 112–118 | I | H | V | L | H | S |
| Nase 104–110 | V | R | Q | G | L | A |
| Nase 13–20 | I | K | A | I | D | G |

[a] Best-fit alignments between Ova 327–332 and sequences of unrelated peptides which are good I-A$^d$ binders were generated as described in the Materials and Methods section

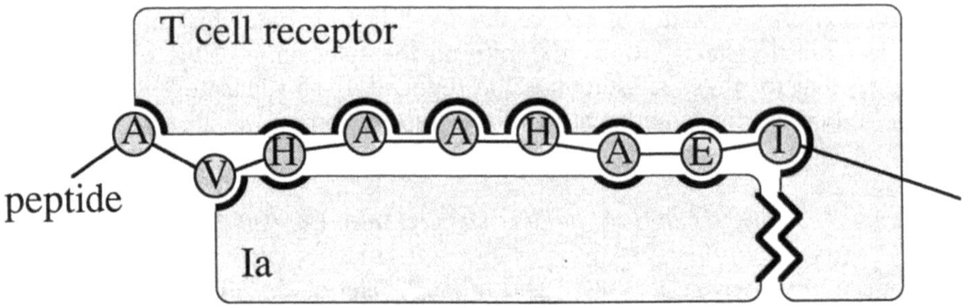


**Fig. 1.** Model of Ova peptide, Ia, and T cell receptor interaction. Putative contact sites with antigen are indicated by *bolded concavities*. The anigen molecule is shown sandwiched in an extended conformation, allowing shared recognition of $H_{328}$, $A_{332}$, and $E_{333}$. All the other residues are shown as either pure T cell or pure Ia-interacting residues. T cell recognition of Ia residues (Ronchese et al. 1987) is also indicated

## Conclusions

The set of experiments summarized in the present report analyze the structure of a previously identified T cell determinant contained in the $IA^d$-restricted ovalbumin peptide (Ova 323–339). Using N- and C-terminal truncated analogs we have defined a core region of six amino acids as critical for $IA^d$ binding (Ova 327–332). We have also been able to show that a number of unrelated peptides that are good $IA^d$ binders contain peptide regions structurally similar to the core region Ova 327–332. We then examined the effect on $IA^d$ binding and T cell activation of a large collection of single substitution analogs of Ova 323–336. The results obtained suggest that as few as two residues may be strongly involved in $IA^d$ binding, while two residues appear to be implicated to a lesser extent. In fact, the interaction Ia–antigen seems to be very permissive, since more than 85% of the substitutions tested appeared to have little or no effect on the binding of Ova 323–336 to purified $IA^d$ molecules. These findings are in keeping with the determinant selection hypothesis and suggest a possible mechanism by which Ia molecules could bind many unrelated peptides.

Furthermore, the set of data described above can be used in an attempt to derive a model for the interaction between antigen, Ia molecules, and T cell receptors (Fig. 1). It has been suggested that immunogenic (DeLisi and Berzofsky 1985; Pincus et al. 1983) peptides may form amphipathic $\alpha$-helical structures with one surface interacting with Ia and another surface with the T cell receptor. This model appears to be inconsistent with our data since T cell and Ia residues do not segregate when the peptide is modeled into an $\alpha$-helix. This lack of segregation also follows from the fact that 9 of the 11 residues studies appeared to be involved in T cell recognition, and finally from the fact that, as discussed above, three of the four possible $IA^d$-interacting residues are also involved in T cell recognition. We suggest that the most economical model to explain the data consists of a trimolecular complex in which the peptide molecule is "sandwiched" between the Ia molecule and the T cell receptor in a two-dimensional planar conformation.

## References

Allen P, Babbitt B, Unanue E (1987) T cell recognition of lysozyme: The biochemical basis of presentation. Immunol Rev 98: 171
Babbitt BP, Matsueda G, Haber E, Unanue E, Allen PM (1985) Binding of immunogenic peptides to Ia histocompatibility molecules. Nature 317: 359
Buus S, Sette A, Colon S, Jenis D, Grey H (1986) Isolation and characterization of antigen–Ia complexes involved in T cell recognition. Cell 47: 1071
Buus S, Sette A, Colon S, Miles C, Grey H (1987a) The relationship between major histocompatibility complex (MHC) restriction and the capacity of Ia to bind immunogenic peptides. Science 235: 1353
Buus S, Sette A, Grey H (1987b) The interaction between protein derived immunogenic peptides and Ia. Immunol Rev 98: 116
Chou P, Fasman G (1978) Empirical predictions on protein conformation. Am Rev Biochem 47: 251
DeLisi C, Berzofsky J (1985) T cell sites tend to be amphipathic structures. Proc Natl Acad Sci USA 82: 7048
Grantham R (1974) Amino acid difference formula to help explain protein evolution. Science 185: 862
Greenwood F, Hunter W, Glover J (1963) The preparation of $^{131}$I-labeled human growth hormone of high specific radioactivity. Biochem J 89: 114
Grey HM, Chesnut R (1985) Antigen processing and presentation to T cells. Immunol Today 6: 101
Guillet JG, Lai MZ, Briner TJ, Buus S, Sette A, Grey HM, Smith JA, Gefter ML (1987) Immunological self-non self discrimination. Science 235: 865
Houghten RA (1985) General method for the rapid solid-phase synthesis of large numbers of peptides: Specificity of antigen–antibody interaction at the level of individual amino acids. Proc Natl Acad Sci USA 82: 5131
Padlan EA (1977) Structural implications of sequence variability in immunoglobulins. Proc Natl Acad Sci USA 74: 2551
Pincus MR, Gerewitz F, Schwartz R, Scheraga H (1983) Correlation between the conformation of cytochrome C peptides and their stimulatory activity in a T-lymphocyte proliferation assay. Proc Natl Acad Sci USA 80: 3297
Ronchese F, Schwartz R, Germain R (1987) Functionally distinct subsites on a class II major histocompatibility complex molecule. Nature 329: 254
Rosenthal AS (1978) Determinant selection and macrophage function in genetic control of immune responsiveness. Immunol Rev 40: 136
Sette A, Buus S, Colon S, Smith J, Miles C, Grey H (1987) Structural characteristics of an antigen required for its interaction with Ia and recognition by T cells. Nature 328: 395
Shimonkevitz R, Colon S, Kappler JW, Marrack P, Grey H (1984) Antigen recognition by H-2 restricted T cells. II. A tryptic ovalbumin peptide that substitutes for processed antigen. J Immunol 133: 2067
Watts TH, McConnell H (1987) Biophysical aspects of antigen recognition by T cells. Am Rev Immunol 5: 461

# Analysis of Specificities Recognized by Monoclonal Antibodies and Cytotoxic T-Lymphocytes Using Chimeric H-2 Molecules

J. F. ABASTADO[1], J. MARYANSKI[2], C. JAULIN[1], and P. KOURILSKY[1]

[1] Unité de Biologie Moléculaire du Gène, U277 de l'Institut National de la Santé et de la Recherche Médicale, UA 535 du Centre National de la Recherche Scientifique, Dpt. d'Immunologie, Institut Pasteur, 25 rue du Dr Roux, 75724 Paris, France

[2] Ludwig Institute for Cancer Research, Lausanne Branch, Chemin des Boveresses, 1066 Epalinges, Switzerland

In order to develop structure-function studies on H-2 molecules, several groups (reviewed by Forman 1987) have taken advantage of the fact that the different domains of H-2 are encoded by separate exons. Chimeric molecules have been built in which a given domain of an H-2 antigen has been replaced by the corresponding domain of another H-2 antigen; this approach, dubbed "exon shuffling," has permitted the assignment to a particular domain of the specificities recognized by monoclonal antibodies (McAbs). However, for the specificities recognized by alloreactive or syngeneic cytotoxic T-lymphocytes (CTLs) this approach has yielded rather limited information: whereas in most cases it could be shown that the speci-

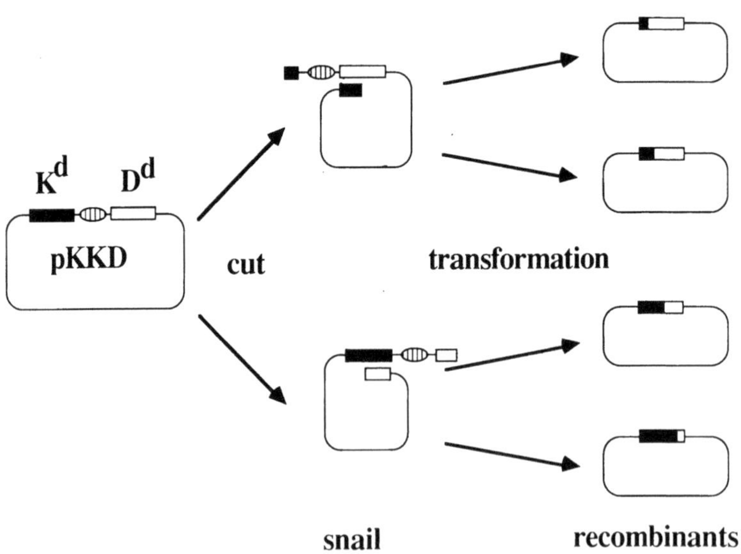


**Fig. 1.** The snail technique. The plasmid pKKD is linearized within the $K^d$ *(top)* or the $D^d$ *(bottom)* sequence and introduced into *E. coli* by transformation. Recircularization occurs by homologous recombination between the two H-2 sequences. Recombinants, which usually represent 80%–90% of the transformants, are identified by the fact that they are kanamycin sensitive

ficity of a given CTL is carried by the first two external domains, shuffling of these two domains usually destroyed the recognition by CTLs.

In order to map more precisely the specificities recognized by McAbs and CTLs, we have developed a new technique to produce a large number of intradomain recombinants between two H-2 antigens (Abastado et al. 1987a). This technique is dubbed the "snail technique" and is depicted in Fig. 1. It is derived from an in vivo recombination technique originally developed by Weber and Weissman. We have cloned on the same plasmid and in the same orientation, the two parental H-2 sequences; these sequences correspond to a full-length cDNA encoding H-2K$^d$ and an incomplete cDNA encoding most of H-2D$^d$. In between, we have inserted a bacterial marker *Km$^r$*, which confers the resistance to kanamycin. The plasmid is linearized with a restriction endonuclease and introduced by transformation into *E. coli.* We have shown that when the double-stranded break is introduced within one of the H-2 sequences, it stimulates the recircularization of the plasmid by homologous recombination between the H-2 sequences so that plasmids carrying a chimeric sequence are obtained. Such plasmids are easily identified as they are kanamycin sensitive. Recombination does not occur at the site of cleavage but at a certain distance from it. In fact the recombination frequency decreases gradually with the distance from the site of cleavage. By varying the site of cleavage, one can change the distribution of the recombination events. Fifty nanograms of plasmid is generally sufficient to produce 500–1000 independent recombinants which are then rapidly analyzed by in situ colony hybridization using oligonucleotide probes specific for H-2K$^d$ or H-2D$^d$.

We have determined the DNA sequence of 22 such recombinants, of which 11 were found to be different. These 11 different recombinants were subcloned in an expression vector carrying the early promoter of the SV40 virus so that the chimeric

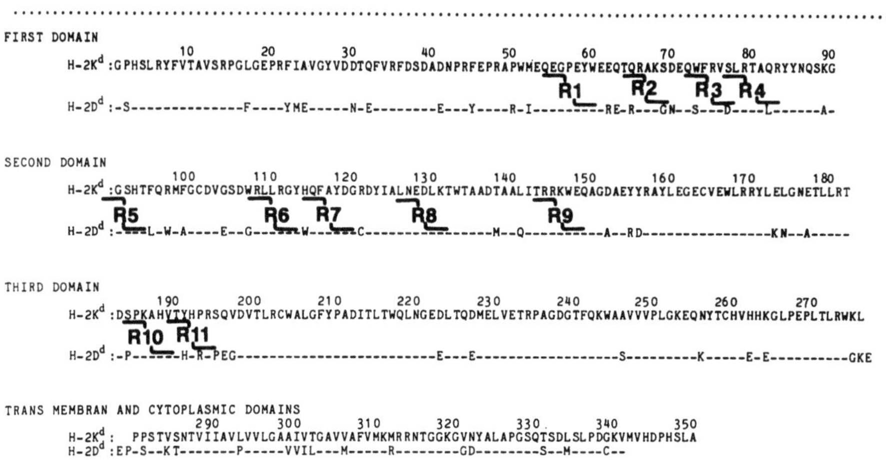


**Fig. 2.** Sequences of the K$^d$/D$^d$ chimeras. The amino acid sequences of the 11 chimeras (R1–R11) are deduced from their nucleotide sequences. The N-terminal part is identical to K$^d$ and the C-terminal part to D$^d$

H-2 sequence could be transcribed from this promoter. These new constructs were then transfected into mouse L cells (Abastado et al. 1987b). All were expressed at the cell surface so that we now have available 11 cell lines expressing chimeric antigen of which the N-terminal part is derived from $K^d$ and the C-terminal part from $D^d$, the part from $K^d$ and that from $D^d$ differing from one recombinant to another (Fig. 2).

We have used these cell lines to characterize McABs specific for $K^d$ or $D^d$. As an example, the pattern of fluorescence of these cell lines with the $D^d$-specific TO112 McAb analyzed with a cytofluorometer is shown in Fig. 3. This McAb recognizes R1–R5 but not R6–R11. This means that the region defined by the recombination points of R5 and R6 is essential for the recognition. By comparing the sequences of the recombinants (Fig. 2), one can see that recombinants that have Leu(L), Trp(W), Ala(A), Glu(E), and Gly(G) residues at positions 95, 97, 99, 104, and 107 respectively are recognized by McAb TO112 and those that do not have these residues are not. By means of this approach, we have analyzed 59 McAbs (Table 1). At least five regions important for the recognition by these McAbs have been identified. These regions are defined by the recombination points of R1 and R2, R4 and R5, R5 and

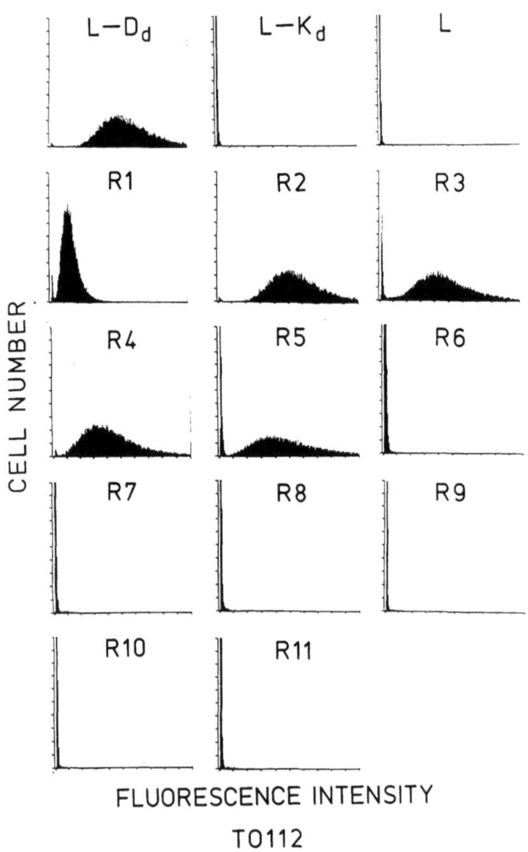


**Fig. 3.** Recognition of the $K^d/D^d$ chimeras by McAb TO112. Untransfected L cells or L cells transfected with H-2$K^d$, H-2$D^d$, or a chimeric $K^d/D^d$ sequence (R1–R11) were stained with McAb TO112. Staining was then analyzed by flow cytofluorometry. *Abscissa*, relative fluorescence intensity (linear amplification); *ordinate*, number of cells

**Table 1.** Tentative assignment of polymorphic residues in $K^d$ and $D^d$ recognized by specific McAbs

| Monoclonal antibodies[a] | Speci-ficity | Chimeras recognized | Chimeras not recognized | Accessible polymorphic residues[b] | Other polymorphic residues |
|---|---|---|---|---|---|
| TO107, TO111, TO116, TO119, TO120, TO128, TO129, TO131, TO133, TO137, TO138, TO139, TO142, TO143, 66-2-4 | $D^d$ | None | R1–R11 | | |
| TO124, TO127, TO132, TO134, TO136, TO141, TO144, 34-421, 27-11-13, 28-8-6, T14c | $D^d$ | R1 | R2–R11 | 62(R), 65(R) | 63(E) |
| TO112, TO113, TO114, TO115, TO117, TO121, TO123, TO126, TO130, 28-11-5, 34-420, 34-5-8, 66-13-6, H81-199-1, T148 | $D^d$ | R1–R5 | R6–R11 | 104(E), 107(G) | 95(L), 97(W), 99(A) |
| TO102, TO103, TO104, TO108, TO109, TO110, TO118, TO125, 66-4-8, 66-13-5, 66-13-19, H147 | $D^d$ | R1–R8 | R9–R11 | 138(M), 141(Q) | |
| TO105, TO106 | $D^d$ | R1–R9 | R10, R11 | 155(R), 173(K), 174(N), 177(A) | 152(A), 156(D) |
| 20-8-4, F35-119-18 | $K^d$ | R5–R11 | R1–R4 | 82(Q), 89(K) | |
| 31-3-4 | $K^d$ | R10, R11 | R1–R9 | 155(Y), 173(E), 174(L), 177(E) | 152(D), 156(Y) |

[a] The McAbs here characterized have been described in Abastado et al. (1987 b)
[b] The polymorphic residues are considered as accessible accoring to Bjorkman et al. (1987); see discussion in text

R6, R8 and R9, and R10 and R11 respectively. They contain, in most cases, a small number of polymorphic residues involved in the recognition by the MyAbs.

Bjorkman et al. (1987) recently described the three-dimensional structure of HLA-A2 determined by X-ray diffraction. Taking into account the high homology between HLA and H-2 sequences, we have assumed, as a working hypothesis, that the three-dimensional structure of an H-2 molecule is not very different and have tried to interpret our data within this framework. All five regions shown to contain residues important for the binding of an McAb contain polymorphic residues accessible to an antibody molecule (Table 1). The residues are located either at the periphery of the molecule (e.g., residues 89, 104, 107, 138, 141, 173, 174, and 177) or within the so-called foreign antigen binding site, with a pointing up side chain (e.g., residues 62, 65 and 155); these accessible residues are the best candidates for the recognition by the McAb. Alternatively, one cannot rule out the possibility that a non-

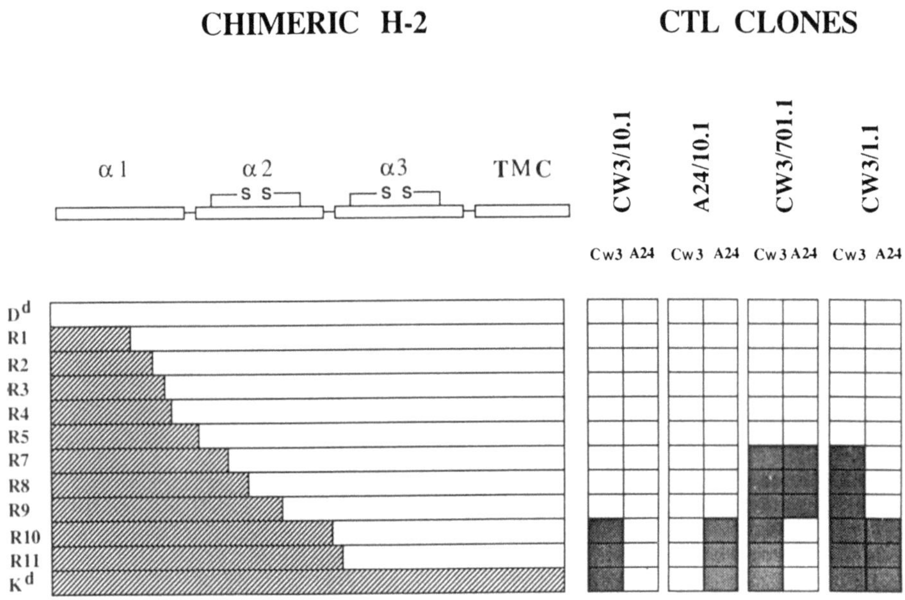


**Fig. 4.** Recognition of the Cw3- and A24-derived peptides by $K^d$-restricted CTL clones on $K^d/D^d$ chimeras. *Left:* Schematic drawing of the parental and chimeric H-2 molecules. The *hatched region* corresponds to the $K^d$-derived portion of the chimera. *Right:* Lysis of the different cell lines in the presence of the Cw3- or A24-derived peptides (residues 170–182) by four CTL clones; *darkened areas* correspond to lysis

accessible polymorphic residue may induce a conformational effect on the histocompatibility molecule. It is also possible that a nonaccessible residue located in the binding site (e.g., residues 63, 152, and 156) and even at the bottom of the groove which constitutes the binding site (residues 95, 97, and 99) may directly interact with the McAb once this has displaced the bound peptide(s) which seems to occupy the site of most or all class I molecules. We are presently constructing new chimeric molecules which should allow us to discriminate between these hypotheses.

To study the specificities recognized by CTLs, we analyzed the capacity of the 11 cell lines expressing chimeric $K^d/D^d$ molecules to present HLA-derived peptides to specific $K^d$-restricted CTL clones (Maryanski et al. 1987). These clones were previously shown to lyse L cells transfected with $K^d$ in the presence of a peptide derived from HLA-A24 or HLA-Cw3 but not untransfected or $D^d$ transfected cells (Maryanski et al. 1986). The results are shown in Fig. 4. At least five of the chimeric H-2 molecules (R7–R11) were shown to be functional, in that they are capable of presenting the specific peptide to the corresponding CTL clone, sometimes very efficiently.

These approaches identify two regions in $K^d$ which are important for the specific recognition by the CTL receptor. These regions are defined by the recombination points of R5 and R7 and those of R9 and R10. Of course, other regions may be identified by studying other CTL clones. According to Bjorkman et al. (1987), the polymorphic residues thus identified are located in the "foreign antigen binding site," either in the large $\alpha$-helix of the second domain or at the bottom of the groove constituted by the first $\beta$-strand of the second domain. No region in the first domain has been identified but this is probably due to the fact that all chimeras tested have a $K^d$-derived N-terminal part and a $D^d$-derived C-terminal part, while all CTL clones are $K^d$ restricted. The construction of reciprocal recombinants ($D^d/K^d$) to test this point is in progress.

In two instances, our results seem to indicate that the T cell recognizes the association of the H-2 molecule and the peptide: Cw3/701.1 recognizes the Cw3 peptide on R9 and R10 and the A24 peptide on R9 but not R10; similarly, Cw3/1.1 recognizes the Cw3 peptide on R9 and R10 and the A24 peptide on R10 but not on R9 (see Fig. 4). This is not compatible with a model in which the restricting element and the foreign antigen would be recognized independently (two or one-and-a-half receptor models, see Pernis and Axel 1985). Our results are compatible with a model in which the restricting element and the peptide would be corecognized by a single site of a single receptor carried by the T cell.

Our data show that the presentation of the peptide by the chimeric H-2 molecule is not sufficient for the recognition by certain CTL clones. For example, the peptide Cw3 is recognized on R7, R8, or R9 by Cw3/701.1 and Cw3/1.1 (Fig. 4), which probably indicates that these chimeras are capable of presenting this peptide; however, the same peptide is not recognized on these H-2 chimeric molecules by Cw3/10.1, which is nevertheless specific for the Cw3 peptide on R10, R11, or $K^d$. We are therefore left with two possibilities: either the T cell receptor recognizes the peptide plus certain polymorphic residues of the MHC presenting molecule or it recognizes a conformation of the peptide induced or stabilized by the restricting element.

We are presently building, by the snail technique, new sets of recombinants. In one set the N-terminal part of the chimeric molecules is now derived from $D^d$ and the C-terminal part from $K^d$, while in another set, on a $D^d$ background, a small number of residues are derived from $K^d$. The accumulation of data using these new chimeric molecules should allow us to characterize the interactions between the H-2 molecule and the receptor of a syngeneic or alloreactive CTL and to pinpoint the residues involved. Interpretation of these data in terms of the three-dimensional structure of a histocompatibility molecule should help in clarifying these points and understanding the molecular basis of MHC restriction.

## References

Abastado JP, Darche S, Godeau F, Cami B, Kourilsky P (1987a) Intramolecular recombination between partially homologous sequences in E. coli and Xenopus laevis oocytes. Proc Natl Acad Sci USA 84: 6496–6500

Abastado JP, Jaulin C, Schutze MP, Langlade-Demoyen P, Plata F, Ozato K, Kourilsky P (1987b) Fine mapping of epitopes by intradomain $K^d/D^d$ recombinants. J Exp Med 166: 327–340

Bjorkman PJ, Saper MA, Samraoui B, Bennett WS, Strominger JL, Wiley DC (1987) The foreign antigen binding site and T cell recognition regions of class I histocompatibility antigens. Nature 329: 512–518

Forman J (1987) Determinants on major histocompatibility complex class I molecules recognized by cytotoxic T lymphocytes. Adv Immunol 41: 135–179

Maryanski JL, Pala P, Corradin G, Jordan BR, Cerottini JC (1986) H-2-restricted cytologic T cells specific for HLA can recognize a synthetic HLA peptide. Nature 324: 578–579
Maryanski J, Abastado JP, Kourilsky P (1987) Specificity of peptide presentation by a set of hybrid mouse class I MHC molecules. Nature 330: 660–662
Pernis B, Axel R (1985) A one and a half receptor model for MHC-restricted antigen recognition by T lymphocytes. Cell 41: 13–16
Weber H, Weissmann C (1983) Formation of genes coding for hybrid proteins by recombination between related, cloned genes in E. coli. Nucleic Acids Res 11: 5661–5669

# The Conformational Flexibility and $\beta_2$-Microglobulin Interaction Dynamics of the H-2K$^b$ Heavy Chain

BRIAN H. BARBER[1], MARGARET H. SMITH[1], HAMISH ALLEN[2], and DAVID B. WILLIAMS[3]

Departments of Immunology[1] and Biochemistry[3], Medical Sciences Building, University of Toronto, Toronto, Canada M5S 1A8 and Biogen Research Corp.[2], Cambridge, MA 02142, USA

## Introduction

The expressed gene products of the class I major histocompatibility complex (MHC) are now recognized as key elements in the mechanism used by cytotoxic T-lymphocytes (CTLs) to detect virus-infected cells (Doherty and Bennick 1981). Although it is widely accepted that CTL antigen receptors can only recognize viral antigens when they are "seen" in association with self class I molecules (i.e., the phenomenon of MHC-restricted recognition; Zinkernagel and Doherty 1979), it has not yet been established how, in molecular terms, the class I molecules "present" viral determinants to CTL receptors. In fact, it is not even clear where the self and nonself components of the target structure intersect within an infected cell. Particularly useful in this regard would be a better understanding of the class I heavy chain dynamics within the cell: the nature of its association with $\beta_2$-microglobulin ($\beta_2$m) postbiosynthesis, the possibility for subsequent dissociation and reassociation with $\beta_2$m or other molecules, as well as the conformational perturbation accompanying the different subunit interactions. Defining the spectrum of interaction sites and conformational states available to class I heavy chains should contribute to establishing which of these are most important with respect to its antigen presentation function.

In this report, we describe the use of rabbit antisera specific for synthetic peptides corresponding to different subregions of the intracytoplasmic portion of the H-2K$^b$ molecule. Using these reagents it has been possible to monitor the presence in different subcellular locations of both the $\beta_2$m-"bound" heavy chains (i.e., the conventional class I dimer) and "free" (i.e., non-$\beta_2$m-associated) heavy chains. The conformational differences between free and bound heavy chains, as well as the role of $\beta_2$m in the transport of heavy chains to the cell surface, are discussed in the context of the functional properties of the class I gene products.

## Materials and Methods

### *Cell Lines*

EL4 (an H-2$^b$ T cell lymphoma) and R1E (an H-2$^k$ thymoma) were obtained from the American Type Culture Collection (Rockville, Maryland) and cultured at 37 °C in 10%

$CO_2$ in RPMI 1640 supplemented with 2–5 mM glutamine, 100 units/ml penicillin, 100 µg/ml streptomycin, and 10% (v/v) fetal calf serum. The R1E cell line (known to be lacking a functional $\beta_2$m gene) was first transfected with a genomic clone of H-$2K^b$ (Allen et al. 1984), and then subsequently transfected with the murine $\beta_2$m gene (Parnes and Seidman 1982), using procedures outlined previously for H-$2D^b$ (Allen et al. 1986).

### *Radiolabeling*

In order to detect cell surface-associated molecules, cells were labeled via lactoperoxidase-catalyzed iodination using $Na^{125}I$ as previously described (Smith et al. 1986). Biosynthetic labeling was carried out using a 10-min exposure of cells at 37 °C to methionine-free medium containing 0.5 mCi/ml $^{35}$S-methionine, followed by a 30-min "chase" with complete medium containing 1 mM methionine and 10% fetal calf serum.

### *Immunoprecipitation*

Subsequent to radiolabeling, cells were solubilized in 0.5% Nonidet P-40 containing 1% aprotinin as a protease inhibitor. Lysates were clarified by high-speed centrifugation and then precleared for nonspecific reactivity by incubation with either normal mouse serum or normal rabbit serum and reaction with a 10% (w/v) suspension of fixed *Staphylococcus aureus.* In some instances these lysates were reacted directly with test antibodies, while in other cases a glycoprotein fraction was prepared by adsorption and elution (with 0.5 M $\alpha$-methyl-mannoside) from columns of lentil lectin (LcH)-Sepharose. The rabbit antipeptide antibodies used to immunoprecipitate class I H-2 antigens are those specific for peptides corresponding to exons 6, 7, and 8 of the H-$2K^b$ molecule (Weiss et al. 1983). Their preparation and characterization have been described elsewhere (Smith et al. 1986). The mouse alloantiserum E-k.b (anti-H-$2K^bD^b$) was obtained from the National Institutes of Health (US) collection, and the monoclonal antibody 20-8-4S (anti-H-$2K^bD^b$) from the American Type Culture Collection.

Immunoprecipitates were analyzed by one-dimensional SDS-PAGE using the Tris-glycine buffer system of Laemmli as previously described. (Allore et al. 1985). Bands were detected using low temperature (−70 °C) autoradiography by exposure of gels dried onto paper to Kodak XAR X-ray film in the presence of Quanta III (Dupont) enhancing screens.

In certain cases, immunoprecipitates were incubated with endoglycosidase H (Endo H, Miles Scientific) prior to electrophoresis. When indicated, immunoprecipitates were solubilized in 1% SDS, precipitated with acetone, and resuspended in an Endo H digestion buffer as previously described (Williams et al. 1985). Half of the sample was reacted with 1.5 mU Endo H, and the other half served as the undigested control.

*Triton X-114 Fractionation of Antigen Prior to Immunoprecipitation*

A modified form of the Triton X-114 (TX-114) partitioning procedure introduced by Bordier (1981) was used to analyze the solubility properties of different class I antigen preparations. Very briefly, cells were lysed on ice in 10 mM, 0.15 M NaCl, pH 7.4 containing 1% aprotinin and 1% TX-114 (Sigma). The lysates were then shifted from 4 °C to 37 °C and after 3 min the detergent-rich ("pellet") phase was separated by centrifugation from the detergent-depleted ("supernatant") phase. The supernatant phase was reextracted twice more by the low temperature addition of further TX-114 and repeating the above procedure. The pellet and supernatant fractions were then resuspended in equal volumes and subjected to immunoprecipitation analysis as indicated above.

## Results

### *1. Cell Surface Glycoproteins from H-$2^b$ Cells Contain a Significant Fraction of H-$2K^b$ Heavy Chains Not Associated with $\beta_2$-Microglobulin*

EL4 tumor cells were labeled by lactoperoxidase-catalyzed iodination and then solubilized with 0.5% Nonidet P-40 (NP-40). Glycoproteins were then purified from this detergent soluble fraction by LcH-Sepharose affinity chromatography and challenged with a panel of different serological reagents specific for class I H-2 antigens. These included mouse alloantisera and monoclonal antibodies specific for determinants on H-$2K^b$ and H-$2D^b$, and the rabbit antipeptide sera raised against synthetic peptides corresponding to exons 6, 7, and 8 of the H-$2K^b$ gene (respectively referred to as antipeptide 6, antipeptide 7, and antipeptide 8). These rabbit antipeptide antisera have been shown to react with specific denatured murine class I heavy chains by Western blotting, and they also immunoprecipitate class I heavy chains from whole cell lysates solubilized in nondenaturing detergents such as NP-40 (Smith et al. 1986). Upon comparing the reactivity profiles of different anti-class I antibodies with the recognition properties of the antipeptide antisera, an unexpected observation was made. When the surface – labeled glycoprotein fraction prepared from EL4 cells was precleared for reactivity with the alloantiserum E-k.b (anti-H-$2K^bD^b$), it was found that antipeptide 8 was still able to react with a significant population of class I heavy chains (Fig. 1). This result was even more surprising when it was observed that the antipeptide 8 immunoprecipitated heavy chains lacked any associated $\beta_2$m. Apparently these "free" class I heavy chains do not possess the H-$2K^b$ or H-$2D^b$ serological determinants seen by the E-k.b alloantiserum. The same result was observed when the glycoproteins were precleared with a rabbit anti-mouse $\beta_2$m antiserum (Smith and Barber, in preparation). In each case, by excising and counting the immunoprecipitated class I heavy chain bands before and after preclearance, the free heavy chain was found to represent approximately 30% of the total heavy chain population. It was also observed that free heavy chains were immunoprecipitated from anti-$\beta_2$m precleared glycoprotein preparations by either antipeptide 6 or antipeptide 7 antisera in a manner analogous to that with antipep-

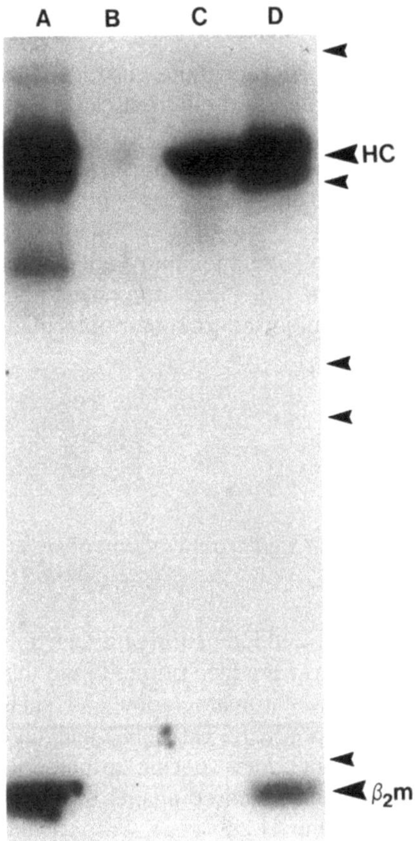


**Fig. 1.** An autoradiogram of the SDS-PAGE patterns for immunoprecipitates formed by reaction of EL4-derived, cell surface-iodinated LcH-purified glycoproteins with *(A)* the alloantiserum E-k.b (anti-H-$2K^bD^b$), *(B)* the reaction with E-k.b after two previous rounds of preclearance with the same alloantiserum, *(C)* the rabbit antipeptide 8 antiserum after two rounds of preclearing with E-k.b, and *(D)* the monoclonal antibody 20-8-4S (anti-H-$2K^bD^b$) without any preclearance. The molecular weight markers indicated by the *smaller arrowheads* to the right of the gel are in descending order 68K, 42K, 29K, 24K, and 15K. The *larger arrowheads* indicate the position of the class I heavy chain *(HC)* and $\beta_2$m

tide 8. A comparison of the two-dimensional gel electrophoresis pattern of H-$2K^b$ heavy chain immunoprecipitated directly with $K^b$ specific monoclonal antibodies, alloantisera, or anti-$\beta_2$m, and that precipitated after preclearance with any of these reagents by antipeptide 8, indicated no detectable difference in their spot patterns (Smith and Barber, in preparation). This result suggests that the difference in the reactivity of $\beta_2$m-bound versus free H-$2K^b$ heavy chains is due to conformational rather than chemical differences. It should also be noted that the same results were observed with LcH glycoprotein preparations from surface labeled C57BL/6 spleen cells (Smith and Barber, in preparation), indicating that this phenomenon is not simply a characteristic of the EL4 tumor cell line.

### *2. Antipeptide 8 Antibodies React with Both $\beta_2$-Microglobulin-Bound and Free Class I Heavy Chains, Whereas Antipeptide 6 and Antipeptide 7 Recognize Only Free Heavy Chains*

In comparing the reactivity profiles of the antipeptide antisera with those of more conventional class I H-2 reagents, a further interesting feature was detected. When

**Fig. 2.** An autoradiogram of the SDS-PAGE pattern for immunoprecipitation of surface-labeled LcH-purified EL4 glycoproteins reacted with *(A)* rabbit antipeptide 6, *(B)* rabbit antipeptide 7, *(C)* rabbit antipeptide 8, and *(D)* the murine alloantiserum E-k.b (anti-H-2$K^bD^b$). The positions of the class I heavy chain *(HC)* and $\beta_2$m are indicated

the ability of antipeptide 6, antipeptide 7, and antipeptide 8 to immunoprecipitate class I H-2 antigens was compared with an alloantiserum such as E-k.b., or anti-$\beta_2$m, antipeptide 8 was observed to immunoprecipitate both dimeric $\beta_2$m-bound and free class I heavy chains, whereas both antipeptide 6 and antipeptide 7 reacted preferentially, if not exclusively, with free heavy chains (Fig. 2). Even on longer exposures of the antipeptide 6 and antipeptide 7 immunoprecipitates depicted in Fig. 2, a corresponding $\beta_2$m band was not detected, whereas the $\beta_2$m band in an antipeptide 8 immunoprecipitate was obvious. The ratio of labeled heavy chain to labeled $\beta_2$m was also consistently higher in the antipeptide 8 immunoprecipitates than in either the alloantiserum or anti-$\beta_2$m precipitates, reflecting the free heavy chain component in addition to the dimeric population among the antipeptide 8 precipitated molecules (Smith and Barber, in preparation). The observation that antipeptide 6 and antipeptide 7 react only with the free heavy chain, and not $\beta_2$m-containing dimers, is most likely a result of conformational rather than primary sequence differences among the population of H-2$K^b$ and H-2$D^b$ heavy chains. Although the potential for differential mRNA splicing does exist in the cytoplasmic region of class I H-2 molecules (Lew et al. 1986), synthetic oligonucleotide hybridization studies of a cDNA library from EL4 has indicated that >90% of the class I molecules contain the sequences corresponding to exons 6 and 7 (D'Agostaro and Barber, unpublished observations). These observations correspond to our previous finding (Smith et al. 1986) that heavy chains immunoprecipitated from EL4 glycoproteins by either antipeptide 6 or antipeptide 7 have two-dimensional gel spot patterns which correspond closely to that observed for E-k.b. (anti-H-2K$^b$, D$^b$). The strong similarity of the electrophoretic patterns for these heavy chains again suggests that the differential reactivity with these sera is due to the conformational properties of these molecules and not the gain or loss of particular epitopes due to chemical or sequence differences.

### *3. Few, if Any, H-2K$^b$ Heavy Chains Reach the Cell Surface in R1E Cells Lacking a Functional $\beta_2$-Microglobulin Gene*

Immunoselection was used to derive a mutant of the C58 (H-2$^k$) thymoma cell line R1.1 which no longer expressed serologically detectable class I determinants (Hyman and Stallings 1976). This mutant, designated R1E, was subsequently shown to possess a defective $\beta_2$m gene (Parnes and Seidman 1982). Taking advantage of this fact, R1E cells were transfected with the H-2K$^b$ gene only, or the H-2K$^b$ gene followed by a further transfection with the murine $\beta_2$m gene, in order to assess the role of $\beta_2$m in the cell surface expression of H-2K$^b$. Since the results described above indicated that H-2K$^b$ heavy chains lacking associated $\beta_2$m could be detected

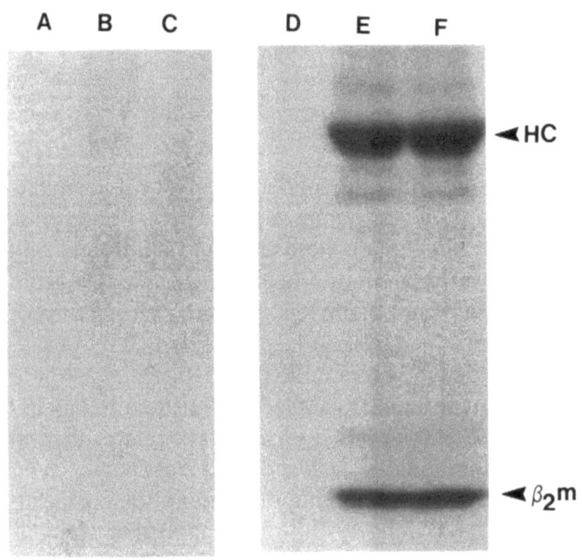


**Fig. 3.** An autoradiogram of SDS-PAGE patterns for immunoprecipitation from two types of cell. The *left panel (A–C)* corresponds to immunoprecipitates from surface-labeled R1E transfected with the H-2K$^b$ gene only, and the *right panel* to surface-labeled R1E sequentially transfected with the H-2K$^b$ and $\beta_2$m genes. NP-40 lysates of each cell population were reacted with preimmune normal rabbit serum *(A* and *D)* and rabbit antipeptide 8 *(B, C, E,* and *F)*. Immunoprecipitates in *lanes C* and *F* were digested with Endo H

among the cell surface-labeled, mature carbohydrate-bearing class I molecules, it was important to determine whether these heavy chains had reached the cell surface on their own, or were generated by subsequent dissociation from $\beta_2$m which had been essential for transit to the cell surface. The data presented in Fig. 3 indicate that the R1E cells transfected with the H-2K$^b$ gene alone express very little if any detectable H-2K$^b$ on the cell surface (as judged by lactoperoxidase-catalyzed iodination and immunoprecipitation of NP-40 lysed cells with antipeptide 8). In contrast to the R1E cells transfected with the K$^b$ only, R1E cells cotransfected with K$^b$ and $\beta_2$m genes express dimeric H-2K$^b$ which can be readily labeled by lactoperoxidase and immunoprecipitated by antipeptide 8 (Fig. 3). The lack of effect of Endo H on the immunoprecipitated heavy chain indicates that the associated carbohydrate moieties have matured from the high mannose to complex form, as expected for the cell surface-associated molecule (Kobata 1979). Cell equivalent quantities of the antipeptide 8 immunoprecipitates have been loaded on the gel in each case, making a comparison of the heavy chain band intensity between tracks B and E in Fig. 3

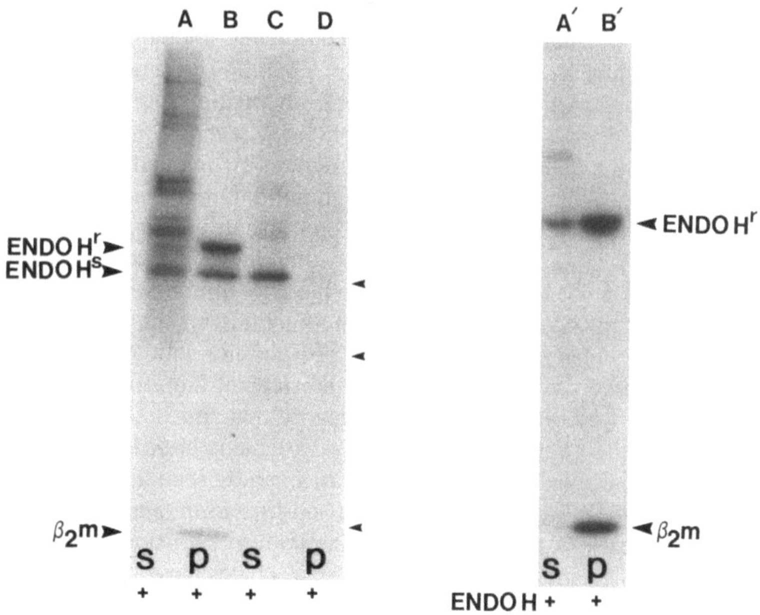


**Fig. 4.** An autoradiogram of SDS-PAGE patterns for immunoprecipitates formed after Triton X-114 fractionation of different whole cell lysates. In the *left-hand panel* are immunoprecipitates formed with antipeptide 8 for $^{35}$S-methionine-labeled R1E cells cotransfected with the H-2K$^b$ and $\beta_2$m genes *(A* and *B),* and $^{35}$S-methionine labeled R1E cells transfected with the H-2K$^b$ gene only *(C* and *D).* Antipeptide 8 immunoprecipitates prepared from the supernatant *(S)* fractions are in *lanes A* and *C,* whereas those from the pellet *(P)* fractions are in *B* and *D.* The molecular weight markers to the *right of this panel* are 46K, 30K, and 14K. In the *right-hand panel* are immunoprecipitates formed with antipeptide 8 for cell surface-iodinated R1E cells cotransfected with H-2K$^b$ and $\beta_2$m genes for both the supernatant *(track A′)* and pellet fractions *(track B′).* The migration positions of Endo H – resistant *(Endo H$^r$)* and Endo H – sensitive *(Endo H$^s$)* heavy chains are indicated, as is that for $\beta_2$m. All immunoprecipitates were reacted with Endo H before electrophoresis, as indicated below each gel track

a reasonable reflection of the differential degree of H-2K$^b$ expression on the cell surface for the single and double transfectants. The lack of cell surface detection is not due to failure of H-2K$^b$ gene expression, or an inability of the antipeptide 8 to react with the H-2K$^b$ synthesized in the absence of $\beta_2$m, because the K$^b$ transfectants, when metabolically labeled with $^{35}$S-methionine, synthesize a heavy chain band of the correct size which can be immunoprecipitated with antipeptide 8 (see Fig. 4).

### *4. Triton X-114 Partitioning Experiments Suggest Differential Conformational Properties for Free and Bound H-2K$^b$ Heavy Chains Both Intracellularly and on the Cell Surface*

The potential for a temperature-dependent phase separation of the detergent Triton X-114 (TX-114) has recently been exploited by a number of groups to study conformational alterations in membrane-associated glycoproteins (Alcaraz et al. 1984; Gething et al. 1986). The method involves solubilization of labeled cells by 1% (w/v) TX-114 at 4 °C (where the detergent is completely soluble), followed by elevation of the temperature to 37 °C, where phase separation occurs and the components in the cell lysate partition into either or both of the "supernatant" (i.e., detergent-depleted) or "pellet" (i.e., detergent-enriched) fractions. Separation of the two fractions by centrifugation allows one to examine by immunoprecipitation the components of the pellet and supernatant under different conditions. The experiment depicted in the left-hand panel of Fig. 4 represents a comparison of Endo H-treated antipeptide 8 immunoprecipitates from the single (K$^b$) and double (K$^b$ plus $\beta_2$m) transfectants pulse-labeled with $^{35}$S-methionine and chased for 30 min. A comparison of the supernatant (S) and pellet (P) fractions for the double transfectant indicates that the majority of class I molecules partition into the pellet fraction. In contrast, the heavy chains in the K$^b$-only transfectant are found predominantly in the supernatant fraction. These results suggest that the detergent solubility (and by implication the conformation) of the heavy chains in each of these two situations may be significantly different. In addition, it can be seen that the heavy chains in the single transfectant are all in the Endo H-sensitive form after 30 min of chase, consistent with a lack of progress towards the mature cell surface form, whereas in the double transfectant at least half of the heavy chains have already been processed to the more mature Endo H-resistant form. These findings are in agreement with the data in Fig. 3 indicating that the lack of functional $\beta_2$m prevented the K$^b$ heavy chain from reaching the cell surface.

The right-hand panel of Fig. 4 indicates the results which are obtained when the same TX-114 partitioning experiment is performed in surface-iodinated cells transfected with both K$^b$ and $\beta_2$m genes. In this situation, which we know permits H-2K$^b$ molecules to reach the cell surface, we again see a partitioning of the labeled heavy chains between the supernatant and pellet fractions (as revealed by antipeptide 8 immunoprecipitation). The expected dimeric form of the H-2K$^b$ molecule, with heavy chain and associated $\beta_2$m, partitions almost exclusively into the pellet fraction (as did the $^{35}$S-methionine-labeled dimer), whereas antipeptide 8 immunoprecipitation of the supernatant fraction reveals a labeled free heavy chain with only trace amounts of $\beta_2$m. Cutting and counting the bands in each track indicated that 24% of the total heavy chain was in the supernatant and 76% in the pellet. Again one sees a distinctly different solubility in the two TX-114 phases of the $\beta_2$m-bound and free H-2K$^b$ heavy chains. This fraction of biosynthetically mature (as revealed by Endo H resistance), cell surface-associated free heavy chain presumably arose by dissociation from $\beta_2$m at some point postmaturation, and may be directly related to the free class I heavy chain population detected in the surface-labeled, LcH-purified glycoprotein fraction (Fig. 1).

## Discussion

The results outlined above highlight the possibility that a non-$\beta_2$m-associated (i.e., free) heavy chain may have a previously unrecognized role to play in the conformational dynamics of class I H-2 antigens. Current dogma would suggest that newly synthesized class I H-2 heavy chains promptly associate with $\beta_2$m in the cisternae of the endoplasmic reticulum, and then remain in the dimeric form thereafter (Maloy and Coligan 1985). The findings presented in this report confirm the need for H-2K$^b$ to associate with $\beta_2$m in order to reach the cell surface, but they also indicate the potential for subsequent dissociation once the cell surface location has been attained. For some time it has been recognized that extracellular $\beta_2$m could be exchanged into dimeric class I molecules expressed on the cell surface (Robinson and Graf 1983; Schmidt et al. 1981), but the extent to which the free heavy chain might be an intermediate in this process was not clear.

If one accepts the possibility that free heavy chains might exist in situ, then it becomes important to consider what implications this might have for the function of class I gene products. A model is presented in Fig. 5 which depicts a possible scenario for the conformational dynamics and functional features of different cell surface-associated forms of class I heavy chain. The first point to emphasize in discussing this representation is the data presented in Fig. 3, which indicated that the presence of a functional $\beta_2$m molecule is a clear requirement for the H-2K$^b$ heavy chains to reach the cell surface. This makes the starting point for newly synthesized class I just reaching the cell surface the dimeric structure depicted on the left. Our data also indicate that it is possible to detect in cell surface-labeled preparations of detergent-solubilized glycoproteins a form of the H-2K$^b$ heavy chain which is not associated with $\beta_2$m (Fig. 1, 2, 4). This we have been able to detect by use of rabbit antisera specific for peptides corresponding to different exon-encoded portions of the intracytoplasmic region of H-2K$^b$. The most effective of these was antipeptide 8, which recognizes the C-terminal region of H-2K$^b$. It is also important to note that the free heavy chain form of H-2K$^b$ was not recognized by various monoclonal antibodies and alloantisera with specificity for H-2K$^b$. Because the serological determinants recognized by these antibodies are associated with the polymorphic regions of the extracellular class I domains (Abastado et al. 1987), we have depicted the dissociation of $\beta_2$m as resulting in a significant conformational change of the $\alpha$1, $\alpha$2, and $\alpha$3 domains. This representation is consistent with previously reported data concerning heavy chain conformational alterations detectable by spectroscopy

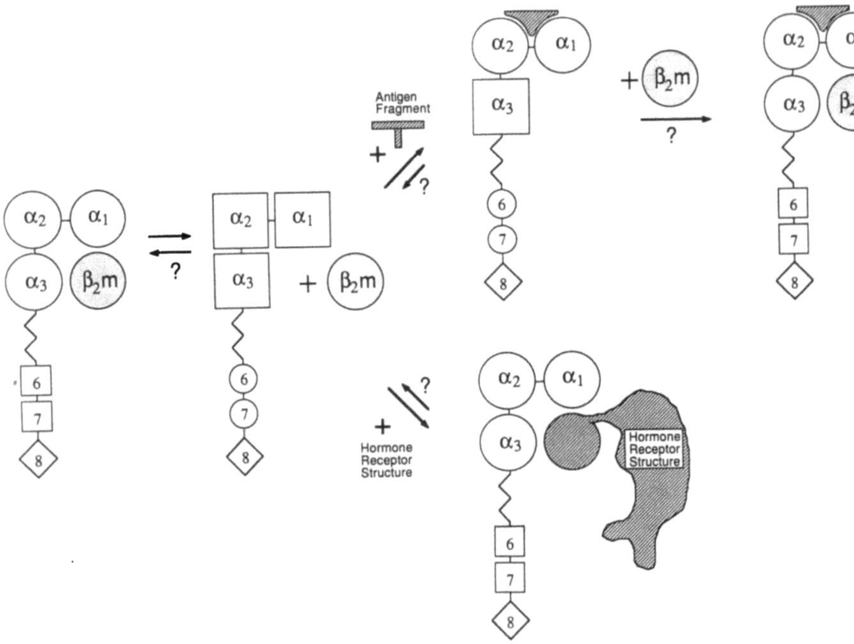


**Fig. 5.** A schematic representation of possible subunit interactions and conformational alterations for the H-2K$^b$ heavy chain. The transition from a *circle* to a *square* (or vice versa) for particular class I domains is intended to reflect a significant conformational change in that region of the molecule. See text for further explanation of the interactions depicted

in papain-solubilized class I molecules dissociated from $\beta_2$m by treatment with guanidinium hydrochloride (Lancet et al. 1979), and with the more recent findings of Allen et al. (1986) in R1E cells transfected with H-2D$^b$ only.

With respect to the intracytoplasmic region of H-2K$^b$ in the transition from the $\beta_2$m-bound to the free state, we know that the epitopes recognized by antipeptide 6 and antipeptide 7 are inaccessible in the dimer, but can be seen in the free H-2K$^b$ heavy chain. Thus the representation in Fig. 5 depicts a conformational change in this region of the molecule as well. These data represent, we believe, the first indication that a conformational alteration in the extracellular domains of the class I molecule can be "transmitted" to the intracytoplasmic region. Thus at least the potential exists for transmembrane signaling to occur via interactions with the extracellular domains of the class I heavy chain. We are not able to tell whether or not this conformational perturbation extends to the exon 8-encoded region of the heavy chain because the antipeptide 8 antiserum contains antibodies able to react with both the $\beta_2$m-bound and free heavy chains.

As indicated above, there are a number of reports documenting the potential to exchange both homologous and heterologous forms of soluble $\beta_2$m into dimeric class I molecules in the surface of mouse and human cells (Robinson and Graf 1983; Schmidt et al. 1981). Thus one might reasonably expect that the dissociation of $\beta_2$m from the heavy chain would be reversible, and that $\beta_2$m could reassociate with the free H-2K$^b$. We have not yet examined this potential with the detergent-solubilized free H-2K$^b$ heavy chains.

Matters of interesting speculation are the conditions which might promote the dissociation of $\beta_2$m from the heavy chain (possibly interactions with other membrane-associated molecules or recycling through acidic intracellular compartments), as well as the functional roles that the free heavy chain might have to play in development, cell physiology, or antigen presentation. Two such possibilities as to the role of the free heavy chains are included in Fig. 5. Lately there have been a number of indications that the class I MHC gene products may associate with the insulin receptor on the cell surface (Fehlmann et al. 1985; Phillips et al. 1986; Due et al. 1986). Although it is not yet clear whether this association involves $\beta_2$m as well as the heavy chain, it does represent one possible role for a free heavy chain intermediate on the cell surface. The conformational properties of the heavy chain in any such interaction with another cell surface receptor structure are clearly a matter of speculation. The representation we have chosen reflects a $\beta_2$m-like interaction which would be consistent with the observed immunoprecipitation of these associated heavy chains with conventional anti-class I reagents.

Although the most completely understood functional role for class I MHC gene products is the presentation of nonself determinants to CTLs, we still lack a molecular understanding of how this presentation process occurs. It has been previously suggested that a free class I heavy chain intermediate may act as an "antigen-capture molecule" in this presentation process (Paabo et al. 1985). Such an interaction, if it were to occur, could be expected to at least locally alter the conformational properties of the heavy chain. It is known from exon-shuffling experiments that the extracellular $\alpha 1$ and $\alpha 2$ domains are important in class I-restricted antigen presentation (Arnold et al. 1984) and that antibodies recognizing serological determinants in this region are able to block CTL recognition of the MHC-restricted complex (Zinkernagel and Rosenthal 1981). Thus our representation is intended to reflect interaction of antigen with this region of the molecule, accompanied by at least partial restoration of the dimeric serological determinants. The conformational status of the rest of the heavy chain is again a matter of speculation, but it is possible that a somewhat altered form of the heavy chain in association with antigen may be contributing to the difficulty of isolating the CTL-recognizable complex of antigen and class I gene product. It is also possible that $\beta_2$m might interact with this intermediate to generate a dimeric class I molecule with a bound antigen fragment.

The anti-intracytoplasmic region peptide antibodies (in particular antipeptide 8) have provided a new perspective on the conformational flexibility and potential dynamics of class I H-2 heavy chains. It still remains, however, to determine whether this new insight will provide us with a better understanding of how these gene products function in the process of antigen presentation to CTL receptors.

## Summary

Peptides synthesized to mimic the exon-encoded sequences within the intracytoplasmic region of the H-2K$^b$ molecule have been used to generate rabbit heteroantisera which react specifically with class I heavy chains. Immunoprecipitation studies with these sera have revealed that cell surface-labeled glycoproteins purified from H-$2^b$ cells by detergent solubilization and lentil lectin affinity chromatography contion a significant fraction (approx. 30%) of H-$2K^b$ lacking an associated $\beta_2$-microglobulin ($\beta_2$-m) molecule. These "free" H-$2K^b$ heavy chains are unreactive with antibodies recognizing conventional H-$2K^b$ serological determinants, but can be immunoprecipitated by the anti-peptide antisera. R1E cells possessing a defective $\beta_2$m gene were transfected with either the $K^b$ gene alone, or both the $K^b$ gene and the $\beta_2$m gene, to examine further the role of $\beta_2$-m in H-$2K^b$ heavy chain expression. The data indicate a requirement for $\beta_2$-m association in order for the H-$2K^b$ heavy chain to reach the cell surface, after which dissociation may occur. A model is presented indicating possible functional roles for the "free" class I heavy chain.

*Acknowledgments.* This work was supported by a Medical Research Council of Canada grant to B.H.B. and a National Cancer Institute of Canada grant to D.B.W.

## References

Abastado J-P, Jaulin C, Schutze M-P, Langlade-Demoyen P, Plata F, Ozata K, Kourilsky P (1987) Fine mapping of epitopes by intradomain $K^d/D^d$ recombinants. J Exp Med 166: 327

Alcaraz G, Kinet J-P, Kumar N, Wank SA, Metzger H (1984) Phase separation of the receptor for immunoglobulin E and its subunits in Triton X-114. J Biol Chem 259: 14922

Allen H, Wraith D, Pala P, Askonas B, Flavell RA (1984) Domain interaction of H-2 class I antigens alter cytotoxic T-cell recognition sites. Nature 309: 279

Allen H, Fraser J, Flyer D, Calvin S, Flavell R (1986) $\beta_2$-microglobulin is not required for cell surface expression of the murine class I histocompatibility antigen H-$2D^b$ or of a truncated H-$2D^b$. Proc Natl Acad Sci USA 83: 7447

Allore RA, Barber BH (1982) Structural studies of purified pig lymph node plasma membrane: association of cytoskeletal components with the plasma membrane and the effect of detergent solubilization. Can J Biochem 60: 57

Arnold B, Burgert N-G, Hamann U, Hammerling G, Kees U, Kvist S (1984) Cytolytic T-cells recognize the two amino-terminal domains of H-2K antigens in tandem in influenza A infected cells. Cell 38: 79

Bordier C (1981) Phase separation of integral membrane proteins in Triton X-114 solution. J Biol Chem 256: 1604

Doherty PC, Bennick J (1981) Monitoring the integrity of self: biology of MHC-restriction of virus-immune T cells. Fed Proc 40: 218

Due C, Simonsen M, Olsson L (1986) The major histocompatibility complex class I heavy chain as a structural subunit of the human cell membrane insulin receptor: Implications for the range of biological functions of histocompatibility antigens. Proc Natl Acad SCI USA 83: 6007

Fehlmann M, Peyron J-F, Samson M, van Obberghen E, Brandenburg D, Brossette N (1985) Molecular association between major histocompatibility complex class I antigens and insulin receptors in mouse liver membranes. Proc Natl Acad Sci USA 82: 8634

Gething M-J, McCammon K, Sambrook J (1986) Expression of wild-type and mutant forms of influenza hemagglutinin: The role of folding in intracellular transport. Cell 46: 939

Hyman R, Stallings V (1976) Analysis of hybrids between an H-$2^+$ TL-lymphoma and an H-$2^+$, $TL^+$ lymphoma and its H-$2^-TL^-$ variant sublime. Immunogenetics 4: 171

Kobata A (1979) Use of endo- and exoglycosidases for structural studies of glycoconjugates. Anal Biochem 100: 1

Lancet D, Parham P, Strominger JL (1979) Heavy chain of HLA-A and HLA-B antigens is conformationally labile: A possible role for $\beta_2$-microglobulin. Proc Natl Acad Sci USA 75: 3844

Lew AM, Lillehoj EP, Cowan EP, Maloy WL, van Schravendjk MR, Coligan JE (1986) Class I genes and molecules: an update. Immunology 57: 3

Maloy WL, Coligan JE (1985) Is $\beta_2$-microglobulin required for MHC class I heavy chain expression? Immunol Today 6: 263

Paabo S, Kampe O, Severinsson L, Anderson M, Fernandez C, Peterson PA (1985) The association between class I transplantation antigens and an adenovirus membrane protein. Prog Allergy 36: 114

Parnes JR, Seidman JG (1982) Structure of wild-type and mutant mouse $\beta_2$-microglobulin genes. Cell 29: 661

Phillips ML, Moule ML, Delovitch TL, Yip CC (1986) Class I histocompatibility antigens and insulin receptors: evidence for interactions. Proc Natl Acad Sci USA 83: 3474

Robinson PJ, Graf L (1983) Dynamic aspects of $\beta_2$-microglobulin replacement on the surface of mouse lymphoma cells. Transplant Proc 15: 2051

Schmidt W, Festenstein H, Ward PJ, Sanderson AR (1981) Interspecies exchange of $\beta_2$-microglobulin and associated MHC and differentiation antigens. Immunogenetics 13: 483

Smlth MH, Parker JMR, Hodges RS, Barber BH (1986) The preparation and characterization of anti-peptide heteroantisera recognizing subregions of the intracytoplasmic domain of class I H-2 antigens. Mol Immunol 23: 1077

Weiss E, Golden L, Zeikut R, Mellor A, Fahrner K, Kvist S, Flavell RA (1983) The DNA sequence of the H-$2K^b$ gene: evidence for gene conversion as a mechanism for the generation of polymorphism in histocompatibility antigens. EMBO J 2: 453

Williams DB, Sweidler SJ, Hart GW (1985) Intracellular transport of membrane glycoproteins: two closely related histocompatibility antigens differ in their rates of transit to the cell surface. J Cell Biol 101: 725

Zinkernagel RM, Doherty PC (1979) MHC-restricted cytotoxic T cells. Adv Immunol 27: 51

Zinkernagel RM, Rosenthal KL (1981) Experiments and speculation on anti-viral specificity of T and B cells. Immunol Rev 58: 131

# Models for MHC-Restricted T Cell Recognition of a Synthetic H-2L$^d$ Peptide

ELIZABETH S. SONG, RICHARD LINSK, and ROBERT S. GOODENOW

Department of Genetics, University of California, Berkeley, CA 94720, USA

## Introduction

Class I and class II major histocompatibility (MHC) antigens are polymorphic cell surface glycoproteins which play central roles in controlling target cell recognition by antigen specific cytotoxic and helper T cells. The class I molecules are expressed ubiquitously and, in general, are involved in the presentation of intracellular and cell surface viral antigens to cytotoxic T-lymphocytes (CTLs). On the other hand, class II molecules are expressed primarily on antigen presenting cells such as macrophages and B cells, where they are involved in the presentation of phagocytized or internalized foreign antigens to helper T cells (Zinkernagel and Doherty 1979; Schwartz 1984). As a rule, T cells exhibit an obligatory requirement of corecognition of antigen and MHC antigen of their own haplotype or "restriction element." Therefore, T cell recognition is said to be "MHC restricted."

Another intriguing feature of T cell recognition is the phenomenon of alloreactivity. It is known that a large percentage of unprimed T cells from one strain show extremely vigorous and rapid reactions against cells derived from other haplotypes. In this a priori recognition of alloantigenic nonself MHC molecules, target antigens may be presented without self restriction elements in apparent violation of MHC restriction. However, many cloned, MHC-restricted, nominal antigen specific T cells exhibit cross-reactivity with allo-MHC antigens, suggesting that the molecular mechanisms underlying these two distinct classes of T cell recognition are somehow intertwined (Ashwell et al. 1986; Sheil et al. 1987). In either case, the specific target structures which allow this cross-reactivity have not been identified. Also, it is still not clear whether MHC molecules are recognized as independent structures by allospecific T cells or whether the relevant T cell target structure consists of a complex between the allo-MHC molecule and some as yet unidentified cellular product or processed peptide. Therefore, it was of interest to determine the functional relationship between these two targets as well as the specific MHC structural features recognized by the T cells.

Although the precise mechanisms underlying MHC restriction remain obscure, recent results suggest that class II molecules present antigen to T cells in the form of processed peptide in association with self-class II molecules (Shimonkevitz et al. 1983; Morrison et al. 1986). Similar mechanisms have been implicated in influenza virus-specific CTLs, recognizing virally derived processed peptides restricted by self-class I molecule (Townsend et al. 1986a, b). Finally, studies of CTLs generated against mouse cell transfectants expressing HLA class I molecules suggest that some examples of allorecognition might involve the processing of the foreign class I antigen and presentation in association with self-MHC products (Maryanski et al. 1986a, b). Implicating the involvement of processed peptide in T cell recognition, the results of Marianski et al. using murine CTLs directed against human HLA peptide in conjunction with murine H-2 do not adequately address the nature of the alloantigenic structure. Therefore, we attempted to generate murine allospecific CTLs capable of recognizing an allogeneic H-2 peptide in the context of self-class I restriction elements.

## Experimental Procedures and Results

### *Generation of CTLs*

1591 is a UV-induced C3H fibrosarcoma which expresses three novel class I molecules not normally present on C3H tissue, in addition to normal H-$2^k$ histocompatibility antigens. These novel genes have been cloned and sequenced (Stauss et al. 1986; Linsk et al. 1986) and the products characterized serologically as well as immunologically using several allospecific and viral specific CTLs (Linsk et al. 1986). One of these genes, designated A149, is nearly identical to the BALB/c H-$2L^d$ gene, and its product, H-$2L^{149}$, differs from H-$2L^d$ by six amino acid substitutions in $\alpha 2$ domain. Despite the homology between them, H-$2L^d$ and H-$2L^{149}$ can be distinguished by several H-$2L^d$ specific CTLs and monoclonal antibodies (Linsk et al. 1986). Utilizing these immunologically distinct but extremely homologous H-2L molecules, we attempted to generate a distinct set of H-$2L^d$ reactive CTLs. Briefly, C3H mice were immunized with 1591 tumor and primed splenocytes were cultured in vitro in the presence of 1591 stimulators expressing H-$2L^{149}$ antigen. After 1 week of growth, cultures were then stimulated with C3H L cell transfectants expressing H-$2L^d$ molecule. Because CTLs require specific antigenic stimulation for growth, only the CTLs directed against the determinants common to both H-$2L^{149}$ and H-$2L^d$, namely the $\alpha 1$ domain, were expected to survive. Since many of the amino acid sequence differences between both H-2L and H-$2^k$ map to the $\alpha 1$ domain, it seemed likely that the majority of these cross-reactive CTLs would be directed against this region. Therefore, we synthesize the peptide whose sequence is derived from the hypervariable regions in the $\alpha 1$ domain (amino acid 61–85) of the H-$2L^d$ molecule. This region was also predicted to form an amphipathic $\alpha$ helix which is presumably a predominant form of peptidic antigens (Novotny and Auffray 1984).

### *Peptide Recognition by Cross-reactive CTLs*

Since the CTLs were selected on the basis of cross-reactivity between two H-2L molecules, we reasoned that these CTLs might be able to recognize the H-$2L^d$ hypervariable peptide – perhaps in association with self restriction elements.

Indeed, we found that these cross-reactive CTLs were able to recognize the 61–85 H-2L$^d$ peptide in association with C3H H-2$^k$ restriction elements. This recognition of the H-2L$^d$ peptide by cross-reactive CTLs is MHC restricted such that only H-2$^k$ restriction elements were able to mediate specific lysis. Analysis using a panel of recombinant inbred, strains suggests that H-2D$^k$ is a predominant restriction element for these CTLs. Also, lysis of syngeneic target cells by these CTLs appears to be dependent on the sequence and conformation of 61–85 H-2L$^d$ peptide, since no killing was observed in the presence of a 61–85 H-2D$^d$ peptide or an analogue of the 61–85 H-2L$^d$ peptide containing D-isomers of glutamine and leucine.

These cross-reactive CTLs are also able to recognize intact H-2L$^d$ molecules expressed on the surface of allogeneic targets, suggesting the structure of the H-2L$^d$ peptide in association with the H-2$^k$ restriction element resembles that presented by the intact H-2L$^d$ molecule itself. This structural mimicry between the intact H-2L$^d$ and H-2L$^d$ peptide in conjunction with an H-2$^k$ restriction element has been confirmed using cross-reactive CTLs grown on untransfected L cells in the presence of soluble H-2L$^d$ peptide. These CTLs, presumably specific to the H-2L$^d$ peptide in conjunction with H-2$^k$ restriction elements, retained the ability to recognize allogeneic targets bearing intact H-2L$^d$ antigen after 8 weeks of growth on soluble H-2L$^d$ peptide only.

## Peptide Recognition by H-2L$^d$ Specific CTLs

The results described above strongly suggest the possible structural conservation between the H-2L$^d$ peptide (in conjunction with H-2$^k$ restriction elements) and the intact H-2L$^d$ antigen borne by allogeneic cells. Therefore, we examined CTL clones – generated against intact H-2L$^d$ antigen – for the recognition of H-2L$^d$ peptide in conjunction with H-2$^k$ restriction elements. Three out of four clones tested were also able to lyse C3H L cells in the presence of H-2L$^d$ peptide and to do so only in conjunction with H-2$^k$ haplotype restriction elements. These data were especially intriguing in that two of the clones were generated against the targets bearing no H-2$^k$ restriction elements.

Together, these results indicate a conserved recognition structure for many H-2L$^d$ reactive CTLs which can be represented not only by intact H-2L$^d$ molecule but also by alloantigenic H-2L$^d$ peptide in conjunction with appropriate restriction elements. Furthermore, it is conceivable that alloreactive CTLs generated against intact alloantigen might, in fact, be directed against an alloantigenic MHC peptide presented by class I molecules.

## Models for T Cell Recognition

Recent crystallographic analysis of the structure of HLA-A2 has revealed that the $\alpha 1$ and $\alpha 2$ domains of the class I molecule comprise one superdomain consisting of two $\alpha$ helices on top of a $\beta$-pleated sheet (Bjorkman et al. 1987 a, b). An antigen

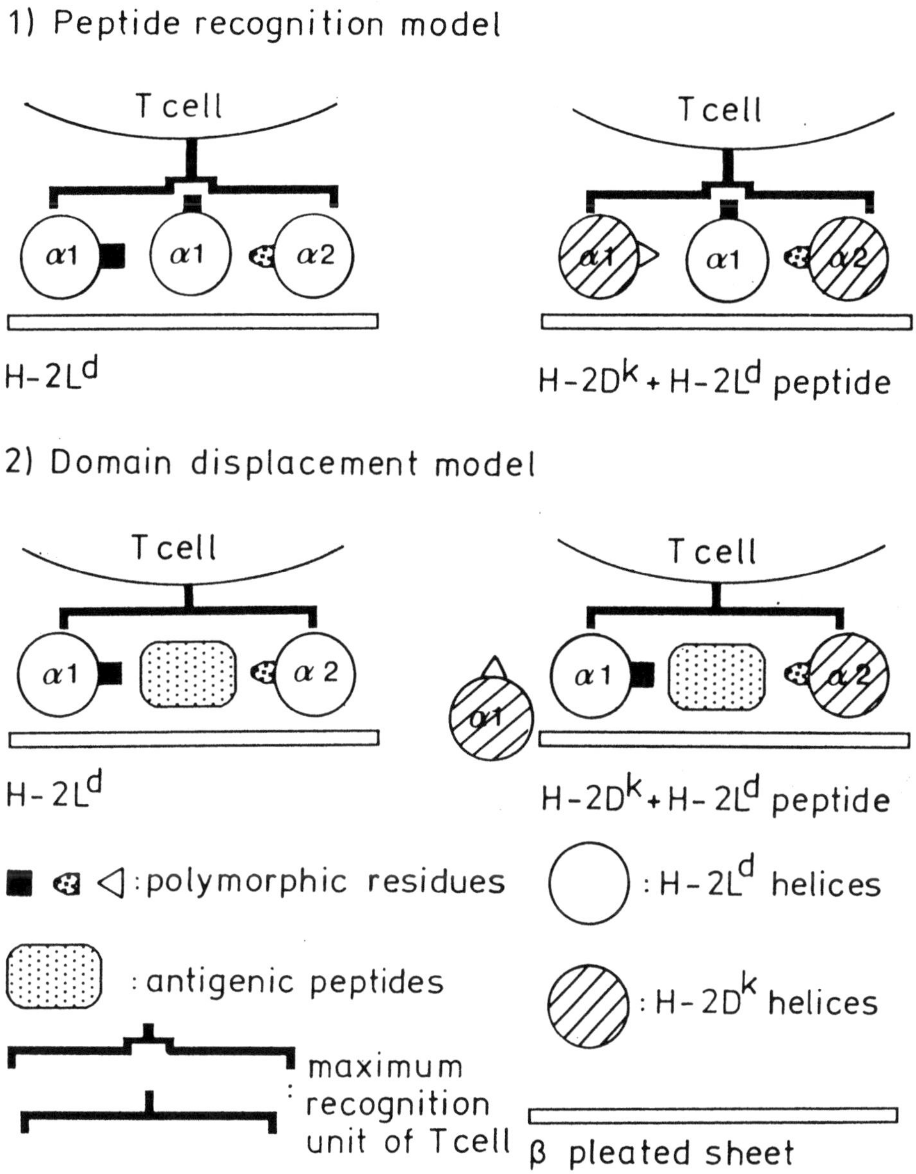


**Fig. 1.** Models for T cell recognition

binding cleft between the two $\alpha$ helices and the $\beta$ pleated sheet provides the foreign antigen binding site(s). In support of this, processed peptides were cocrystallized and actually observed in the crystals of this class I molecule. One of the most interesting features of this structure is that the majority of amino acid residues on the helices which are polymorphic among various class I molecules are oriented toward the cleft and therefore appear unavailable for direct interaction with the T cell receptor. Instead, these polymorphic residues seem to be important in antigen binding. However, our results suggest that an alloantigenic peptide can reconstitute the antigenic structure such that at least some of the polymorphic residues are directly recognized by the T cell receptor. In light of these observations, we envision two models for alloantigenic recognition to explain the spatial constraints between the T cell recognition of the H-2L$^d$ $\alpha$1 helix on intact H-2L$^d$ versus that reconstituted on H-2D$^k$ (see Fig. 1).

### *Peptide Recognition Model*

In this model, the alloreactive CTLs recognize an allogeneic peptide presented in the antigen binding cleft. In fact, some fraction of the H-2L$^d$ molecules could present their own peptide (i.e., H-2L$^d$ $\alpha$1 helix). Alternatively, H-2D$^k$ restriction elements could be uniquely suited for the presentation of this H-2L$^d$ peptide in the same conformation as that seen on the H-2L$^d$ molecule. Therefore, the polymorphic residues common to H-2L$^d$ and H-2D$^k$ yet distinct from other class I molecules would be crucial for the presentation of the H-2L$^d$ peptide. This model would predict that H-2L$^d$ alloreactive CTLs would not be able to recognize isolated H-2L$^d$ molecules in the absence of processed peptide. One experimental approach which has the potential to test this model is to examine whether the CTLs that we have used are able to recognize purified H-2L$^d$ reconstituted into lipid membranes. Several recent reports employing this approach have concluded that some T cells might in fact be able to recognize isolated class I antigens (Brian and McConnell 1984; Blackman et al. 1986; Goldberg and Mescher 1986). However, even with the purified membranes, it would be very difficult to ensure that the class I preparation is indeed free of any contaminating peptide.

### *Domain Displacement Model*

Residues mapping to the $\alpha$ helix might, in fact, constitute part of polymorphic structure of the intact H-2L$^d$ molecule recognized by the alloreactive CTLs. If so, the peptide derived from the H-2L$^d$ helix might be able to displace the homologous $\alpha$ helix of the H-2D$^k$ restriction element and therefore reconstitute the antigenic determinant borne by intact H-2L$^d$ molecule. This model is especially attractive if alloantigenic CTLs recognize conformational determinants comprising the $\alpha$1 helix from H-2L$^d$ together with other parts of class I molecule, such as regions in $\alpha$2 helix common to H-2L$^d$ and H-2D$^k$. In this case, domain displacement could ensure the proper spatial relationship among components of the determinants recognized. In agreement with this hypothesis, limited sequence homology has recently been detected between several peptide antigens presented by I-E$^d$ class II molecule and $E_\beta{}^d$ molecule (Guillet et al. 1987). Such homology could explain the binding of certain peptides if homologous segments of the MHC structure are displaced by ligands.

Both of the models discussed above assume that structurally homologous antigenic determinants are exhibited by both H-2L$^d$ and the H-2D$^k$ restriction element in association with H-2L$^d$ peptide. However, the extent of the structural homology required is difficult to estimate, even though the 61–85 H-2L$^d$ peptide presumably takes an $\alpha$ helical conformation. The avidity of the T cell and target cell interaction depends not only on the affinity of the individual T cell receptor and antigen interaction but also on the overall avidity represented by the number of target cell MHC antigens exhibiting the appropriate determinants. Nonetheless, the observation that antigenic cross-reactivity can be detected using CTLs derived by different protocols argues strongly that the epitopes borne by the H-2L$^d$ peptide are structurally homologous to an intact H-2L$^d$.

## Implications

The observation described in this report might be especially relevant to the phenomenon of T cell cross-reactivity. It has previously been observed that T cells specific for nominal antigen in association with self-MHC restriction elements often exhibit cross-reactive recognition with allogeneic MHC products (Sredni and Schwartz 1980; Braciale et al. 1981; Naquet et al. 1983; Schwartz 1984). This cross-reactivity is nonrandom and probably reflects some underlying homology among various target molecules (Ashwell et al. 1986). In light of these studies and our observations, it is reasonable to assume that the allogeneic cross-reactivity observed for other, non-MHC peptide antigens might also reflect some level of structural sequence homology between the target molecules. It is conceivable that T cell repertoire is generated in order to recognize nominal antigen in the context of self-MHC molecule and this repertoire is fortuitously cross-reactive with foreign MHC molecules due to relatively conserved MHC frameworks. In this context, allograft rejection would also be MHC restricted in its recognition of alloantigen in the context of self-MHC molecules. It is also clear that the pattern of T cell MHC specificity largely depends on the MHC products expressed in the developmental environment of the thymus. In particular, a variety of experiments involving chimeric mice suggest that CTLs preferentially recognize antigen/MHC complexes resembling "altered self-H-2" (Bevan 1977; Bevan and Hunig 1981; Hunig and Bevan 1982). However, it was not possible to determine from these experiments what the molecular nature of an altered self-MHC molecule might be. The realization that many T cells recognize processed peptides in association with MHC products provides a possible explanation. Since some of the thymic MHC molecules responsible for T cell education are undoubtedly also engaged in the presentation of self-MHC peptides, it is conceivable that thymic selection for the recognition of altered self-MHC molecules might, in fact generate a T cell repertoire specific for altered self-MHC peptides.

## References


Ashwell JD, Chen C, Schwartz RH (1986) High frequency and nonrandom distribution of alloreactivity in T cell clones selected for recognition of foreign antigen in association with self class II molecules. J Immunol 136: 389–395

Bevan MJ (1977) In a radiation chimera, host H-2 antigens determine immune responsiveness of donor cytotoxic cells. Nature 269: 417–418

Bevan MJ (1984) High determinant density may explain the phenomenon of alloreactivity. Immunol Today 5: 128–131

Bevan MJ, Hunig TR (1981) T cells respond preferentially to antigens that are similar to self H-2. Proc Natl Acad Sci USA 78: 1843–1847

Bjorkman PJ, Saper MA, Samraoui B, Bennett WS, Strominger JL, Wiley DC (1987 a) Structure of the human class I histocompatibility antigen, HLA-A2. Nature 329: 506–512

Bjorkman PJ, Saper MA, Samraoui B, Bennett WS, Strominger JL, Wiley DC (1987 b) The foreign antigen binding site and T cell recognition regions of class I histocompatibility antigens. Nature 329: 512–518

Blackman M, Yagüe J, Kubo R, Gay D, Coleclough C, Palmer E, Kappler J, Marrack P (1986) The T cell repertoire may be biased in favor of MHC recognition. Cell 47: 349–357

Braciale T, Andrew M, Braciale V (1981) Simultaneous expression of H-2-restricted and alloreactive recognition by a cloned line of influenza virus-specific cytotoxic T lymphocytes. J Exp Med 153: 1371–1376

Brian A, McConnell H (1984) Allogeneic stimulation of cytotoxic T cells by supported planar membranes. Proc Natl Acad Sci USA 81: 6159–6163

Goldstein S, Mescher M (1986) Cell-sized, supported artificial membranes (pseudocytes): responses of precursor cytotoxic T lymphocytes to class I MHC proteins. J Immunol 137: 3383–3392

Guillet J-G, Ming-Zong L, Briner TJ, Buus S, Sette A, Grey HM, Smith JA, Gefter ML (1987) Immunological self, nonself discrimination. Science 235: 865–870

Hunig TR, Bevan MJ (1981) Specificity of T-cell clones illustrates altered self hypothesis. Nature 294: 460–462

Hunig TR, Bevan MJ (1982) Antigen recognition by cloned cytotoxic T lymphocytes follows rules predicted by the altered-self hypothesis. J Exp Med 155: 111–125

Klein J (1987) Natural history of the major histocompatibility complex. Wiley, New York

Linsk R, Vogel J, Stauss H, Forman J, Goodenow RS (1986) Structure and function of three novel MHC class I antigens derived from a C3H ultraviolet-induced fibrosarcoma. J Exp Med 164: 794–813

Maryanski JL, Accolla RS, Jordan B (1986 a) H-2-restricted recognition of closed HLA class I gene products expressed in mouse cells. J Immunol 136: 4340–4347

Maryanski JL, Pala P, Corradin G, Jordan BR, Cerottini J-C (1986 b) H-2 restricted cytolytic T cells specific for HLA can recognize a synthetic HLA peptide. Nature 324: 578–579

Morrison LA, Lukacher AE, Braciale VL, Fan DP, Braciale TJ (1986) Differences in antigen presentation to MHC class I- and class II-restricted influenza virus-specific cytolytic T lymphocyte clones. J Exp Med 163: 903–921

Naquet P, Pierres A, Pierre M (1983) Dissection of the poly (glu60 ala30 tyr10) (GAT)-specific T-cell repertoire in H-2I$^k$ mice. I. GAT plus self-I-A$^k$-reactive T-cell clones can recognize alloreactive and/or restriction determinants on nonself-Ia molecules. Immunogenetics 18: 478–488

Novotny J, Auffray C (1984) A program for prediction of protein secondary structure from nucleotide sequence data: application to histocompatibility antigens. Nucleic Acids Res 12: 243–255

Schwartz RH (1984) The role of gene products of the major histocompatibility complex in T-cell activation. In: Paul WE (ed) Fundamental immunology. Raven, New York, pp 379–438

Sheil JM, Bevan MJ, Lefrancois L (1987) Characterization of dual-reactive H-2K$^b$-restricted anti-vesicular stomatitis virus and alloreactive cytotoxic T cells. J Immunol 138: 3654–3660
Shimonkevitz R, Kappler J, Marrack P, Grey H (1983) Antigen recognition by H-2-restricted T cells. I. Cell-free antigen processing. J Exp Med 158: 303–316
Sredni B, Schwartz RH (1980) Alloreactivity of an antigen-specific T cell clone. Nature 287: 855–857
Stauss HJ, Linsk R, Fischer A, Watts S, Banasiak D, Haberman A, Clark I et al. (1986) Isolation of the MHC genes encoding the tumour-specific class I antigens expressed on a murine fibrosarcoma. J Immunogenet 13: 101–111
Townsend ARM, Rothbard J, Gotch FM, Bahadur G, Wraith D, McMichael AJ (1986a) The epitopes of influenza nucleoprotein recognized by cytotoxic T lymphocytes can be defined with short synthetic peptides. Cell 44: 959–968
Townsend ARM, Bastin J, Gould K, Brownlee GG (1986b) Cytotoxic T lymphocytes recognize influenza haemagglutinin that lacks a signal sequence. Nature 324: 575–577
Zinkernagel R, Doherty P (1979) MHC-restricted cytotoxic T cells: Studies on the biological role of polymorphic major transplantation antigen determining T-cell restriction-specificity, function, and responsiveness. Adv Immunol 27: 51–177

# Endorphin Receptors and HLA Class I Molecules

F. H. J. Claas[1], W. Verduyn[1], M. Mommaas-Kienhuis[2], W. M. A. Verhoeven[3], J. M. van Ree[4], and J. J. van Rood[1]

[1] Department of Immunohaematology and Bloodbank, University Hospital Leiden, Leiden, The Netherlands
[2] Department of Dermatology, University Hospital Leiden, Leiden, The Netherlands
[3] Department of Biological Psychiatry, University of Utrecht, Utrecht, The Netherlands
[4] Rudolf Magnus Institute of Pharmacology, University of Utrecht, Utrecht, The Netherlands

The HLA system, the major histocompatibility complex (MHC) in man, consists of a set of linked genes on chromosome 6, which code for different classes of highly polymorphic cell membrane molecules. The biological function of the HLA-A, -B, -C or class I antigens, which are expressed on almost all nucleated cells, mainly concerns immune recognition. Cytotoxic T-lymphocytes can recognize virus-infected cells only in the context of HLA class I molecules, a phenomenon called MHC restriction (Zinkernagel and Doherty 1974). However, MHC class I antigens may also be involved in nonimmunological recognition processes (Dausset and Contu 1980).

Genetic differences in MHC class I antigens have been correlated to variation in the size of hormone-dependent tissues like testis and vesicular gland (Ivanyi 1978). One could speculate that these correlations are due to an involvement of MHC class I molecules in hormone-receptor interactions. Such interactions have recently been found for MHC class I products and the binding of insulin (Fehlmann et al. 1985; Simonsen et al. 1985; Philips et al. 1986; Kittur et al. 1987), glucagon (Simonsen and Olsson 1983), epidermal growth factor (Schreiber et al. 1984), and endoperoxide-thromboxane (Curry et al. 1984) to their respective receptors. These results seem to confirm the hypothesis originally suggested by Svejgaard and Ryder in 1976, that MHC antigens may interfere in the interaction between nonimmunological ligands and their receptors in the cell membrane. Also our findings concerning the interactions between $\gamma$-type endorphins and HLA class I antigens may be in line with this hypothesis.

Gamma-type endorphins, i.e., des-Tyr[1]-$\gamma$-endorphin (DT$\gamma$E) and desenkephalin-$\gamma$-endorphin (DE$\gamma$E) induce behavioral effects in rats, which are in some aspects comparable with those exerted by neuroleptics (De Wied et al. 1978). Subsequent clinical studies showed that $\gamma$-type endorphins decreased psychotic symptomatology in a number of schizophrenic patients (Verhoeven et al. 1979; Van Ree et al. 1982). Since the frequency of HLA-B15, -Bw22, and -B13 was increased in those patients who responded to treatment with $\gamma$-type endorphins, genetic factors coded for within the HLA region may be associated with the response of patients to treatment with $\gamma$-type endorphins (De Jongh et al. 1982; Claas et al. 1986).

We then tested whether HLA class I antigens are involved in the binding of $\gamma$-type endorphins to their receptors. The protocol we used was very similar to the one described for the studies on the interaction between penicillin and the HLA class I alloantigens (Claas et al. 1982). Isolated lymphocytes were preincubated in a

**Table 1.** Interference of DTγE in the reactivity of HLA alloantisera with their corresponding antigens in complement-dependent lymphocytotoxicity

| Molecular concentration of DTγE | Anti-A2 | Anti-B15 |
|---|---|---|
| $6 \times 10^{-7}$ | 0 | 0 |
| $6 \times 10^{-6}$ | 0 | 0 |
| $3 \times 10^{-5}$ | 0 | 20[a] |
| $1 \times 10^{-4}$ | 0 | 40 |
| $3 \times 10^{-4}$ | 0 | 60 |
| $6 \times 10^{-4}$ | 0 | 60 |

[a] Percentage inhibition of the cytotoxicity reaction after preincubation of the lymphocytes from a healthy volunteer (HLA: A2, Aw32, B15, B40) with graded concentrations of DTγE

solution of γ-type endorphins and after washing the binding of anti-HLA alloantibodies was tested in a complement-dependent cytotoxicity assay.

The percentage lysis of the individual HLA-typing sera was determined against both untreated lymphocytes and γ-endorphin-treated lymphocytes of the same donors. In some cases inhibition of complement-dependent cytotoxicity was found; this depended on the concentration of DTγE used and the specificity of the target antigens (Table 1). A preferential inhibition was found for alloantisera recognizing HLA-B15, -Bw22, and -B13, although other HLA-A or HLA-B antigens were partially masked by the incubation with endorphin as well (Claas and van Rood 1985; Claas et al. 1986). It was striking that the antigens, of which the frequency was increased among the schizophrenic patients who responded to treatment with γ-type endorphins, showed the highest level of inhibition in the in vitro studies with DTγE. This correlation between the in vitro inhibition of complement-dependent cytotoxicity by DTγE and the clinical improvement after treatment with DTγE or DEγE in the 64 patients tested so far, suggests a functional relationship between the two phenomena. However, no explanation for this correlation is available at the moment.

Next, we started to study the mechanism responsible for the observed in vitro interaction between DTγE and HLA class I molecules (Table 2). One of the possibilities is that the HLA class I molecules are necessary for the binding of DTγE to lymphoid cells either because they function as the receptor for DTγE or because they are part of a compound receptor (Simonsen and Olsson 1983). Therefore we compared the binding of radioactive endorphin (DEγE) to Daudi, a cell line lacking expression of HLA class I molecules, with that to an HLA-B15 positive EBV-transformed B cell line. As can be seen from Table 3, no difference in binding was observed, showing that the presence of HLA class I molecules is not a prerequisite for binding of γ-endorphins to cells.

A second possibility could be that DTγE binds to its specific receptor without the involvement of HLA molecules in the binding process, but the HLA class I molecules have a function in the signal transduction into the cell. This might induce a conformational change of the HLA molecules, which could explain the inhibition studies. However, at the moment we do not have a tool to test this hypothesis as we found

**Table 2.** Protocols to study the possible mechanisms involved in the interaction between $\gamma$-type endorphins and HLA class I molecules

| Possible mechanisms | Assays |
|---|---|
| 1. HLA involved in DT$\gamma$E binding | Radioactive binding studies |
| 2. HLA involved in the signal transduction triggered by DT$\gamma$E binding to its specific receptor | cAMP or $Ca^{2+}$ levels |
| 3. HLA acts as transport molecule for the DT$\gamma$E receptor complex | Electron microscopy |

**Table 3.** Replaceable binding of $^3$H-DE$\gamma$E to Daudi and a HLA-B15 positive B cell line

| $^3$H-DE$\gamma$E added | Daudi | B15$^+$ EBV line |
|---|---|---|
| $2 \times 10^{-6}$ M | 0.05[a] | 0.08 |
| $4 \times 10^{-6}$ M | 0.45 | 0.30 |
| $6 \times 10^{-6}$ M | 0.75 | 0.60 |
| $8 \times 10^{-6}$ M | 1.06 | 1.01 |

[a] DE$\gamma$E bound to $10^6$ cells ( $\times 10^{13}$ M)

no evidence for an increase or decrease of the levels of cAMP or $Ca^{2+}$ after binding of DT$\gamma$E to lymphoid cells.

As a third hypothesis we are testing whether the HLA class I molecules are functioning as transport molecules, as suggested for adenovirus glycoprotein E19 (Paäbo et al. 1985). After binding of DT$\gamma$E to its specific receptor, the HLA class I molecules may be involved in the transport of this DT$\gamma$E-receptor complex into the cell. An assay for the ultrastructural visualization of this interaction has recently been developed by one of us (Mommaas et al., in preparation). In brief, DT$\gamma$E is conjugated to 10-nm gold particles whereas HLA class I molecules are visualized indirectly using monoclonal antibodies against HLA class I conjugated to 5-nm gold particles. After incubation of the cells with these preparations, binding, internalization, and colocalization of DT$\gamma$E and HLA class I molecules can be studied using electron microscopy.

Preliminary data indeed show evidence for involvement of the HLA class I molecules in the internalization of DT$\gamma$E. The nature of this interaction has to be studied more extensively, however, before conclusions can be drawn.

*Acknowledgments.* We wish to thank Ria Castelli-Visser and Marian Wijsman for excellent technical assistance and Ingrid Curiël for the preparation of the manuscript.

## References

Claas FHJ, van Rood JJ (1985) The interaction of drugs and endogenous substancés with HLA class-I antigens. Prog Allergy 36: 135–150

Claas FHJ, Runia-van Nieuwkoop R, van de Berge W, van Rood JJ (1982) Interaction of penicillin with HLA-A and -B antigens. Hum Immunol 5: 83–90

Claas FHJ, van Ree JM, Verhoeven WMA, van der Poel JJ, Verduyn W, de Wied D, van Rood JJ (1986) The interaction between gamma-type endorphins and HLA class I antigens. Hum Immunol 15: 347–356

Curry RA, Messner RP, Johnson GJ (1984) Inhibition of platelet aggregation by monoclonal antibody reactive with $\beta_2$-microglobulin chain of HLA complex. Science 224: 509–511

Dausset J, Contu L (1980) Is the MHC a general self-recognition system playing a major unifying role in an organism? Hum Immunol 1: 5–14

De Jongh BM, Verhoeven WMA, van Ree JM, de Wied D, van Rood JJ (1982) HLA and the response to treatment with $\gamma$-type endorphins in schizophrenia. J Immunogenet 9: 381–388

De Wied D, Kovacs GL, Bohus B, van Ree JM, Greven HM (1978) Neuroleptic activity of the neuropeptide $\beta$-LPH 62–77 (Des-Tyr$^1$-$\gamma$-endorphin; DT$\gamma$E). Eur J Pharmacol 49: 427–436

Fehlmann M, Peyron JF, Samson M, van Obberghen E, Brandenburg D, Brossette N (1985) Molecular association between major histocompatibility complex class I antigens and insulin receptors in mouse liver membranes. Proc Natl Acad Sci USA 82: 8634–8637

Ivanyi P (1978) Some aspects of the H-2 system, the major histocompatibility system in the mouse. Proc R Soc Lond [Biol] 202: 117–158

Kittur D, Shimizu Y, DeMars R, Edidin M (1987) Insulin binding to human B lymphoblasts is a function of HLA haplotype. Proc Natl Acad Sci USA 84: 1351–1355

Paäbo S, Kämpe O, Severinsson L, Andersson M, Fernandez C, Peterson PA (1985) The association between class I transplantation antigens and an adenovirus membrane protein. Prog Allergy 36: 114–134

Philips ML, Moule ML, Delovitch TL, Yip CC (1986) Class I histocompatibility antigens and insulin receptors: evidence for interactions. Proc Natl Acad Sci USA 83: 3474–3478

Schreiber AB, Schlesinger J, Edidin M (1984) Interaction between major histocompatibility complex antigens and epidermal growth factor receptors on human cells. J Cell Biol 98: 725–731

Simonsen M, Olsson L (1983) Possible roles of compound membrane receptors in the immune system. Ann Inst Pasteur Immunol 134D: 85–92

Simonsen M, Skjødt K, Crone M, Sanderson A, Fujita-Yamaguchi Y, Due C, Rønne E, Linnet K, Olsson L (1985) Compound receptors in the cell membrane: ruminations from the borderland of immunology and physiology. Prog Allergy 36: 151–176

Svejgaard A, Ryder LP (1976) Interaction of HLA molecules with non-immunological ligands as an explanation of HLA and disease associations. Lancet 2: 547–549

Van Ree JM, Verhoeven WMA, de Wied D, van Praag HM (1982) The use of synthetic peptides from $\gamma$-type endorphins in mentally ill patients. Ann NY Acad Sci 398: 478–485

Verhoeven WMA, van Praag HM, van Ree JM, de Wied D (1979) Improvement of schizophrenic patients by treatment with (des-Tyr$^1$)-$\gamma$-endorphin (DT$\gamma$E). Arch Gen Psychiatry 36: 294–298

Zinkernagel RM, Doherty PC (1974) Immunological surveillance against altered self components by sensitized T lymphocytes in lymphocytic choriomeningitis. Nature 251: 547–549

# MHC Class I-Insulin Receptor Interaction in the Cell Membrane

M. LAURIE PHILLIPS, MARGARET L. MOULE, and CECIL C. YIP

Banting and Best Department of Medical Research, University of Toronto, Toronto, Ontario M5G 1L6, Canada

## Introduction

This forum deals with the interactions between major histocompatibility complex (MHC) class I heavy chains ($CI_H$) and other proteins. Most familiar to immunologists are the noncovalent association of class I heavy chain with $\beta_2$-microglobulin ($\beta_2$m) and the binding of peptide antigens (Ags) to MHC class I molecules. The latter interaction, which results in MHC-restricted Ag presentation, is commonly referred to as MHC+X, signifying the expression of new antigenic determinants in the complex. Several other manuscripts in this volume deal with changes in antigenic determinants on class I heavy chains generated when a given $CI_H$ associates with $\beta_2$m of different species. These interactions are considered to represent special cases of MHC+X. MHC class I Ags also interact with cell surface proteins (MHC+Y). The notion that MHC+Y are biologically relevant complexes is not new (reviewed in Meruelo and Edidin 1980; Dausset and Contu 1980; Ivanyi 1978; Edidin 1983). It is derived from observations that particular MHC haplotypes are associated with specific diseases and with variations in physiological responses. It is also invoked as an explanation for the MHC-restricted recognition of "minor" transplantation antigens. In this manuscript we will specifically discuss the possibility that MHC class I Ags are subunits of a compound insulin receptor on cell membranes.

Simonsen and Olssen (1983) have defined a "compound receptor" as a molecular complex which attains its final, biologically active structure by rearrangement and assembly in the membrane of several active subunits. Implicit in their definition is the notion that the interaction between the subunits must have structural consequences which are of biological importance to either or both components of the compound receptor. In order for an interaction between class I Ags and insulin receptors to satisfy this definition, two conditions must be met. First, there must be a physical proximity of the insulin receptor subunits and MHC class I Ags on cell membranes, and second, MHC class I Ags must influence the physiological function of insulin receptors or vice versa.

## The Structure of Insulin Receptors

Insulin receptors of a variety of tissues consist of at least two major glycoproteins, including a 125- to 130-kd insulin binding subunit and a 90- to 95-kd subunit with protein kinase activity (Yip et al. 1980; Massague et al. 1980; Kasuga et al. 1982). A 40- to 45-kd subunit and, under special experimental circumstances, an 85-kd subunit of unknown function have also been detected (Yip and Moule 1983; Massague and Czech 1980). Massague et al. (1980) have concluded that the insulin receptor is a heterodimer of two 125-kd and two 95-kd subunits, considering the 40- to 45-kd component to be a proteolytic fragment of the 95-kd subunit. Yip and Moule (1983) have proposed an alternate model of the insulin receptor composed of one 130-kd, one 90-kd, one 85-kd, and two 40- to 45-kd subunits. The similarity in $M_r$ between the 45-kd MHC class I heavy chain and the putative 40- to 45-kd insulin receptor subunit suggests that the two may in fact be identical.

## Evidence for a Physical Proximity of Insulin Receptors and MHC Class I Molecules on Cell Membranes

Evidence for the physical association of MHC class I molecules and insulin receptors is derived primarily from two types of experiments, those which demonstrate specific coprecipitation and those which show cocapping of the two components.

### *Coprecipitation of Insulin Receptors and MHC Class I Ags*

Several laboratories have demonstrated that photoaffinity-labeled insulin receptors can be immunoprecipitated by antibodies (Abs) to MHC $CI_H$. Haplotype specific precipitations have been reported for H-2 k, b, and s mouse strains using monoclonal antibodies (McAbs) to H-2K and H-2D antigens (Chvatko et al. 1983; Brossette et al. 1984; Phillips et al. 1986; Fehlmann et al. 1985; Yip et al. 1986). An example of haplotype specific precipitation of insulin receptors by anti-class I McAbs is shown in Fig. 1. Abs to MHC class II Ags were not found to precipitate insulin receptors in these experiments. Similarly, Abs to human HLA class I but not class II Ags have been shown to immunoprecipitate photoaffinity-labeled insulin receptors from two human lymphoblastoid cell lines (Samson et al. 1986) and a human hepatoma cell line (Cousin et al. 1987). The ability of an anti-HLA class I McAb to precipitate purified insulin receptors following reaction with $^{125}$I-labeled insulin has also been observed (Due et al. 1986). Pretreatment of cells with a bifunctional chemical cross-linker appears to enhance the precipitation of insulin receptors by anti-class I McAbs (Samson et al. 1986) but is clearly not required. This coprecipitation is not induced by photoaffinity labeling of insulin receptors (Phillips et al. 1986) and cannot be explained on the basis of nonspecific coprecipitation of membrane proteins (Phillips et al. 1986; Fehlmann et al. 1985; Samson et al. 1986; Cousin et al. 1987). Preclearing experiments using McAbs to class I MHC Ags and to insulin receptors initially showed that 25%–30% of insulin receptors on cell membranes may be asso-

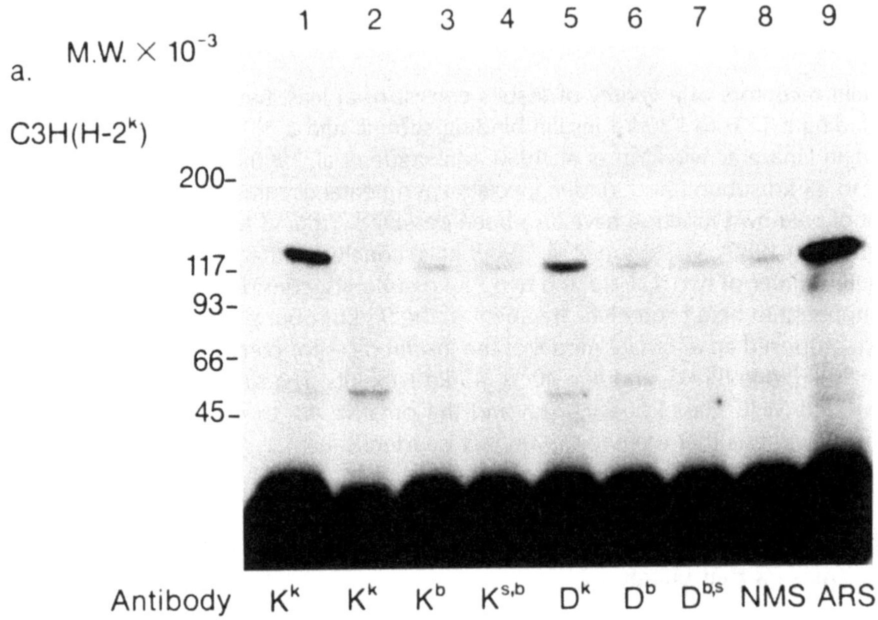


**Fig. 1.** Liver plasma membrane fractions were prepared from C3H mice and photoaffinity labeled with $N^{eB29}$-monoazidobenzoyl-insulin as previously described (Phillips et al. 1986). Control membranes were labeled in the presence of 67 μg unlabeled native insulin per ml (+). Solubilized membrane proteins were precipitated using the following antibodies: 11.4 ($K^k$, anti-H-2$K^k$; *lanes 1, 2*); 28-13-3 ($K^b$, anti-H-2$K^b$; *lane 3*); 34-1-2 ($K^{b,\ s}$, anti-H-2$K^{b,\ s}$, *lane 4*); 15-5-5S ($D^k$, anti-H-$D^k$, *lane 5*); 28-14-8 ($D^b$, anti-H-2$D^b$, *lane 6*); 20-8-4 ($D^{b,\ s}$, anti-H-2$D^{b,\ s}$, *lane 7*); NMS *(lane 8)*; or anti-insulin receptor serum (ARS, *lane 9*). 130-kd subunit; 90-kd subunit; 45-kd subunit

ciated with class I Ags (Brossette et al. 1984; Phillips et al. 1986), although this appears to vary with the age of the animals examined (Yip et al. 1986). Recent experiments in our laboratory suggest that this figure may be as high as 75% in some mouse liver plasma membrane fractions (Table 1), and Samson et al. (1986) have reported 90% of insulin receptors on human B lymphoblastoid cell lines to be associated with HLA class I molecules.

### *Cocapping of Insulin Receptors and MHC Class I Ags on Cells*

The ability to cocap insulin receptors and class I Ags on the cell surface also suggests a selective physical association between these proteins (Due et al. 1986; Simonsen and Olsson 1983). The binding of FITC-labeled insulin and FITC-labeled anti-insulin receptor Abs to U-937 cells is inhibited by capping of MHC class I Ags with Abs to class I heavy chains or $\beta_2$m but not by McAbs to class II Ags (Due et al. 1986). Simonsen et al. (1985) have also reported that the binding of $^{125}$I-labeled insulin to the human monocytoid cell line U-937 can be inhibited 70% by an anti-class I MHC specific McAb (PA 2.6) but not by Abs to $\beta_2$m. In addition they found

**Table 1.** Immunoprecipitation of subcellular membrane fractions: % association of insulin receptors and MHC class I Ags[a]

| Membrane fraction | Experiment 1 | Experiment 2 |
|---|---|---|
| Golgi light | 4.6 | 3.3 |
| Golgi intermediate | 1.4 | 3.9 |
| Golgi heavy | 6.7 | 4.6 |
| Plasma membrane (smooth vesicles) | 31.9 | 29.3 |
| Plasma membrane (smooth membrane) | 37.8 | 75.8 |

[a] Hepatocyte membranes from C3H mice were prepared and characterized as previously described by Posner et al. (1985). The membranes were labeled with $B_{29}$MABI and the solubilized membrane proteins were immunoprecipitated with 11.4 McAb, ARS, or NMS. Samples were analyzed by SDS-PAGE followed by autoradiography (Phillips et al. 1986). Individual insulin receptor subunit bands were cut from gels and counted. NMS values were subtracted from those of 11.4 and ARS. Percent association was calculated as the ratio of CPM from 11.4 to ARS for the 130-kd subunit of the insulin receptor

that pretreatment of the cells with insulin inhibited subsequent binding of PA 2.6 (Simonsen et al. 1985). Since the latter experiments were performed at 4 °C, where no capping is expected to occur, the insulin presumably inhibits binding of the McAb to the MHC Ag.

While coprecipitation and cocapping experiments suggest that the 45-kd insulin receptor protein originally identified by photoaffinity labeling (Yip and Moule 1983; Massague and Czech 1980) is an MHC class I heavy chain, to date there is only one report which addresses this question directly. Due et al. (1986) immunoprecipitated the 130-kd, 90-kd, and 45-kd insulin receptor bands from U937 cells using an McAb specific for the $\alpha$-subunit of the insulin receptor. The 45-kd protein was subsequently shown by immunoblot analysis to react with an anti-HLA-A3 McAb.

### *Association of $\beta_2$m* with Insulin Receptors

Simonsen and Olssen (1983) have proposed that the molecular mechanism of class I MHC-insulin receptor interaction involves the competitive and reversible replacement of $\beta_2$m by the 90-kd or 130-kd insulin receptor subunit. They found that insulin binding is reduced in the presence of excess $\beta_2$m (which presumably would dissociate free $CI_H$ from insulin receptor) (Simonsen and Olsson 1983; Simonsen et al. 1985). Experiments by the same workers, however, show inhibition of insulin binding following capping of cell surface proteins by antibodies to $\beta_2$m (Due et al. 1986). In our own studies (Phillips et al. 1986) and those of Fehlmann and his colleagues (Samson et al. 1986; Cousin et al. 1987), photoaffinity-labeled insulin receptors could be immunoprecipitated by Abs to $\beta_2$m, suggesting that $\beta_2$m is present in the receptor complex. Clearly this question requires further examination.

## Evidence for a Physiological Influence of MHC Class I Molecules on Insulin Action

The influence of MHC Ags on three aspects of insulin receptor function – insulin binding an affinity, insulin receptor cycling, and receptor tyrosine kinase activity – has been examined. In addition to the data reviewed below, some recent experiments by Goodenow and his co-workers on the effects of H-2L antigens on insulin binding in vitro and on insulin action in vivo are described in this forum (see Goodenow et al., this volume).

### *Effect of MHC Class I Ags on Insulin Binding*

Petersen et al. (1985) have reported significant differences in the high affinity binding constant of insulin to isolated hepatocytes among H-2 congeneic mice. Although there may be several explanations for this observation, one possibility is that the insulin binding subunit of the receptors is influenced by the polymorphism of the $CI_H$ with which it associates. Kittur et al. (1987) examined the effect of specific HLA antigens on insulin binding. They measured insulin binding among different radiation-induced HLA-loss mutants of an Epstein-Barr virus-transformed human B cell clone (HLA-A1, -A2, HLA-B8, -B5, HLA-C). No significant differences were found between the low affinity sites of different mutants. They found that lines expressing HLA-B5, alone or in combination with other class I Ags, bound insulin with low affinity ($0.7–3.3 \times 10^8 M^{-1}$) and appeared to carry only a single type of binding site while lines lacking HLA-B5 and positive for other HLA-A or -B antigen bound insulin with high affinity ($4.7–20 \times 10^8 M^{-1}$) and appeared to carry two types of insulin binding site. HLA-C Ag did not appear to affect insulin binding.

Edidin and his co-workers (Kittur et al. 1987) have reported that a mutant line expressing no HLA antigens still bound insulin, suggesting that class I Ags are not obligate parts of the receptor. By contrast, Due et al. (1986) have observed that an MHC class I positive murine T-lymphoma line, R1, binds $^{125}$I-labeled insulin while a class I negative subclone (R1e) does not, although both lines bind anti-insulin receptor McAbs. Similarly, the HLA-class I negative Burkitt lymphoma line, Daudi, does not bind insulin with high affinity (Due et al. 1985). However, high affinity insulin binding has been recovered following transfection of this cell line with the murine H-2L$^d$ class I gene (see Goodenow et al., this volume).

### *Internalization of MHC Class I Ags and Insulin Receptors*

To determine whether class I Ags influence insulin receptor cycling or vice versa, Cousin et al. (1987) examined their respective rates of internalization. They reported an initial rapid internalization of insulin receptors ($t_{1/2} = 2–3$ min) in the presence of insulin. By contrast, MHC class I antigens were internalized at a much slower rate ($t_{1/2} = 2$ h). The presence of insulin had no effect on internalization of class I Ags, and treatment of cells with Abs to class I Ags followed by second Abs had no effect on the rate of internalization of insulin. They concluded that class I MHC Ags and insulin receptors are independently internalized. If, however, the percentage of MHC class I Ags associated with insulin receptors is small relative to the total number of MHC class I Ags, then the internalization of receptor-associated class I Ags may be difficult to discern.

Our laboratory has examined the association of class I Ags and insulin receptors in subcellular membrane fractions. We find that while 40%–75% of insulin receptors in plasma membranes are associated with MHC class I Ags, only 1%–6% of insulin receptors present in Golgi membrane fractions are associated with class I Ags (Table 1). Thus, the association of the two components is also not likely to play a role in the transport of newly synthesized receptors to the cell surface.

### *Effects of MHC Class I Ags on Receptor Tyrosine Kinase Activity*

Recently our laboratory has asked whether MHC class I association could be involved in the regulation of insulin tyrosine kinase activity. Two experimental approaches were used. First, insulin receptors were immunoprecipitated with anti-insulin receptor Abs (ARS), normal mouse serum (NMS), or anti-class I McAbs. The precipitates were then allowed to undergo phosphorylation in the presence or absence of insulin and were subsequently analyzed for insulin receptor autophosphorylation. Alternatively, insulin-stimulated autophosphorylation of the receptor

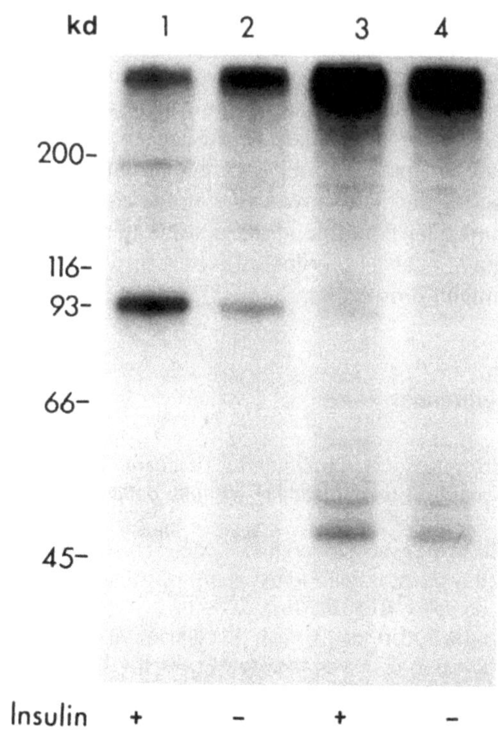


**Fig. 2.** Solubilized liver plasma membrane fractions from C3H mice were incubated in the presence (+) or absence of insulin and phosphorylated with $^{32}P$-ATP and $Mn^{2+}$. The reaction was stopped and the samples were immunoprecipitated with ARS *(lanes 1 and 2)* or 11.4 (anti-H-2K$^k$) *(lanes 3 and 4)*. Samples were then analyzed by SDS-PAGE and autoradiography

was performed prior to immunoprecipitation with the same Ab panel (Fig. 2). In both cases insulin-stimulated autophosphorylation of the 90-kd insulin receptor protein was detected in the anti-insulin receptor precipitates (Fig. 2, lanes 1 and 2) but not in samples precipitated with anti-class I McAbs (Fig. 2, lanes 3 and 4) or control sera. Control experiments have shown that this is not a result of phosphatase activity in the anti-class I McAbs or a result of loss of class I Ags from the receptor complex following receptor autophosphorylation. Our present data suggest that the population of insulin receptors associated with class I Ags has a reduced capacity to undergo autophosphorylation.

## Summary

Considerable evidence has accumulated for a physical association between class I Ags and insulin receptors. It is likely that this represents only a single example of a large number of nonimmunological interactions of MHC class I molecules and other proteins. Similar associations have been reported for other peptide hormone receptors, including epidermal growth factor receptors (Schreiber et al. 1984; Meruelo and Edidin 1979) and glucagon receptors (Lafuse and Edidin 1980). Such associations are not indiscriminate, however, since negative results have been found for transferrin receptors (Due et al. 1986) and for atrial natriuretic factor receptors (Phillips et al. 1986). The biological importance of the association between insulin receptors and class I Ags is not completely understood. However, there is accumulating evidence that class I Ags influence both insulin binding and insulin action. Furthermore, there is evidence to suggest that different polymorphic forms of the MHC $Cl_H$ can confer a degree of polymorphism on the receptor complex. This concept has implications for variability in insulin receptor-ligand interaction and the etiology of receptor-related diseases. At present there is no evidence to suggest that insulin receptor proteins influence the immunological functions of MHC Ags. It is interesting to speculate, however, that the association between insulin receptors and class I molecules could generate unique immunological determinants on the class I heavy chain, alter its binding capacity for foreign antigens, or influence class I restricted antigen presentation.

## References

Brossette N, van Obberghen E, Fehlmann M (1984) Interaction between insulin receptors and major histocompatibility complex antigens in mouse liver membranes. Diabetologia 27: 74–76

Chvatko Y, van Obberghen E, Kiger N, Fehlmann M (1983) Immunoprecipitation of insulin receptors by antibodies against class I antigens of the murine H-2 major histocompatibility complex. FEBS Lett 163: 207–211

Cousin JL, Samson M, Pilch P, Fehlmann M (1987) Internalization of insulin receptors and HLA antigens in human hepatoma cells. Biochem J 242: 403–410

Dausset J, Contu L (1980) Is the MHC a general self-recognition system playing a major unifying role in an organism. Hum Immunol 1: 5–17

Due C, Linnet K, Langeland-Johansen N, Olsson L (1985) Analysis of insulin receptors on heterogeneous eukaryotic cell populations with fluorochrome-conjugated insulin and fluorescence-activated cell sorter. Advantages and limitations to the $^{125}$I-labeled insulin methodology. Diabetologia 28: 749–755

Due C, Simonsen M, Olsson L (1986) The major histocompatibility complex class I heavy chain as a structural subunit of the human cell membrane insulin receptor: Implications for the range of biological functions of histocompatibility antigens. Proc Natl Acad Sci USA 83: 6007–6011

Edidin M (1983) MHC antigens and non-immune functions. Immunol Today 4: 269–270

Fehlmann M, Peyron JF, Samson M, van Obberghen E, Brandenburg D, Brossette N (1985) Molecular association between major histocompatibility complex class I antigens and insulin receptors in mouse liver membranes. Proc Natl Acad Sci USA 82: 8634–8637

Ivanyi P (1978) Some aspects of the H-2 system, the major histocompatibility system in the mouse. Proc R Soc Lond [Biol] 201: 117–158

Kasuga M, Karlsson FA, Kahn CR (1982) Insulin stimulates the phosphorylation of the 95,000-dalton subunit of its own receptor. Science 215: 185–187

Kittur D, Shimizu Y, DeMars R, Edidin M (1987) Insulin binding to human B lymphoblasts is a function of HLA haplotype. Proc Natl Acad Sci USA 84: 1351–1355

Lafuse W, Edidin M (1980) Influence of the mouse major histocompatibility complex, H-2, on adenylate cyclase activity and on glucagon binding to liver cell membranes. Biochemistry 19: 49–54

Massague J, Czech MP (1980) Multiple redox forms of the insulin receptor in native liver membranes. Diabetes 29: 945–947

Massague J, Pilch PF, Czech MP (1980) Electrophoretic resolution of three major insulin receptor structures with unique subunit stoichiometries. Proc Natl Acad Sci USA 77: 7137–7141

Meruelo D, Edidin M (1980) The biological function of the major histocompatibility complex: hypothesis. Contemp Top Immunobiol 9: 231–253

Petersen KG, Heidenreich S, Pilot PR, Kerp L (1985) Insulin secretion in vitro and insulin binding to isolated hepatocytes in congeneic mice with different H-2 complexes. Diabetes 34: 256–259

Phillips ML, Moule ML, Delovitch TL, Yip CC (1986) Class I histocompatibility antigens and insulin receptors: Evidence for interactions. Proc Natl Acad Sci USA 83: 3474–3478

Posner BI, Khan MN, Bergeron JJM (1985) Peptide hormone receptors in intracellular structure from rat liver. Methods Enzymol 109: 219–231

Samson M, Cousin JL, Fehlmann M (1986) Cross-linking of insulin receptors to MHC antigens in human B lymphocytes: Evidence for selective molecular interactions. J Immunol 137: 2293–2298

Schreiber AB, Schlessinger J, Edidin M (1984) Interaction between major histocompatibility complex antigens and epidermal growth factor receptors on human cells. J Cell Biol 98: 725–731

Simonsen M, Olsson L (1983) Possible roles of compound membrane receptors in the immune system. Ann Inst Pasteur Immunol 134D: 85–92

Simonsen M, Skyodt K, Crone M, Sanderson A, Fujita-Yamaguchi Y, Due C, Ronne E, Linnet K, Olsson L (1985) Compound receptors in the cell membrane: Ruminations from the borderland of immunology and physiology. Prog Allergy 36: 151–176

Yip CC, Moule ML (1983) Structure of the insulin receptor in rat adipocytes. The three interconvertible redox forms. Diabetes 32: 760–767

Yip CC, Yeung CWT, Moule ML (1980) Photoaffinity labeling of insulin receptor proteins of liver plasma membrane preparations. Biochemistry 19: 70–76

Yip C, Moule M, Hirstwood C, Cinader B, Phillips ML (1986) Class I histocompatibility antigens and insulin receptors: Evidence for age-dependent interactions. In: Jaworski MA (ed) The immunology of diabetes mellitus. Elsevier, Amsterdam, p 215

# Soluble Classical Class I MHC Antigens in Solution in the Body Fluids

P. B. Singh[1], R. E. Brown[2], and B. J. Roser[1]

[1] Department of Immunology, AFRC Institute of Animal Physiology and Genetics Research, Babraham Research Station, Babraham Hall, Cambridge, CB2 4AT, England
[2] Department of Psychology, Dalhousie University, Halifax, Nova Scotia B3H 4J1, Canada

## Introduction

Classical class I transplantation antigens (HLA-A, -B, -C in man; H-2K, H-2D in mouse) of the major histocompatibility complex (MHC) are membrane-bound glycoproteins that act as restricting elements in antigen recognition by T-lymphocytes (Klein 1986). However, there have been reports of classical class I transplantation antigens that have been found in true solution in the body fluids of mouse (Callahan et al. 1975, 1976), man (van Rood et al. 1970; Vincent and Revillard 1979), and rat (Spencer and Fabre 1987; Singh et al. 1987a). The function of these molecules has not been defined but suggestions ranging from there being no function for them at all (Gussow and Ploegh 1987) to a role for them in inducing tolerance to self (Kamada et al. 1981; Kress et al. 1983; Cosman et al. 1982) have been put forward.

We now show that classical, polymorphic class I molecules are found in the circulation of the rat and are then excreted in the urine, and that urine taken from normal donors which differ only at the class I MHC locus can be distinguished between by detector animals. Thus, the family of class I glycoproteins which are markers of tissue uniqueness in the membrane-bound form are also excreted into the environment, where they may act as the recognition signals that regulate social interactions between individuals within a species.

## Materials and Methods

### *Animals*

Rats of the DA ($RT1^{av1}$), PVG ($RT1^{c}$), PVG-$RT1^{u}$, and (PVG × DA)$F_1$ hybrid strains were bred in our own conventional animal house or in a specific pathogen-free (SPF) unit. Some PVG-$RT1^{u}$ animals were obtained from Sir William Dunn School of Pathology, University of Oxford. Hooded Lister rats were obtained from the Psychology Department at Cambridge University.

For behavioral experiments, rats were housed in groups of two or three in opaque plastic cages (NKP Ltd.) on a reversed 12:12 L:D cycle with lights off at 6.00 hours at a temperature of 20° ± 2 °C. Food (Labsure irradiated diets CRM NUTS) and water were freely available.

*Neonatally Tolerant Rats*

Rats were rendered tolerant according to a previously published protocol (Roser and Dorsch 1979). Briefly, DA or PVG neonates were injected intravascularly with 75 µl of a 50% v/v suspension of (PVG × DA)$F_1$ hybrid bone marrow cells within 24 h of birth. As adults, these rats failed to reject skin grafts from donors homozygous for the unshared antigens of the $F_1$ for more than 50 days. Their lymph node cells were shown to be unreactive against donor antigens in popliteal lymph node graft-versus-host assays (Ford 1978) before they were used in experiments.

*Enzyme-Linked Immunoabsorbent Assays*

Enzyme-linked immunoabsorbent assays (ELISAs) were carried out in rigid, nonsterile, flat-bottomed, 96-well polystyrene plates (Nunc Immunoplates II, Gibco Ltd., Middlesex, UK). To coat individual wells, 200 µl of a 20 µg/ml solution of purified monoclonal anti-class I antibody in PBS/0.1% azide, pH 7.4, was used. The plates were covered and kept at 4 °C until use. On the day of use, excess antibody was removed and wells postcoated up to the brim with PBS/azide containing 10% v/v fetal calf serum (FCS) for 1 h at 4 °C to block nonspecific protein adsorption sites. After 1 h the plates were washed twice with PBS + 0.5% w/v Tween 20 (Sigma, Poole, England) (PBST). Two hundred microliters of a solution containing the class I MHC antigen was then added to the appropriate wells and incubated overnight at 4 °C. After three washes with PBST and three washes with PBSR + 2% v/v FCS, 200 µl of a second noncompetitive anti-class I monoclonal antibody coupled to biotin was added at 8 µg/ml and incubated for 1 h. The plate was again washed three times with PBST and three times with PBST + 2% v/v FCS before addition of 200 µl of a 4 µg/ml concentration of Stepavidin-coupled horseradish peroxidase (Miles Laboratories Ltd., Slough, UK). The plates were then incubated for 20 min at 4 °C. After washing twice with PBST + 2% v/v FCS and twice under tap water for approximately 15 s each time, 200 µl of the substrate 3,3′,5,5′-tetramethyl benzidine at 100 µg/ml was added. The reaction was stopped after 5 min with 50 µl of 2 M $H_2SO_4$ and the plates read in a Titerik Multiscan ELISA plate reader (Flow Labs. Ltd., Irvine, Scotland).

In the assay for detection of the $A^{av1}$ class I MHC antigens of the DA strain the capture antibody on the plate was JY3/109 (haplotype specificity, *av1* +, *c* −, *f* −, *h* −, *l* −, *n* −, *o* +, *u* −, *b* +), while the second biotin-labeled antibody was JY1/116 (haplotype specificity, *av1* +, *c* −, *f* −, *h* −, *l* −, *n* −, *o* −, *u* −, *b* +). For the detection of the $A^c$ antigens of the PVG strain, the capture antibody was YR5/310 (haplotype specificity, *av1* −, *c* +, *f* −, *h* +, *l* −, *n* −, *o* −, *u* −), with YR5/12 (haplotype specificity, *av1* −, *c* +, *f* −, *k* −, *l* −, *n* +, *o* −, *u* −) as the biotin-labeled second stage. All are rat IgG alloantibodies.

*Affinity Chromatography*

Class I molecules were extracted from liver cell membranes or lymph by affinity chromatography with specific monoclonal anti-class I antibodies covalently coupled to cyanogen bromide-activated Sepharose 4B (Pharmacia, Uppsala, Sweden). To prepare the membrane-bound form, DA strain rat livers were homogenized in a buffer which consisted of 0.05 M TrisHCl pH 7.2, 0.25 M sucrose, 25 mM KCl, 5 mM $MgCl_2$, and 1 mM phenylmethylsulfonyl fluoride (Sigma, Poole, UK) to inhibit proteolysis. Crude membranes were obtained by centrifugation at $10^3 \times g$ for 20 min to remove nuclei and large debris and at $1.05 \times 10^5 \times g$ for 1 h over a 40% w/w sucrose cushion to sediment the membrane fraction. This was then made up to 10 mg/ml with 0.02 M Tris/HCl pH 7.8 containing 0.5% w/v *n*-octylglucopyranoside (Sigma).

The solubilization was carried out for 1 h at 4 °C, stirring gently throughout. The material was then centrifuged at $1.05 \times 10^5 \times g$ to remove debris. The RT1.$A^{av1}$-containing supernatant underwent affinity chromatography according to Parham (1979), on a JY1/116 anti-RT1.$A^{av1}$ affinity column which was eluted with 0.015 M carbonate buffer pH 11.0 containing 0.5% w/v *n*-octylglucopyranoside. The eluted fractions were concentrated using a CX30 concentrator (Millipore Corp. Bedford, Mass. USA) before running on SDS-PAGE.

To purify the soluble RT1.$A^{av1}$ molecules, lymph was spun at $1.05 \times 10^5 \times g$ for 1 h so that all the lipid and lipoproteins floated and could be removed. The opalescent lymph was loaded onto the JY1/116 affinity column and washed as above, except there was no detergent in the buffers. RT1.$A^{av1}$ was eluted with 0.5% w/v *n*-octylglucopyranoside in 0.015 M carbonate buffer pH 11.0, concentrated, and run on SDS-PAGE as for the solubilized membrane material.

To purify soluble RT1.$A^{av1}$ molecules from urine, 500 ml urine collected from eight DA animals was spun at $105000 \times g$ to remove any material that may block the columns and then treated as in the case of the lymph.

### *Sodium Dodecylsulfate–Polyacrylamide Gel Electrophoresis (SDS-PAGE)*

SDS-PAGE analysis of proteins purified by affinity chromatography was undertaken according to Laemmli (1970) and bands stained with Coomassie blue.

### *High Pressure Liquid Chromatography (HPLC)–Gel Filtration*

Under nondissociating conditions, 200 μl of PVG serum and 200 μl of purified RT1.$A^c$ molecules from DA urine were fractionated on an LKB TSK G3000SDW column (LKB Produckter, Bromma, Sweden) running at 0.5 ml/min in PBS/azide.

### *Scatchard Analysis*

The anti-$A^{av1}$ antibody JY1/116 and the anti-$A^c$ antibody YR5/12 were iodinated using the lactoperoxidase method (Heusser et al. 1973) at a ratio of 1 μCi to 1 μg purified IgG. These antibodies were then used to measure the number of antigenic sites on the relevant rat erythrocytes. Titrated amounts of antibody were incubated with 25 μl of a 10% v/v RBC suspension for 1 h at 4 °C and the cells separated from unbound antibody by centrifugation on a 200 µl mixture of di-*n*-butylphthalate (8 parts) and di-nonyl phthalate (2 parts) (BDH Chemicals Ltd., Poole, UK). The tubes were frozen with solid $CO_2$ and then cut at the midpoint of the frozen oil column. The RBC pellet and the floating supernatant were counted separately in a Packard model 5210 $\gamma$ counter (Packard Instrument Co., Illinois, USA) to give a precise measure of bound and unbound antibody. The antibody affinity and number of antigenic sites were then calculated by the method of Scatchard.

### *Saturation Analysis*

One-tenth of the amount used in Scatchard analysis (Scatchard 1949), i.e., 25 µl of a 1% v/v RBC suspension was then used in an antibody-binding inhibition assay using 25 µl of neat serum of the relevant type in 12 replicates. The three reactants, $^{125}$I-labeled antibody, serum containing antigen, and RBCs, were incubated for 1 h on a shaker at 4 °C to reach equilibrium. The cells were washed three times with PBS azide + 2% v/v FCS, harvested onto glass fiber paper, washed again using a Skatron harvester (Flow Laboratories, Irvine, Scotland), and counted in a Packard $\gamma$-counter. Results were processed according to the formulae described by Ekins (1974). Serum in which the amount of class I molecule was known from saturation analysis was aliquoted and stored at −20 °C. This was always run in ELISA as a positive calibration control. The amount of antigen in unknown samples run in these assays could then be estimated accurately by measurement of the horizontal displacement of regression lines drawn through the linear portions of the ELISA titration curves.

### *Half-Life Measurement*

Four DA rats rendered neonatally tolerant of PVG antigens were bled of 3 ml of blood from the tail vein. They were then immediately injected i.v. with 5 ml of PVG serum. After 10 min the first blood sample was taken. A blood sample was then taken from one pair of rats 2 h later and from the other pair after another 2 h. The pairs were then bled alternately at 4-h intervals.

### *Habituation–Dishabituation Tests*

PVG-$RT1^u$ subjects were given a 15-min habituation to the test arena (an opaque plastic cage measuring 29.6 × 23.6 × 14.6 cm with a ½-in. square wire mesh top which rose another 5.6 cm above the cage rim) and were then tested once per week in a habituation–dishabituation test in which nine sequential 2-min odor presentations were given. For these presentations, 0.1 ml of liquid was placed on a 7-cm diameter disk of Whatman No. 1 filter paper which was taped to the mesh top of the cage, with the center of the disk 13.5 cm from the floor of the arena. The odorized disk was replaced every 2 min, by removing the top of the cage and replacing it with another top carrying a new odorized disk. Tops and arenas were washed in 70% ethanol after each test and filter papers were used only once.

On the first three 2-min tests, water was placed on the filter paper, so the subject had a 6-min habituation to the test procedure before the first urine sample was presented for the next three 2-min trials. On trial 7, the second urine sample was presented for three trials. Donor urine was collected from ten individual PVG and ten individual PVG.R1 males the same age as the subjects in Urimax metabolic cages for 6 h during the day, and frozen in aliquots until used. Each subject was given two urine samples from known individuals.

The time spent rearing on the hind legs and sniffing, with the nose within 1 cm of the disk throughout each 2-min period, was recorded by an observer using a stopwatch. Each of the rats was tested in three habituation–dishabituation tests. The observer was "blind" to the order of testing odorant samples, which were prepared by an assistant according to a random list.

### *Nitrogen-Purged Urine*

Oxygen-free compressed nitrogen (BOC Ltd., Guildford, Surrey, UK) was bubbled through a reservoir of distilled water and a 0.2-μm sterilizing filter (Flow Labs.) before bubbling through 5 ml of each urine sample for 48 h at a flow rate of 50 ml per minute.

### *Dialysis of Urine*

Five milliliters of urine was dialyzed at +4 °C against 5 liters of PBS in a visking tubing bag (size 8/32, Scientific Instrument Centre, London, UK) with a molecular weight cut off of 12 000.

### *Statistical Methods*

Analysis of the behavioral data was done using separate randomized block (repeated measures) analyses of variance for each experiment. Post hoc analyses were done using Newman-Keuls tests (Kirk 1968).

## Results

### *Classical Class I MHC Antigens in Solution in the Circulation*

From two panels of monoclonal antibodies (McAbs), two-site ELISA assays were developed that could detect A-region soluble class I transplantation antigens in both the $RT1^{av1}$ and the $RT1^{c}$ rat strains.

The ELISA was used to detect soluble class I molecules in serum and lymph (Fig. 1 a). Lymph has twice the concentration of soluble class I antigens as serum. Analysis of the MHC molecules, recovered from lymph by affinity chromatography in SDS-PAGE, revealed a heterodimeric molecule with a heavy chain of 39 kd and a

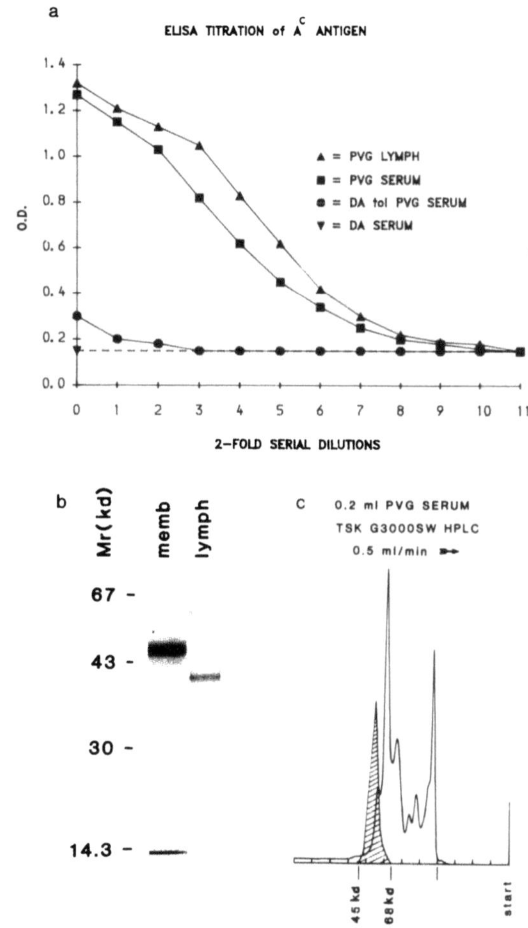


**Fig. 1a–c.** Detection and characteristics of class I molecules in serum and lymph. **a** ELISA detections of the RT1.A$^c$ antigen in central lymph (▲) and serum (■). The level of donor PVG antigen in the serum of DA rats made tolerant of the (PVG × DA)$F_1$ hybrid strain (●) is low but readily detectable because of the constant, small background (—) in this assay (▼, control DA serum). **b** SDS-PAGE analysis of membrane-bound and soluble class I molecules. Membrane-bound A$^{av1}$ molecules from DA liver have a heavy chain $M_r$ of 47 kd and a $\beta_2$-microglobulin light chain of 13 kd. The lymph molecules has a smaller heavy chain ($M_r$ 39 kd). **c** HPLC gel filtration of DA serum. The serum RT1.A$^{av1}$ molecules, detected by ELISA, eluted as a sharp peak *(shaded area)* at the trailing edge of the serum protein peaks with an apparent $M_r$ between that of BSA (68 kd) and ovalbumin (43 kd) (Singh et al. 1987a)

light chain of 13 kd typical of $\beta_2$-microglobulin (Fig. 1 b). The body fluids of the rat therefore contain authentic classical polymorphic class I antigens in solution, and fractionation of the serum molecule on HPLC–gel filtration (Fig. 1 c) under nondissociating conditions shows these molecules to be true monomers and not molecular aggregates released from formed elements in the blood brought about by the process of clotting.

*Turnover and Urinary Excretion of Serum Class I Molecules*

In order to measure accurately the molar concentration of soluble class I MHC antigens in the serum, the method of saturation analysis was used (Ekins 1974). This method depends on a calibrated source of these molecules which was provided by determination of the number of molecules on the erythrocyte surface by Scatchard analysis (using the anti-RT1A$^c$ McAb, YR5/12, 4500 sites per PVG erythrocyte, $K = 1.5 \times 10^8\ M^{-1}$; using the anti-RT1A$^{av1}$ antibody JY1/116, 10800 sites per DA erythrocyte, $K = 1 \times 10^8\ M^{-1}$). These erythrocytes were then used as targets in an equilibrium competition assay to measure the concentration of soluble molecules in the sera of two strains of rat (Singh et al. 1987a). The concentration of the RT1.A$^c$ molecule of the PVG strain was found to be 379 ng/nl ($K = 7.5 \times 10^7\ M^{-1}$), and that of the RT1.A$^{av1}$ molecule of the DA strain, 347 ng/ml ($K = 4.5 \times 10^8\ M^{-1}$). The concentration of the different lymph draining from lymphoid tissue is about twice this level (Fig. 1a).

Previous experiments (Roser et al. 1983) had shown that within 24 h after liver grafting, the concentration of the donor class I MHC antigen in the host circulation was near maximal. This implied that the soluble molecule in the body fluids may be rapidly produced and therefore may also turn over rapidly. Accordingly, we measured the half-life of the serum molecule. This was done by capitilizing on the facts that neonatally tolerant rats contain only trace amounts of donor antigen in their circulation (Fig. 1a) and exhibit no reactivity against donor MHC antigens (Roser and Dorsch 1982). DA (RT1$^{av1}$) rats tolerant of the PVG (RT1$^c$) MHC were injected with

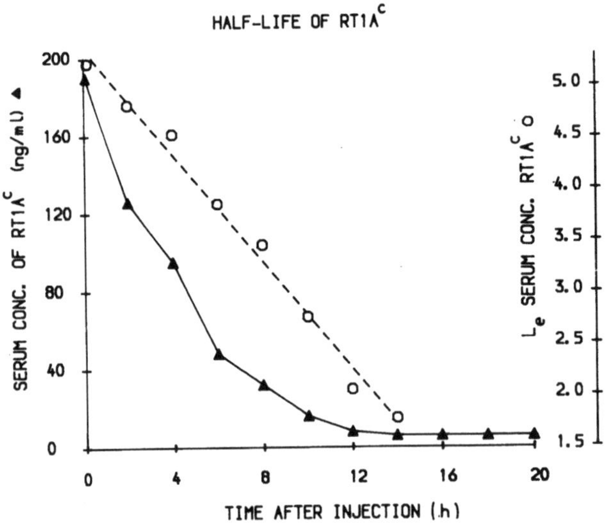


**Fig. 2.** Half-life of serum class I molecules. Four DA rats rendered neonatally tolerant of PVG antigens were bled of 3 ml of blood and injected i.v. with 5 ml of PVG serum. Analysis of the serum for the RT1.A$^c$ antigens by ELISA showed that the molecules disappeared from the recipients' circulation exponentially with a gradient *(λ)* on a semilogarithmic plot of −0.2608. From classical pharmokinetics: $t_{1/2} = \frac{\log_{e^2}}{-\lambda} = 2.66$ h

the large amounts of donor (PVG) serum and the level of donor antigens in their blood was measured at frequent intervals. These experiments showed that the decay of donor molecules was exponential, with a half-life of 2.7 h (Fig. 2).

The short half-life of the molecules in the circulation suggested that they were either rapidly degraded or excreted. Examination of urine showed that these molecules were readily detectable and calibration of the ELISA with the serum molecule showed that the concentration of the class I molecules in the urine was quite variable and ranged from 40 to 190 ng/ml. Class I molecules purified from urine by affinity chromatography showed that they not only possess the typical 39-kd heavy chain and 13-kd light chain associated with the serum class I molecule but also a major protein band at 30-kd suggesting fragmentation of class I molecules is occurring in the urine (Fig. 3 a). This was confirmed on fractionation of class I antigen, extracted from the urine, by HPLC–gel filtration and assay of the fractions by ELISA (Fig. 3 b). This showed that the antigenic material emerges from the HPLC column as a broad band of activity from about 50 kd to 15 kd $M_r$. This is in sharp contrast to the behavior of the class I molecule in serum, which emerged from the HPLC column as a well defined narrow peak between 68 kd and 43 kd (Fig. 1 c).

### *Olfactory Detection of MHC Class I Associated Urinary Odors*

The ready detectability of polymorphic class I MHC antigens of the RT1.A$^c$ and RT1.A$^{av1}$ types and their degradation products in the urine of normal rats by ELISA prompted us to see whether rats could detect urinary odors associated with these molecules. In these experiments we used the habituation–dishabituation method (Sundberg et al. 1982) in which untrained detector animals were habituated to the odor of urine samples from donors of one strain and then tested for dishabituation when exposed to the urine from donors of another strain.

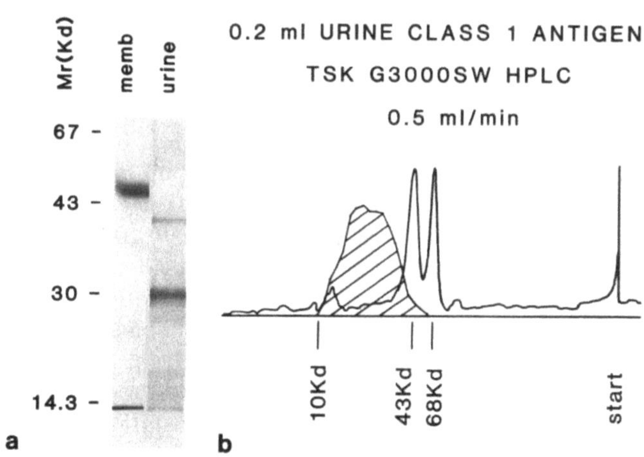


**Fig. 3 a, b.** Detection and characteristics of soluble class I molecules in urine. **a** SDS-PAGE analysis of RT1.A$^{av1}$ molecule in urine. The gel is run under reducing conditions. **b** HPLC–gel filtration of urine class I molecules under nondissociating conditions (Singh et al. 1987 a)

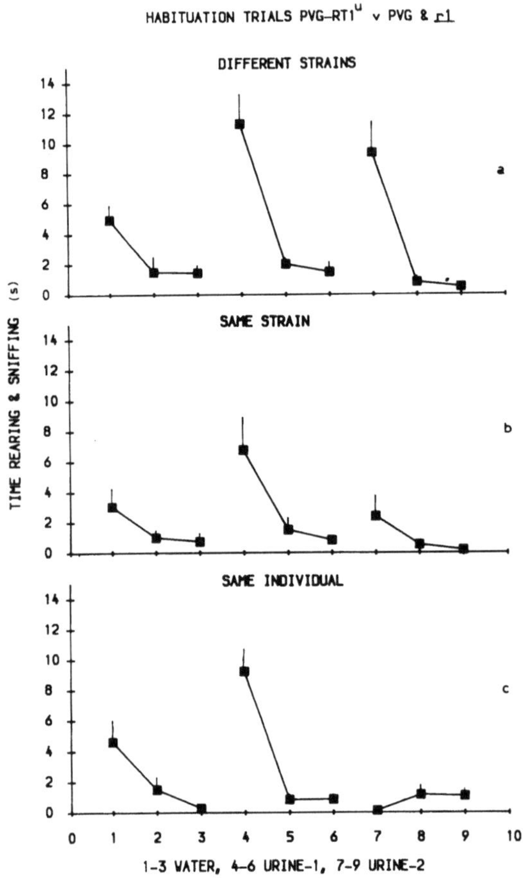


**Fig. 4 a–c.** PVG-RT1$^{u}$ strain of rats can distinguish between urine sample taken from animals that differ only at the RT1.A classical class I genetic locus. **a** The urine samples were from different strains and, after the three tests with water, the first (PVG) urine sample was investigated significantly more than the third water sample. The response habituated rapidly and when the second (PVG.R1) urine sample was presented, the subject dishabituated. Response to the second odor habituated rapidly. **b** Urine samples from two individual males of the same strain (PVG-1, PVG-2) were presented. There was no significant dishabituation. **c** Two urine samples from the same individual. There was no significant dishabituation. The figure shows mean time (±SEM) rearing and sniffing at odors on each 2-min presentation (Singh et al. 1987 a)

Figure 4 a shows that PVG-*RT1*$^{u}$ rats could readily discriminate between the odor of PVG urine containing the A$^{c}$ molecule and PVG.R1 urine containing the A$^{av1}$ molecule. There were significant differences among the nine means for all eight subjects ($F = 24.249$; $df = 8{,}104$; $P < 0.001$) and post hoc tests showed that significantly more time was spent investigating odors on tests 4 and 7 than on any other test ($P < 0.01$). In Fig. 4 b there were significant differences among the nine means over all eight subjects ($F = 17.730$; $df = 8{,}120$; $P < 0.001$); more time was spent sniffing the odor on test 4 than any other test ($P < 0.01$). Thus individuals of the PVG strain cannot be distinguished between as evidenced by the lack of dishabituation on trial 7. The odor is a stable phenotype since urine collected from the same individual, but at different times, cannot be discriminated between. There were significant differences among the nine means for alle eight subjects ($F = 21.053$; $df = 8{,}120$; $P < 0.001$), and the sniffing on test 4 was significantly more than on all other tests ($P < 0.01$). Time sniffing on test 7 did not differ from that on test 6.

*Is the odoriferous component in the urine the class I molecules themselves or the remainder urine with the class I molecules removed?* The most direct mechanism to explain the MHC discrimination by the detector animals would be that the class I molecules themselves were the odorants.

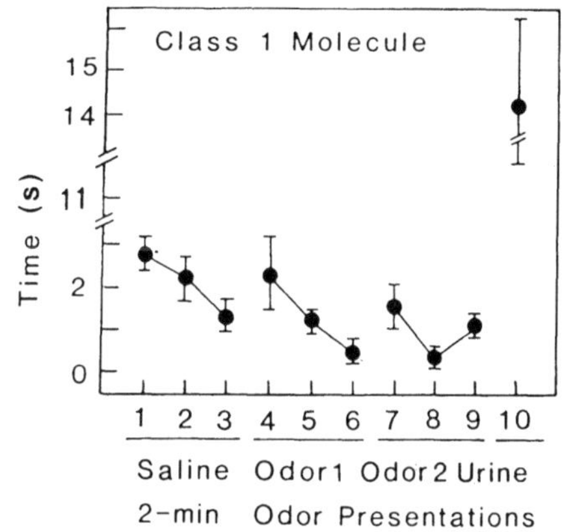


**Fig. 5.** Mean time (±SEM) rearing and sniffing at PBS containing A-region class I molecules purified from PVG and PVG.R1 urine

Figure 5 shows that A-region class I molecules purified from PVG (RT1.A$^{c}$) and PVG.R1 (RT1.A$^{av1}$) urine by monoclonal antibody affinity chromatography could not be used by 12 hooded Lister subjects to discriminate between the two strains. The only odor which was sniffed more than the others was the urine presented on trial 10 ($F = 26.56$; $df = 9{,}99$; $P < 0.001$). The response to the class I molecule on trials 4 and 7 was no greater than that to saline. The remainder could be used by another 12 subjects to discriminate between the two strains (Fig. 6). There were significant differences in sniffing time over the nine odor presentations ($F = 27.24$; $df = 8{,}88$; $P < 0.001$), with more sniffing on trials 4 and 7 than on any other trial ($P < 0.01$).

*Are the molecules that confer the MHC-related odor volatile?* Figure 7 shows that nitrogen-purged urine cannot be used by PVG-*RT1*$^{u}$ detector animals to discriminate between the PVG and PVG.R1 strains. There were significant differences in snif-

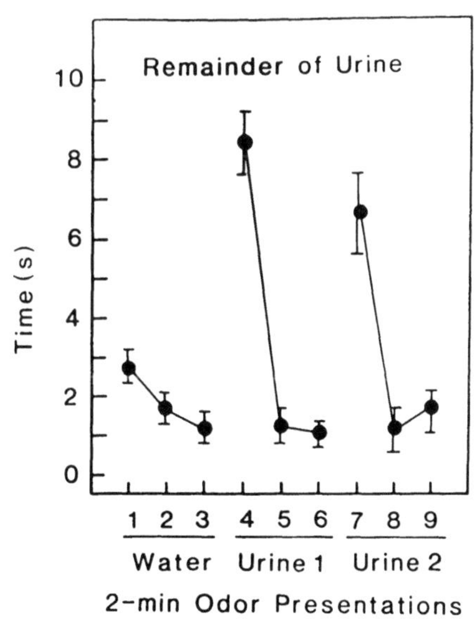


**Fig. 6.** Mean time (±SEM) rearing and sniffing at the remainder PVG or PVG.R1 urine after removal of the A region class I molecules

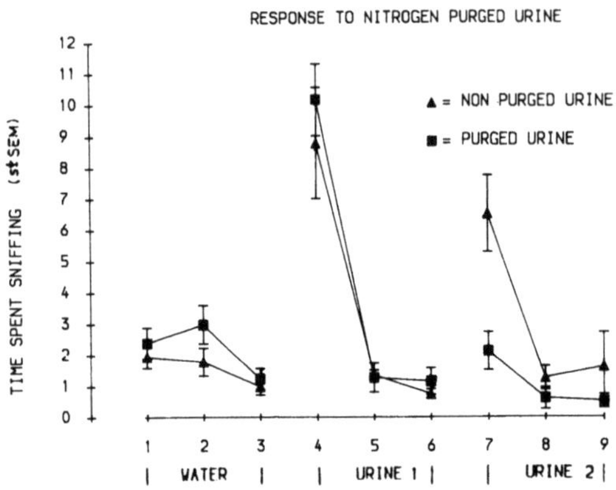


**Fig. 7.** Mean time (±SEM) rearing and sniffing at PVG or PVG.R1 urine after removal of volatiles by nitrogen passage for 48 h

fing times over the nine odor presentations for all eight subjects ($F = 29.25$; $df = 7{,}56$; $P < 0.001$). The urine remained attractive since more time was spent sniffing odors on trial 4 than on any other trial ($P < 0.01$). For the untreated control urine there were significant differences in sniffing times over the nine presentations ($F = 12.145$; $df = 7{,}56$; $P < 0.001$). More time was spent sniffing odors on trials 4 and 7 than on any other trial ($P < 0.01$).

However, dialysis experiments have shown that the odiferous components in the urine are associated with a high molecular weight fraction.

Using dialysis tubing of 12000 dalton cut-off, 5 ml urine was dialyzed against 5 liters of PBS for 48 h; the retentate was shown, using the habituation–dishabituation protocol, to enable detector animals to discriminate between PVG and PVG.R1 rat strains. There were significant differences over all nine odor presentations for all eight subjects ($F = 7.934$; $df = 7{,}56$; $P < 0.001$) and more time was spent sniffing odors 4 and 7 than any other odors ($P < 0.01$) although the dishabituation responses were diminished to 4 s. The dialysate could not be used to discriminate between PVG and PVG.R1 urine. The response to the dialysates was no different than that to PBS alone.

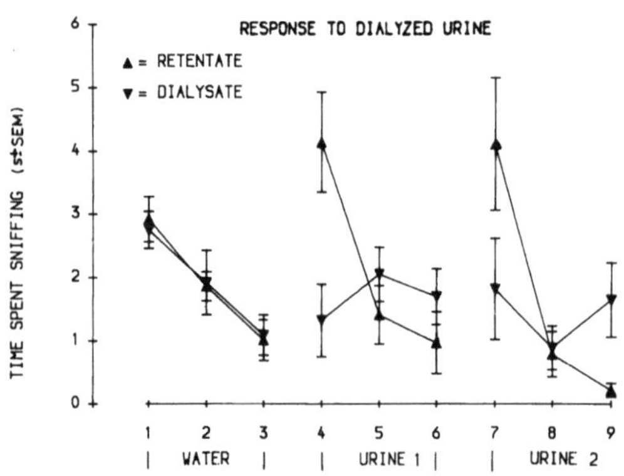


**Fig. 8.** Mean time (±SEM) rearing and sniffing at the retentates and dialysates after dialysis of PVG and PVG.R1 urine

## Discussion

Classical class I transplantation antigens have previously only been studied as membrane-bound molecules that have been studied in the context of graft rejection and antigen presentation (Klein 1986). Therefore, there is at present no conceptual framework that accommodates the presence of soluble forms of any class I molecules. However, it appears that a number of distinct forms of such molecules clearly exist. This study has confirmed the presence of such molecules in the sera of the rat (Fig. 1) which have a short half-life in the circulation (Fig. 2) and are excreted in the urine (Fig. 3) via the kidneys, where over 90% of the molecules undergo degradation (Singh et al. 1987 b).

The excretion of classical class I transplantation and degradation products in the urine seemed to us important since the precision of olfactory discrimination in both mice and rats has been shown to include the ability to identify urine samples from congeneic animals that differ only at the MHC (Beauchamp et al. 1985; Yamazaki et al. 1979). In fact it has been possible to show that trained mice can distinguish the H-2K mutant *bm1* mouse from the nonmutant C57BL/6 mouse (Yamazaki et al. 1983), implying the direct involvement of class I MHC gene products rather than products of unknown MHC-linked genes.

Behavioral experiments showed that untrained PVG-*RT1*$^{u}$ rats could discriminate between urine odors of male rats of the PVG and PVG.R1 strains as evidenced by the pronounced dishabituation on trial 7 (Fig. 4 a). Individual PVG males are not discriminable by their urine odors (Fig. 4 b).

The immunological function of classical class I molecules as restricting elements has led to the suggestion that the MHC-related odor is governed by the action of immune-response gene effects of the class I genetic locus on commensal bacterial flora of the skin, urinary tract, or gut (Howard 1977). Individual MHC haplotypes would then be associated with unique flora. However, two observations mitigate against this idea. Yamazaki et al. (1985) found that in radiation chimeras of the $F_1$ into parent type, urinary odors are of the $F_1$ donor. Since it has been shown that class I restricted immune responsiveness is heavily biased toward the recipient parental MHC type (Bevan and Fink 1978; Zinkernagel and Doherty 1979), it would be expected that the immune response phenotype would remain parental in these animals. In addition, it has been shown that mice (Yamazaki et al. 1982) that are congeneic with respect to the nonclassical Qa:T1a class I loci can be distinguished between by odor. These molecules have no known Ir gene function, though they are found in the circulation of both mice (Maloy et al. 1984; Kress et al. 1983) and rats (Spencer and Fabre 1987). Urine collected from germ-free animals will be assayed in the habituation–dishabituation system to test the hypothesis that bacterial flora are necessary for the production of the MHC-related odors.

Intact class I MHC antigens purified from urine by affinity chromatography are not the odoriferous component in the urine (Fig. 5). This result could, perhaps, be predicted since these molecules are large glycoproteins and therefore essentially nonvolatile. Mechanisms have been described for the detection of (Wysocki et al. 1980), and behavioral responses to (Johns et al. 1978), nonvolatile urinary molecules by means of the vomeronasal organ; however, there is clear evidence that volatile odors are also used by rodents to identify MHC type of urine odors (Fig. 7). The remainder of the urine after removal of the class I molecules can be used to discriminate the two strains of rat (Fig. 6). The odoriferous component therefore resides with this fraction and not with the class I molecules themselves.

There are two possible mechanisms which may explain the role of soluble class I MHC antigens themselves in conferring the MHC-associated odors found in the urine. Firstly, the odoriferous component may not be the intact class I molecules, but rather small volatile fragments of the MHC molecule. These fragments would have been too small to be removed by the affinity chromatography and would therefore remain in the urine. Secondly, the odorants could be small volatiles that associate with the class I molecules in an allele-specific way. The class I MHC antigens in this case act as a carrier which transports small volatiles from a common pool in the serum and releases them in the urine, where, having dissociated from the carrier, they give rise to the odor profile which is specific for the class I molecule.

The fact that rats fail to discriminate between PVG and PVG.R1 by smelling their sera (since all sera contain the mixture volatiles) (Brown et al. 1987) is consistent with the latter mechanism. Moreover, dialysis experiments (Fig. 8) have shown that the odoriferous component in the urine is associated with a high molecular weight fraction, since the retentate inside the dialysis tubing can be used to discriminate between the PVG and PVG.R1 rat strains. If the molecules were freely dialyzable a dilution of 1:1000 would render them ineffective in this assay, since dilution of neat urine by this factor cannot be used to discriminate between these strains (unpublished result).

Further experiments to investigate the role of these molecules in conferring the MHC-associated odors are projected. Purified $A^{av1}$ MHC class I antigens will be injected into PVG ($RT1^{c}$) animals and the removal of the antigen from the serum and appearance in the urine will be monitored by the ELISA. This should bring into the urine the $A^{av1}$ specific odor by either association of small volatiles with these molecules, or by its degradation to small fragments which then act as specific odorants.

## Summary

Soluble, monomeric classical class I transplantation antigens have been found in the rat at a serum concentration of 350–380 ng/ml. They have a short half-life in the circulation of 2.7 h, and are then excreted in the urine, where the majority are degraded into smaller fragments. Urine collected from recombinant congeneic inbred rat strains which differ only at the classical class I locus RT1.A and thus excrete different allelomorphs of these molecules in their urine, can be distinguished between using the habituation–dishabituation assay.

Intact class I molecules are not the odoriferous component in the urine though the remainder of the urine after removal of class I molecules by affinity chromatography can be used to discriminate between these rat strains. Purging of the urine of volatiles by nitrogen passage for 48 h leaves the urine attractive to detector animals but does not allow urine collected from the two strains to be distinguished between. Dialysis of the urine shows that the high molecular weight fraction > 12000 can be used to discriminate between these two strains.

## References


Beauchamp GK, Yamazaki K, Boyse EA (1985) The chemosensory recognition of genetic individuality. Sci Am 253: 66–72

Bevan MJ, Fink PJ (1978) The influence of thymus H-2 antigens on the specificity of maturing killer and helper cells. Immunol Rev 42: 3–19

Brown RE, Singh PB, Roser BJ (1987) The major histocompatibility complex and the chemosensory recognition of individuality in rats. Physiol Behav 40: 65–73

Callahan GN, Ferrone S, Allison JP, Reisfeld RA (1975) Detection of H-2 antigens in serum. Transplantation 20: 431–433

Callahan GN, Ferrone S, Poulik MD, Reisfeld RA, Klein J (1976) Characterisation of Ia antigens in mouse serum. J Immunol 1351–1355

Cosman D, Khoury G, Jay G (1982) Three classes of mouse H-2 messenger RNA distinguished by analysis of cDNA clones. Nature 281: 603–605

Ekins RP (1974) Basic principles and theory. Br Med Bull 30: 3–11

Ford WL (1978) The preparation and labelling of lymphocytes. In: Weir DM (ed) Handbook of experimental immunology. Blackwell, Oxford, pp 23.1–23.22

Gussow D, Ploegh H (1987) Soluble class I antigens: a conundrum with no solution. Immunol Today 8: 220–222

Heusser C, Boesman M, Nordin JH, Isliker H (1973) Effect of chemical and enzymatic radioiodination on in vitro human C1q activities. J Immunol 110: 820–828

Howard JC (1977) H-2 and mating preferences. Nature 266: 406–407

Johns MA, Feder HH, Komisaruk BR, Mayer AD (1978) Urine-induced reflex ovulation of anovulatory rats may be a vomeronasal effect. Nature 272: 446–448

Kamada K, Davies HS, Roser BJ (1981) Reversal of transplantation on immunity by liver grafting. Nature 292: 840–842

Kirk RE (1968) Experimental design: Procedures for the behavioral sciences. Brooks/Cole, Belmont

Klein J (1986) The natural history of the major histocompatibility complex. Wiley and Sons, New York

Kress M, Cosman D, Khoury A, Jay G (1983) Secretion of a transplantation-related antigen. Cell 34: 189–196

Laemmli UK (1970) Cleavage of structural proteins during the assembly of the head bacteriophage T4. Nature 227: 680–685

Maloy WL, Coligan JE, Barra Y, Jay G (1984) Detection of a secreted form of the murine H-21 class I antigen with an antibody against its predicted carboxyl terminus. Proc Natl Acad Sci USA 81: 1216–1220

Parham P (1979) Purification of immunologically active HLA-A and -B antigens by a series of monoclonal antibody columns. J Biol Chem 254: 8709–8712

Roser BJ, Dorsch SE (1979) The cellular basis of transplantation tolerance in the rat. Immunol Rev 46: 55–86

Roser BJ, Kamada K, Zimmerman F, Davies HS (1983) Immunosuppressive effect of experimental liver allografts in liver transplantation. In: Calne RY (ed) Grune and Stratton, New York, pp 35–54

Scatchard G (1949) The attractions of proteins for small molecules and ions. Ann NY Acad Sci 51: 660–672

Singh PB, Brown RE, Roser B (1987a) MHC antigens in urine as olfactory recognition cues. Nature 327: 161–164

Singh PB, Brown RE, Roser B (1988) Classical class I transplantation antigens in solution in the body fluids; individuality signals to the environment. J Exp Med vol 168, pp 195–211

Spencer SC, Fabre JW (1987) Water-soluble form of RT1.A class I MHC molecules in the kidney and liver of the rat. Immunogenetics 25: 91–98

Sundberg H, Doring K, Novikov S, Ursin H (1982) A method for studying responses and habituation to odours in rats. Behav Neural Biol 34: 113–119

Van Rood JJ, van Leeuwen A, van Santen MCT (1970) Anti HL-A2 inhibitor in normal human serum. Nature 226: 366–367

Vincent C, Revillard JP (1979) Characterization of molecules bearing HLA determinants in serum and urine. Transplant Proc 11: 1301–1302

Wysocki CJ, Wellington JL, Beauchamp GK (1980) Access of urinary non-volatiles to the mammalian vomeronasal organ. Science 207: 781–782

Yamazaki K, Yamaguchi M, Baranoski L, Bard J, Boyse EA, Thomas L (1979) Recognition among mice: Evidence from the use of Y maze differentially scented by congeneic mice of different major histocompatibility types. J Exp Med 150: 755–760

Yamazaki K, Beauchamp GK, Bard J, Thomas L, Boyse EA (1982) Chemosensory recognition of phenotypes determined by the *T1a* and H-2K regions of chromosome 17 of the mouse. Proc Natl Acad Sci USA 79: 7828–7831

Yamazaki K, Beauchamp GK, Egorov IK, Bard J, Thomas L, Boyse EA (1983) Sensory distinction between $H\text{-}2^{b}$ and $H\text{-}2^{bm1}$ mutant mice. Proc Natl Acad Sci USA 80: 5685–5688

Yamazaki K, Beauchamp GK, Thomas L, Boyse EA (1985) The hematopoietic system is a source of odorants that distinguish major histocompatibility types. J Exp Med 162: 1377–1380

Zinkernagel RM, Doherty PC (1979) MHC-restricted cytotoxic T cells: Studies on the biological role of polymorphic major transplantation antigens determining T-cell restriction-specificity, function and responsiveness. Adv Immunol 27: 51–177

# Is There Room for a Neutralistic View on MHC Polymorphism?

L. P. de Waal

Central Laboratory of the Netherlands Red Cross Blood Transfusion Service and Laboratory for Experimental and Clinical Immunology, University of Amsterdam, Amsterdam, The Netherlands

The most remarkable feature of the major histocompatibility complex (MHC) is its enormous polymorphism. In man, almost 100 HLA alleles received official recognition from the WHO HLA Nomenclature Committee (Bull WHO 1985). Yet the biological significance of this extensive polymorphism remains obscure.

The most common explanation is that MHC polymorphism secures that for every foreign antigen MHC alleles exist in the population, providing the context in which an antigen is recognized by T cells. Individuals lacking the appropriate MHC alleles cannot generate an antigen-specific T cell response and, as a result, will be negatively selected. The population as a whole will survive despite the loss of nonresponder individuals (Klein and Figueroa 1981; Nagy et al. 1981). The biological significance of MHC antigens should come from functional studies. Indeed, in vitro studies on the interaction between antigen-presenting cells and T cells indicate that MHC alleles differ in their capacity to provide an appropriate context for antigen-specific T cell activation (Scott et al. 1983 a, b; Van Eden et al. 1984; Bruserud et al. 1985; Hensen and Elferink 1984; De Waal et al. 1983). In this way, antigen-specific T cell responder and nonresponder individuals can be identified. However, until now very few data have been put forward to support selection at the population level as the basis for MHC polymorphism.

In man, it has been shown that the HLA allele frequency of the descendents of a group of Dutch colonists in Surinam, after suffering from a typhoid fever and a yellow fever epidemic, differed from the HLA allele frequency in the normal Dutch population (De Vries et al. 1979). From animal studies, there are two lines of evidence for the in vivo involvement of MHC in disease susceptibility. First, there is the influence of the chicken-B complex on resistance to Marek's disease and, second, the influence of the mouse H-2 system on the survival after Sendai virus infection (Briles et al. 1977; Parker et al. 1978). In the latter example, in vivo disease susceptibility seems to correlate with in vitro T cell nonresponsiveness (Kast et al. 1986). However, especially in an outbred population such as the human species, there are hardly any indications for HLA association with infectious diseases (Pollack and Rich 1985; Tiwari and Terasaki 1985). Moreover, diseases associated with HLA alleles almost never interfere with reproduction and, therefore, cannot play a major role in selection (Tiwari and Terasaki 1985).

Another argument against a strong selective influence of MHC polymorphism is presented by the Syrian hamster and the Norway rat and by some mouse populations where a very low degree of MHC polymorphism for an extended period does not seem to harm the animals in any way (Klein 1987). If MHC polymorphism per se is not submitted to strong selective pressure, what then is the function of polymorphic MHC antigens?

In my view, the major biological role for MHC antigens is to ascertain tolerance induction for self-antigens. It has been demonstrated in the mouse that neonatal tolerance induction for an H-2 allele influences the T cell repertoire in such a way that nonresponsiveness against certain combinations of self+X develops (Müllbacher 1981; Schwartz and Doherty 1982). Furthermore, there is abundant evidence that the T cell repertoire for self+X overlaps with the allo-MHC repertoire (Burakoff et al. 1978; Von Boehmer et al. 1979). During T cell ontogeny, tolerance is induced for self-MHC molecules alone and for associations of other self molecules with self-MHC (Matzinger 1981). Recently, evidence was presented for clonal elimination as the mechanism for inducing T cell tolerance to self-MHC antigens (Kappler et al. 1987). As a consequence, tolerance induction for self-MHC and for association of other self molecules with self-MHC antigens will result in the development of holes in the T cell repertoire (Klein and Figueroa 1981; Nagy et al. 1981). If epitopes created by the association of self molecules with self-MHC antigen A cross-react with epitopes formed by the association of foreign antigen with A, the result is T cell nonresponsiveness for the foreign antigen unless T cells, recognizing the foreign antigen in the context of self-MHC antigen B, are present. Therefore, the expression of a balanced number of different HLA gene products is needed in individuals to compensate for the holes in the T cell repertoire induced by these same molecules. Thus, the induction of tolerance for self-antigens on the one hand and the expression of a number of different MHC molecules on the other result in the generation of a T cell repertoire that is relatively complete at the level of the individual. This means that individuals in an outbred population can generate an immune response against almost every foreign antigen, despite the fact that the T cell repertoires of different individuals need not be identical.

Thus far, antigen-specific T cell nonresponsiveness for foreign antigens in vitro correlates in general poorly with the development of disease. The relativity of MHC-controlled T cell nonresponsiveness has been demonstrated by the use of different immunization schedules or by the use of dendritic cells as antigen-presenting cells to abolish nonresponsiveness (Müllbacher and Brenan 1980: Boog et al. 1985; Kast et al. 1988). Furthermore, it can be assumed that the importance of antigen processing of foreign antigens encountered in vivo has been underestimated. It has now been shown that not only helper T cells recognize intracellularly generated peptide fragments of antigens (Guillet et al. 1987), but that the same holds for cytotoxic T-lymphocytes (Germain 1986). This would imply that after processing, antigen can be presented in an immunogenic form to the immune system, and that this process, depending on the antigen, does not always efficiently take place in in vitro cultures (Pilarski and Mohamed 1987). The consequence would be that in vivo almost all MHC alleles can provide the appropriate context for T cell recognition of foreign antigen.

The absence of strong environmental pressure will allow the existence of the observed MHC polymorphism. The generation of such an apparently superfluous number of MHC alleles is merely the reflection of the accumulation of mutations which have become fixed in the population. This process probably did not start with a single allele at each of the functional MHC loci, but already started in earlier species, and is largely controlled by random genetic drift (Klein 1987).

In conclusion, we would like to propose that the major biological role of MHC antigens is tolerance induction for self-antigens and not the genetic control of infectious diseases.

*Acknowledgments.* I would like to thank Dr. P. Ivanyi and Dr. W. M. Kast for valuable suggestions and stimulating discussion.

## References


Boog CJP, Kast WM, Timmers HTM, Boes J, de Waal LP, Melief CJM (1985) Abolition of specific immune response defect by immunisation with dendritic cells. Nature 318: 59–62

Briles WE, Stone HA, Cole RK (1977) Marek's disease: effects of *B* histocompatibility alloalleles in resistant and susceptible chicken lines. Science 195: 193–195

Bruserud O, Stenersen M, Thorsby E (1985) T-lymphocyte responses to Coxsackie B4 and mumps virus. II. Immunoregulation by HLA-DR3- and -DR4-associated restriction elements. Tissue Antigens 26: 179–192

Bull WHO 63 (1985)

Burakoff SJ, Finberg R, Glimcher L, Lemonnier F, Benacerraf B, Cantor H (1978) The biological significance of alloreactivity. The ontogeny of T-cell sets specific for alloantigens and modified self antigens. J Exp Med 148: 1414–1422

De Vries RRP, Meera Khan P, Bernini LF, van Loghem E, van Rood JJ (1979) Genetic control of survival in epidemics? J Immunogenet 6: 271–287

De Waal LP, Kast WM, Melvold RW, Melief CJM (1983) Regulation of the cytotoxic T-lymphocyte response against Sendai virus analyzed with H-2 mutants. J Immunol 130: 1090–1096

Germain RN (1986) The ins and outs of antigen processing and presentation. Nature 322: 687–689

Guillet, JG, Lai MZ, Briner TJ, Burns S, Setto A, Grey H, Smith JA, Geften ML (1987) Immunological self, nonself discrimination. Science 235: 865–870

Hensen EJ, Elferink D (1984) The immune response to (T,G)-A-L and GAT in man: an association of nonresponsiveness to (T,G)-A-L with HLA-DRw8. Hum Immunol 10: 113–127

Kappler JW, Roehm N, Marrack P (1987) T-cell tolerance by clonal elimination in the thymus. Cell 49: 273–280

Kast WM, Bronkhorst AM, de Waal LP, Melief CJM (1986) Cooperation between cytotoxic and helper T lymphocytes in protection against lethal Sendai virus infection. J Exp Med 164: 723–738

Kast WM, Boog CJP, Roep BO, Voordouw AC, Melief CJM (1988) Failure or success in the restauration of virus-specific cytotoxic T-lymphocyte response defects in class-I H-2 mutant mice by dendritic cells. J Immunol 140: 3186–3193

Klein J (1987) Origin of MHC polymorphism: the trans-species hypothesis. Hum Immunol 19: 155–162

Klein J, Figueroa F (1981) Polymorphism of the mouse H-2 loci. Immunol Rev 60: 23–57

Matzinger P (1981) A one-receptor view of T-cell behaviour. Nature 292: 497–501

Müllbacher A (1981) Natural tolerance: a model for Ir-gene effects in the cytotoxic T-cell response to H-Y. Transplantation 32: 58–60

Müllbacher A, Brenan M (1980) Cytotoxic T-cell response to H-Y in "non-responder" CBA mice. Nature 285: 34–36

Nagy ZA, Baxevanis CN, Ishii N, Klein J (1981) Ia antigens as restriction molecules in Ir-gene-controlled T-cell proliferation. Immunol Rev 60: 59–83
Parker JC, Whiteman MD, Richter CB (1978) Susceptibility of inbred and outbred mouse strains to Sendai virus and prevalence of infection in laboratory rodents. Infect Immun 19: 123
Pilarski LM, Mohamed Z (1987) Role of antigen presenting cells in the cytotoxic T-cell response to minor histocompatibility antigens (MIHA). I. In vitro function of cells pulsed with MIHA in vivo. Scand J Immunol 26: 11–19
Pollack MS, Rich RR (1985) The HLA complex and the pathogenesis of infectious diseases. J Infect Dis 151: 1–8
Schwartz DH, Doherty PC (1982) Induction of neonatal tolerance to H-$2^k$ in B6 mice does not allow the emergence of T cells specific for H-$2^k$ plus vaccinia virus. J Exp Med 156: 810–821
Scott H, Fausa O, Thorsby E (1983a) T-lymphocyte activation by a gluten fraction, glyc-Gli. Scand J Immunol 18: 185–191
Scott H, Hirschberg H, Thorsby E (1983b) HLA-DR3- and HLA-DR7-restricted T-cell hyporesponsiveness to gluten antigen: a clue to the aetiology of coeliac disease? Scand J Immunol 18: 163–167
Svejgaard A, Platz P, Ryder LP (1982) HLA and disease 1982 – a survey. Immunol Rev 70: 193–218
Tiwari JL, Terasaki PI (1985) HLA and disease associations. Springer, Berlin Heidelberg New York
Van Eden W, Elferink BG, de Vries RRP, Leiker DL, van Roos JJ (1984) Low T-lymphocyte responsiveness to *Mycobacterium leprae* antigens in association with HLA-DR3. Clin Exp Immunol 55: 140–148
Von Boehmer H, Hengartner H, Nabholz M, Lenhardt W, Schreier MH, Haas W (1979) Fine specificity of a continuously growing killer-cell clone specific for H-Y antigen. Eur J Immunol 9: 592–597

# Subject Index

MIX
Papier aus verantwortungsvollen Quellen
Paper from responsible sources
FSC® C105338

If you have any concerns about our products,
you can contact us on
**ProductSafety@springernature.com**

In case Publisher is established outside the EU,
the EU authorized representative is:
**Springer Nature Customer Service Center GmbH**
**Europaplatz 3, 69115 Heidelberg, Germany**

Printed by Libri Plureos GmbH
in Hamburg, Germany